85

Advances in Biochemical Engineering/Biotechnology

Series Editor: T. Scheper

Springer-Verlag Berlin Heidelberg GmbH

Biotechnology in India II

Volume Editors: T.K. Ghose · P. Ghosh

With contributions by
S. Chand, T.K. Ghose, P. Ghosh, S.K. Gupta, B.B. Lohray,
K. Mazumdar-Shaw, P. Mishra, S. Nath, A.K. Panda,
S. Suryanarayan

Springer

Advances in Biochemical Engineering/Biotechnology reviews actual trends in modern biotechnology. Its aim is to cover all aspects of this interdisciplinary technology where knowledge, methods and expertise are required for chemistry, biochemistry, microbiology, genetics, chemical engineering and computer science. Special volumes are dedicated to selected topics which focus on new biotechnological products and new processes for their synthesis and purification. They give the state-of-the-art of a topic in a comprehensive way thus being a valuable source for the next 3–5 years. It also discusses new discoveries and applications.

In general, special volumes are edited by well known guest editors. The series editor and publisher will however always be pleased to receive suggestions and supplementary information. Manuscripts are accepted in English.

In references Advances in Biochemical Engineering/Biotechnology is abbreviated as *Adv Biochem Engin/Biotechnol* as a journal.
Visit the ABE home page at
http://www.springerlink.com/series/abe/

ISSN 0724-6145
ISBN 978-3-662-14590-6 ISBN 978-3-540-36466-5 (eBook)
DOI 10.1007/978-3-540-36466-5

Library of Congress Catalog Card Number 72-152360

http://www.springer.de

Originally published by Springer-Verlag Berlin Heidelberg New York in 2003
Softcover reprint of the hardcover 1st edition 2003

Typesetting: Fotosatz-Service Köhler GmbH, Würzburg
Cover: KünkelLopka GmbH, Heidelberg; design & production GmbH, Heidelberg

Printed on acid-free paper 02/3020mh – 5 4 3 2 1 0

Series Editor

Professor Dr. T. Scheper

Institute of Technical Chemistry
University of Hannover
Callinstraße 3
30167 Hannover, Germany
E-mail: scheper@iftc.uni-hannover.de

Volume Editors

Professor Tarun K. Ghose

281 SFS Apartments
Hauz Khas
New Delhi 110 016, India
E-mail: tarunghose@attbi.com

Professor Purnendu Ghosh

Birla Institute of Scientific Research
Statue Circle
Jaipur-302 001, India
E-mail: pghosh@bisrjaipur.com

Editorial Board

Advances in Biochemical Engineering/Biotechnology also Available Electronically

For all customers with a standing order for Advances in Biochemical Engineering/Biotechnology we offer the electronic form via SpringerLink free of charge. Please contact your librarian who can receive a password for free access to the full articles. By registration at:

http://www.springer.de/series/abe/reg_form.htm

If you do not have a standard order you can nevertheless browse through the table of contents of the volumes and the abstracts of each article at:

http://www.springerlink.com/series/abe/

There you will find also information about the

- Editorial Board
- Aims and Scope
- Instructions for Authors

Preface

The biotechnology business in India with an increase from USD 500 million in 1997 and reaching an estimated USD 1 billion next year – health related products accounting for 60%, agro and veterinary products together 15%, and contract R&D, reagents, devices and supplies adding up to the remaining 25% of which the diagnostics share was about 10% of the total – surely presented an encouraging picture even five years ago. While volumes have increased, the pattern has not. According to a report, prepared by McKinsey & Co, India's Pharmaceutical industry including domestic and export sales and contract services totals nearly USD 5 billion. Furthermore, the company optimistically projects the growth to a factor of five-fold only if both the industry and the government are able to put in place achievable solutions that must take care of the formidable obstacles preventing further growth. If this assessment is correct, then the established transformation made by IT growth should also provide the confidence required by the high expectations for biotechnology which have arisen in the country in recent years. Some contributors to this are overenthusiastic – these are bureaucrats, some retired scientists and of course the complacent politicians who have the least knowledge of what the new biotechnology is all about. However, there are clear indications of biotechnology growth demonstrated by a few but rapidly expanding biotech companies such as Biocon Ltd, Shantha Biotech (P) Ltd, Dr. Reddy's Laboratory, Bharat Biotech Ltd, Ranbuxy Ltd, Lupin, Torrent, RPG Life Sciences, United Breweries, BCG Laboratory, Godrej Biotech, Zydus-Cadila, Bengal Immunity, EID Parry, Wockhardt and several other small but fast growing units. In addition, there are dozens of joint ventures among these companies and other biotechnological start-ups with each other or with multinational corporations having a presence in India to sell products produced abroad or those from joint R&D operations in India.

Biotechnology entrepreneurship in India is severely handicapped by the lack of critical service sectors such as expression hosts, vectors, genome libraries, large scale sterile fluid transport systems, animal cell bioreactors and their components for clean and prolonged performance and an almost total absence of a biosystems–engineering database for downstream purification of drugs and therapeutic proteins. There are often complaints that government agencies are the only sources for funding research by academics with little or nothing coming from industry. While the statement is basically true, Indian industries, like others almost anywhere, have to have clear-cut goals and because they are not in a position to control the distribution of tax money nor are they charitable institutions, they can only invest funds in new ideas based on good scientific logic and those having a reasonable chance of success.

Cutting edge research with high investment possibilities such as physiomics, metabolic engineering and pathway management, search for or synthesis of new genes for useful applications, in vitro synthesis of short chain peptides, cell-surface engineering, product design models, separation of large molecules like DNA, RNA, etc, to name a few have yet to become established in India. Difficulties of obtaining very large funding, absence of sophisticated analytical infrastructures and indeed top scholars continuing to look for better opportunities elsewhere continue to handicap project leaders and universities. Despite these difficulties, the fast growing biopharma industry is making gains in international business with comparatively priced life-saving drugs that are challenging products of multinational pharmaceutical companies operating outside India.

Parliament has intensively debated the problems Indian biotechnology is faced with in order to be able to match its ability in IPR, TRIPS and GATT Agreements applicable to GM agro and transgenic animal and medical products, protection of the nation's eco balance and the unique biodiversity endowment. It defined its services and obligations to the people. However, except for the availability of high quality human resources, few of the legislative steps taken so far can match with those existing in most parts of the developed globe. In the midst of innumerable problems faced by the largest section of the Indian population, the scientific contributions made to lift life sciences from their observed empirical domain to the level of exactness is very clearly visible in the quality of students graduating in these sciences with the support of biochemical engineering. It is certainly an excellent achievement. Notwithstanding, one source predicts the slow but steady progress with a lower number of highly trained people moving to the USA and Europe and the rate of development in the rich nations slowing down, India's growth rate will reach a high social peak in about twenty years when most certainly teaching will not be conducted in the classroom but in cyber space. Given India's strong IT base, the information cross over appears to be the obvious choice. The Department of Biotechnology plans to award USD 120 million for supercomputer networking a dozen centers to allow research with qualified access to the recently discovered genome sequences and its extensions into proteomics and physiomics data bases. The usual hype associated with anything new in biotechnology is also visible in genome science. Most of those involved have inadequate understanding of what products are needed and how must they be designed and where the huge amount of money has to come from to bring the drug to the market place following approval.

These two volumes with fourteen chapters give accounts of several areas of active research pursuits and many project supports provided by the DBT. Two important sectors of significant importance namely, bioinformatics and politics of management and protection of GM products appearing for human consumption could not be reported. The science and engineering of bioinformatics covering areas like genomics, proteomics and physiomics are new and expensive for biotechnology applications. However, after the supercomputer network connecting a dozen centers in the country becomes fully functional, contributions in these areas at least in few places are expected to be seen. The other area somewhat nebulous in nature relates to the IPR in biotechnology and its legal jurisdictions exercising control over GM agro crops and transgenic animal and plant products for human consumption. Even amongst the active participants, there exist important gaps of knowledge and general understanding of this technology is poor causing the debates to generate more heat than light. A forum for professional interactions between legislators, lawyers, arbitrators, judges and social science activists is currently absent. Until this is in place, knowledge-based and useful contributions with balanced views will have to wait.

The editors were pleased to see that biochemical engineering training and research pioneered and promoted at the IIT Delhi over the last 25 years with bioethanol as one of its substantive, graduate research programs must have helped formulate the recently declared national policy on the use of bioethanol as a transport fuel. This is a big step forward in support of India's advances in biotechnology. The editors are grateful to all authors for submitting the reviewed and edited final manuscripts. The senior editor (TKG) would like to give special thanks to the Managing Editor, Prof. T. Scheper for prompt acknowledgement of the properly edited final manuscripts.

Jaipur, May 2003

Tarun K. Ghose
Purnendu Ghosh

Contents

Contents of Volume 84

Biotechnology in India I

Volume Editors: T. K. Ghose · P. Ghosh

Adv Biochem Engin/Biotechnol (2003) 85: 1–27
DOI 10.1007/b11043

Bioethanol in India: Recent Past and Emerging Future

Purnendu Ghosh[1] · Tarun K. Ghose[2]

[1] Birla Institute of Scientific Research, Statue Circle, Jaipur – 302001, India.
E-mail: pghosh@bisrjaipur.com
[2] 281 SFS Apartments, Hauz Khas, New Delhi – 110016, India.
E-mail: tarunghose@attbi.com

There is renewed interest in bioethanol technology in view of its large potential as a transportation fuel. Bioethanol production based on lignocellulosic biomass, being the technology of the future, has been examined. The major issue is the production of ethanol at a competitive price. Biomass-based ethanol technologies are still evolving and the commercialization of this technology has to overcome various bottlenecks. Keeping this perspective in view, bioethanol technologies are analyzed in terms of feedstock availability, pretreatment strategies, efficient hydrolytic agents, availability of recombinant ethanologens and process economics with a focus on Indian research efforts. It provides indicators for research priorities to achieve these objectives.

Keywords. Bioethanol, Lignocellulose, Cellulases, Pretreatment, Ethanologens, Vacuum cycling

Abbreviations

Ad	Alcohol dehydrogenase
BERC	Biochemical Engineering Research Centre
DST	Department of Science and Technology
ED	Entner-Duodoroff
EMP	Embden-Meyerhoff-Parnas
FPU	Filter paper unit
IIChE	Indian Institute of Chemical Engineers
IIT	Indian Institute of Technology
IUPAC	International Union of Pure and Applied Chemistry
LCB	Lignocellulosic biomass
NREL	National Renewable Energy Laboratory
Pd	Pyruvate decarboxylase
Pp	Pentose phosphate
RFA	Renewable Fuel Association
SSCF	Simultaneous saccharification and cofermentation
SSF	Simultaneous saccharification and fermentation
Ta	Transaldolase
Tc	Tetracycline
Tk	Transketolase
Xd	Xylose dehydrogenase
Xi	Xylose isomerase
Xk	Xylulokinase
Xr	Xylose reductase
Yac	Yeast artificial chromosome

1 Introduction

India produces 1.3 billion liters of ethanol from cane molasses against a installed annual capacity of 3.2 billion liters [1]. Ethanol is known as a beverage and has been widely used as a constituent of Ayurvedic medicines since the vedic times. The first distillery in the country was set up in 1805 at Cawnpore (Kanpur).

Besides a substantial consumption for potable purposes ethanol is also used as a feedstock for the chemical industry; nearly 50% of the total production is consumed by the chemical industry in India. The bulk/specialty chemicals manufactured in India from ethanol include acetic acid, acetic anhydride, vinyl acetate monomer, pyridine, picoline and derivatives, ethyl acetate, monoethylene glycol, amines, ethoxylates etc. The market size of the alcohol-based chemical industry in India is about US $ 1.2 billion [2].

Until the late 1980s India was only country in the world where the price of cane molasses was kept under government control and it varied between 12 and 15 times less than what one could buy from the world market. This policy had hindered rather than helped the development of efficient alcohol technology applicable to the conditions of the then developing world. It had also prevented consideration of other comparable substrates for their conversion into alcohol.

Global experience of market forces reveals that, finally, nothing survives under protection and the tax-payers must pay for the policy. More than two decades ago it was suggested at the 23rd IIChE Annual Symposium in Delhi that one of the most important criteria for selecting a substrate for its conversion into a large volume energy product like ethanol must be priced on the basis of its available energy content [3]. It was argued that if the price of molasses be brought up to the level where it could compete with other substrates available in quantities in India, most importantly the cellulose of cane baggase or rice straw, a way could be worked out to bring a technology in place. It was mentioned that, while the sugar from the cane juice is directly fermented into alcohol, an in situ utilization of the cellulose content of baggase could be a good way of utilizing the most energy-rich components of sugarcane stock (Table 1).

The per capita fuel demand in India is increasing with the improvement of living standards and consequent mobility. The compounded annual growth rate in the consumption of transport fuel is expected to be around 7.5% in the near future. The consumption pattern of petrol and diesel in the four metro cities in India for the year 2000–2001 is given in Table 2 [4]. The demand for petrol and diesel in the country is expected to reach 16.9 and 99.5 billion liters, respectively, by 2006–2007 [5]. India's requirements for these two high-energy fuels are largely met through imports of crude oil and indigenous refining capacities. In the year 2000–2001, India imported 70% of its annual crude requirement of 107 million tonnes; costing foreign exchange of over US $ 17.5 billion [6]. The use of ethanol in the transport sector can greatly reduce oil imports.

Table 1. Sugar cane to bioethanol mass balance [3]. Basis 1000 kg cane stock

Constituents (kg)		Sugar (kg)		Ethanol (kg) (based on 95% conversion)
		Direct	via saccharification	
Pith	(800)			
α-Cellulose	(46.1)	–	50.7	24.6
Hemicellulose	(25.9)	–	28.5	8.4
Lignin	(22.0)	-	–	–
Sugar	(112.0)	112	–	54.3
Water	(594.0)	–	–	–
Sub total:		112	79.2	87.3
Rind	(200)			
α-Cellulose	(41.6)	–	45.8	22.1
Hemicellulose	(32.7)	–	36.0	10.6
Lignin	(49.1)	–	–	–
Wax, Ash etc.	(2.9)	–	–	–
Water	(74.0)	–	–	–
Sub total:		112	81.8	32.7
Total:		112	161.0	120.0

Table 2. Petrol/diesel consumption pattern in select metro cities in India, 2000–2001 [4]

City	Petrol Consumption (million liters)	Diesel Consumption (million liters)
Delhi	750	500
Kolkata	200	150
Mumbai	180	350
Bangalore	220	200

It is believed that, due to its properties as a comparatively cleaner burning fuel, octane booster and fuel extender, the use of ethanol in the transportation sector will be very large in the near future. The use of petrol blended with ethanol is a standard practice in Brazil. Until the mid 1990s nearly 95% of cars in Brazil ran on pure ethanol. When fuel prices dropped, the country turned to 20–24% gasoline blends. Although addition of ethanol increases the evaporation rate of volatile organic compounds, some analysts believe that the reduction of carbon monoxide emissions attributed to ethanol blends effectively offsets the volume loss due to increased volatility. The existing Indian specification allows for mixing of ethanol with petrol up to 5%. This specification needs to be changed to at least 10%.

The transportation sector is a major contributor of greenhouse gas emissions. The total vehicle population in India is about 53 million, and it contributes over 70% of the total carbon monoxide emission. It is desirable for a country like India to use ethanol because of the significant impact it will make in lowering the greenhouse gas emissions.

The world ethanol production was 33.5 billion liters in 2001 (Table 3) [7]. Brazil is the leading producer where ethanol production in 2001 was 11.9 billion liters. USA is the second largest producer of ethanol in the world. According to the estimates of the Renewable Fuel Association (RFA), by the end of 2003 the US ethanol production will reach 13.2 billion liters, and thus may surpass Brazil and become the largest ethanol producing country.

The three major classes of feedstocks for ethanol production are sugars (e.g., molasses, cane juice), starches (e.g., corn, wheat, cassava) and lignocelluloses (e.g., rice straw and bagasse-like agricultural residues, wood, energy crops). Sugar from cane is the primary feedstock for ethanol production in Brazil. In the USA, sugar is derived from starch, mainly corn. Cane molasses, a byproduct of the sugar industry, is the main source of ethanol in India. Sugarcane juice is not presently used in the country for ethanol production. The sugarcane yield in India is

Table 3. World ethanol production, 2001 [7]

Region	Production (billion liters)
Europe	6.3
Americas	20.5
Asia, Oceania and Africa	6.7
World	33.5

Table 4. Potential availability of ethanol from molasses in India [5, 8]

Year	Sugar production (MMT)	Molasses production (MMT)	Ethanol production (million liters)	
			Potential	Actual
1998–99	15.5	7.1	1597	1320
1999–00	18.2	8.0	1800	1380
2000–01	18.5	8.5	1980	1442
2006–07	24.7	11.3	2540	–

70 tonnes per hectare. The yield of molasses is related to sugarcane production. Molasses production varies from 4–4.5% of the cane crushed. The potential of ethanol production from molasses in India is projected to be over 2.5 billion liters (Table 4) [5, 8]. Considering that the demand in India for petrol and diesel will be about 116.4 billion liters in 2006–2007, a 10% ethanol supplementation will require about 12 billion liters of ethanol annually only for transportation sector. Thus, the current rate of ethanol production from molasses, after meeting the requirements of potable and alcohol-based chemical industries, will not be able to meet the ethanol demand of the transportation sector unless additional resources, such as agricultural residues, are used to enhance ethanol production.

Besides seizing an opportunity to transform the country's transportation system, bioethanol addresses a number of serious environmental problems created by the use of either petrol or diesel. In the following sections we will present the case of India's R & D efforts in the search of a new bioethanol technology during the years 1974–1989. From the data generated principally at IIT, Delhi's biochemical engineering research activities gave both hope and expectations of the future of its research. The major issue is the availability of bioethanol at a competitive price. The sugar and starch-based bioethanol technologies have achieved technological success under the specific conditions of the availability of raw materials and are now being commercially used. Bioethanol production based on lignocellulose, being the technology of the future, has been examined in some detail.

2 Lignocellulose Bioethanol Technology

There is one basic advantage of cellulose over native sugar derived from sugarcane in terms of yields of alcohol from these sources, namely:

$$C_6H_{12}O_6 \rightarrow 2C_2H_5OH + 2CO_2 \ (b = 0.511)$$
(glucose)

$$n(C_6H_{10}O_5) \xrightarrow{nH_2O} 2nC_2H_5OH + 2nCO_2 \ (b = 0.568)$$
(cellulose)
(b = Stoichiometric coefficient)

This means that enzymatically saccharified cellulose yields 11.1% more ethanol than glucose. The story is different in the case of acid saccharification of cellulose because, besides ethanol, many other degradation products of biomass appear in the product [3]. The possibility of a high-rate continuous enzymatic saccharification of high concentrations (~30%) of cellulose suspension in buffer yielding glucose syrup (16–18%) conducted in a membrane bioreactor at the US Army Natick Development Center, Massachusetts, was conclusively established [9]. This was achieved even when employing low activity cellulase enzymes expressed by a first generation fungal mutant of *Trichoderma viride* 9123 [10]. This pioneering work led many people to believe that enzymatically derived glucose would one day constitute a cheap substrate for a large number of fermentation products including bioethanol.

In the coming forty to fifty years, it is believed that biomass will be the largest source of energy in the world. This seems a reality because of the large availability of a variety of biomass as well as continued research efforts. The broad category of biomass includes wood, agricultural crops and residues, municipal solid wastes, animal wastes and energy crops. Depending on the biomass characteristics and local availability, various end-uses of biomass have been envisaged. They include the use of biomass as a fuel, producer of heat, generator of electricity or as a feedstock for a variety of chemicals, pharmaceuticals, dyes, paints, detergents etc. The conversion technologies being developed to achieve these end-uses include gasification, combustion, anaerobic digestion, acid/enzyme hydrolysis and fermentation etc. Biomass resources are abundant and have multiple application potential. It is believed that humankind can "meet all of its food, organic material, and transportation energy needs from biomass while coproducing substantial quantities of power and still leaving abundant land for nature" [11]. Among the various competing uses, bioethanol from lignocellulosic biomass appears to have near-term economic potential. Moreover, converting lignocellulosic biomass to bioethanol results in environmental benefits. The potential air pollutant emission benefits resulting from curtailed burning could be significant, and could far outweigh any new production and transportation-related emissions resulting from a bioethanol production facility [12].

The availability of select crop residues (rice straw, wheat straw and bagasse) in India was more than 420 million tonnes in 1997 (Table 5) [13]. These materials are not currently used to derive desired economic and environmental benefits and they could become important resource bases for bioethanol production.

Table 5. Amount of crop residue in India, 1997 [13]

	Residue to product ratio (RPR)	Amount (MMT)
Rice straw	1.75	220
Wheat straw	1.75	121
Bagasse	0.29	80

Rice straw is a potential feedstock for bioethanol production in India, being the major rice producer in the world. Farmers generally burn rice straw but due to concerns for air quality, this practice will not be encouraged in the future. Rice straw can also be tilled back into the soil. But this disposal method, due to the problems of disease and weed infestations, can damage future crops. It is therefore advisable for the rice growers to use the straw for bioethanol production rather than burn or till it [12].

Regardless of the source, lignocellulosic biomass contains cellulose, hemicellulose and lignin; typically 35–55% cellulose, 20–40% hemicellulose and 10–25% lignin. Cellulose consists of β-1,4-linked glucose polymer; hemicellulose contains complex polymers of pentoses and hexoses and lignin is a complex heterogeneous polymer of phenylpropanoid units. In the lignocellulose bioethanol technology, cellulose and hemicellulose are hydrolyzed to sugars – mainly glucose, xylose, arabinose, galactose and mannose – using acids and or enzymes. The resultant sugars are converted to ethanol using yeast or bacteria.

Both dilute and concentrated acid options are available for cellulose hydrolysis. Although acid hydrolysis processes are effective for both hardwoods and softwoods, their major disadvantage is the formation of by-products and reversion compounds, and also the need for an expensive reactor. These provide opportunity for the further development of enzymatic process for cellulose hydrolysis. According to the projections of NREL [14] future cost reductions could be 3–4 times greater for enzyme-based process than for the acid process for cellulose hydrolysis.

In the enzymatic process, the lignocellulosic biomass is first pretreated in order to increase its accessibility for the cellulolytic enzymes. Lignin, a valuable coproduct, is available in the process. The value of lignin depends on the biomass source and the process by which it is recovered. Lignin, depending on its quality, can be processed into high-value specialty products such as electrically conducting polymers, plasticizers, and phenolic resin extenders which may be used as binders in the production of plywood or fiberboard.

The enzyme-based technology for ethanol production from lignocellulosics thus consists of biomass pretreatment, production and application of cellulase enzyme, conversion of sugars derived from cellulose and hemicellulose to ethanol by ethanologens and the recovery of ethanol.

3
Pretreatment

The hydrolysis of cellulose and hemicellulose to sugars is the key step in the lignocellulose bioethanol process technology. Native cellulose is a recalcitrant substrate for enzymatic hydrolysis. Since the rate and extent of hydrolysis of natural cellulose is very slow, it needs to be pretreated for its effective hydrolysis to sugars by enzymes. Typically, hydrolysis yields in the absence of pretreatment are less than 20% of theoretical yields, whereas yields after pretreatment often exceed 90% of the theoretical. Since the characteristics of the treated lignocellulose such as crystallinity, lignin and hemicellulose content and specific surface area are all related to its hydrolysis, the rationale for pretreatment has been to separate the individual components of lignocellulose with minimum losses, concomitant with an increase in surface area and a decrease in crystallinity. Various mechanical, physical, chemical and biological approaches, either singly or in combination, have been attempted to meet these objectives [15].

Mechanical pretreatment methods include grinding, milling and extrusion. Ball milling is an effective means of pretreatment; it reduces particle size and disrupts the crystalline structure. Vibratory ball milling appears to be more effective than ordinary ball milling in translating energy input into size reduction and alteration of cellulose crystalline structure [16]. The differential two-roller mill has been used for the pretreatment of a variety of cellulosic substrates [17]. The two-roller mill approach uses less energy compared to ball mill for an equivalent susceptibility towards enzymatic hydrolysis. The mechanical pretreatment methods utilize shearing and impacting forces to yield a fine substrate possessing lower crystallinity, thus enhancing its susceptibility to enzyme action. It is not clear if the increase in cellulose reactivity due to milling is because of the decrease in particle size (and thus increased available surface area) or due to the decrease in crystallinity. Although mechanical approaches increase cellulose reactivity towards enzymatic hydrolysis, these are unattractive both energy- and cost-wise [18].

Thermal treatments, autohydrolysis and steam explosion, involve water or steam to upgrade cellulose saccharification. Autohydrolysis, which involves heating in water at 170–200°C, removes hemicellulose from lignocellulosic materials. The autohydrolysis reaction involves the formation of acetic acid from acetyl groups located in the hemicellulose. The acid formed catalyzes the hydrolysis of hemicellulose as well as the breakdown of the lignin-cellulose matrix [19]. In the steam explosion process lignocellulose is subjected to high-pressure steam for a specified period, followed by the sudden release of pressure. Pretreatment processes earlier practiced operated at higher temperatures (220–270°C) and short residence times (40–90 s), whereas later developments recommended lower temperatures (190–200°C) and longer residence times (10 min). Steam explosion at lower temperatures is preferred, as at higher temperatures the material becomes progressively darker because of the production of toxic phenolic compounds from lignin pyrolysis. Lignin melts at elevated temperatures and it has been observed by several investigators that lignin does not return to its original form when it condenses upon cooling [14]. A continuous steam explosion process has been found to be more effective than a batch treatment, because of the possibility of exercising precise control of the operating conditions and the efficient utilization of steam. A typical continuous process may consume 0.7 kg of steam per kg of lignocellulose processed.

Chemical pretreatment methods have been extensively used for removing lignin and structurally modifying lignocellulose. These methods involve the use of alkali (sodium hydroxide, ammonia), acids (sulfuric, hydrochloric), gases (chlorine dioxide, nitrogen dioxide, sulfur dioxide), oxidizing agents (hydrogen peroxide, ozone), cellulose solvents (cadoxen, CMCS), lignin solvents (ethanol, butanol) etc. Dilute sulfuric acid (0.5%) at 190°C is an effective agent for the pretreatment of lignocellulosic materials for subsequent enzymatic hydrolysis. Dilute acid treatment does not remove lignin from the substrate, but it modifies the lignin-carbohydrate linkage to a great extent. The increased susceptibility to enzyme attack is mainly attributed to hemicellulose removal and the reduction in the degree of polymerization of cellulose during the treatment. Alkalis such as NaOH and ammonia increase the biodegradability of lignocellulosic materials [20]. The treatment of lignocellulose by dilute alkali causes swelling, decreases the degree of polymerization and crystallinity, separates lignin and disrupts the lignin structure. Swelling of cellulose, due to the saponification of intermolecular ester bonds, promotes enzyme penetration. It is possible to decrease the requirement of alkali by presoaking; a significant increase in sugar production by presoaking followed by heat treatment of cellulosic substrate has been observed [20]. Liquid ammonia, under moderate pressure (15 bar) and temperature (50–80°C), has also been used to increase cellulose reactivity towards enzymatic saccharification [21]. The ammonia freeze-explosion approach disrupts the lignin seal and possibly decrystallizes the cellulose.

In IIT, Delhi's investigations, several organic solvents were evaluated for their effectiveness in removing lignin from wood and agricultural residues [22]. These include ethanol, butanol, ethylamine, phenol, acetone, ethylene glycol. The comparison of activation energy for lignin separation by different pretreatment agents indicated a minimum activation energy with aqueous solvents. Among the various solvent systems, aqueous ethanol and aqueous butanol have been found to be most effective towards delignification. Various organic and inorganic compounds (e.g., mineral acids, aluminum chloride, aluminum sulfate, ferric chloride, maleic acid, oxalic acid, salicylic acid and aromatic acids) have been found to be effective catalysts in the delignification process. The solvent process brings native lignin into the solvent phase. The solvent process provides a good separation of the major biomass components and unaltered lignin is available in the process as a valuable coproduct. The process is, however, energy intensive due to high energy expended in the recovery of the solvents. This technology is perhaps more suited to produce high-grade cellulose.

Solvents such as cadoxen and CMCS possess the ability to dissolve cellulose. The dissolution of cellulose by these solvents and its subsequent regeneration is one way of cellulose pretreatment [23]. CMCS, a non-toxic aqueous alkaline complex of iron and sodium tartarate, dissolves up to 4% cellulose at higher temperature. Cellulose can be regenerated by adding excess water. Cadoxen, an aqueous solution of ethylenediamine and cadmium oxide, readily dissolves cellulose. The addition of NaOH to cadoxen increases the solubility of cellulose. A three-step process consisting of hemicellulose separation (by acid or alkali), celluose dissolution and cellulose regeneration has been developed, but its attractiveness is limited because of the toxicity of cadmium compounds and difficulties in the recovery of the solvent.

An ideal pretreatment process is expected to produce a reactive fiber which yields pentoses in non-degraded form, exhibits no significant inhibition of fermentation, requires little or no size reduction, and can be effectively operated at a high solid/liquid ratio. The search is on for developing a near-ideal process as the pretreatment technology is intimately linked to further processing of biomass, e.g., hydrolysis and fermentation.

4 Cellulases

The important parameters in the hydrolysis such as the yield and concentration of sugars, the duration of hydrolysis and the enzyme loading are all interrelated, and obviously depend on both the characteristics of cellulose and those of the enzyme cellulases. In the economic exploitation of cellulose ethanol technology, efforts to increase the activities and productivities of the cellulase enzyme thus play a significant role.

The process of cellulose breakdown to sugars involves participation of three major classes of cellulases: (i) endoglucanases, which randomly split cellulose polymer, (ii) exoglucanases, which include glucanhydrolases that preferentially liberate glucose or glucose dimer units (cellobiose) from cellulose chain ends and cellobiohydrolases that preferentially liberate cellobiose from the end of the cellulose chain, and (iii) β-glucosidase, which catalyzes the hydrolysis of cellobiose and soluble cellodextrins to glucose. *Trichoderma reesei*, the most studied fungus for cellulases, has been shown to possess two genetically distinct endoglucanases, two cellobiohydrolases and one β-glucosidase. These enzymes work synergistically, which results in efficient decrystallization and hydrolysis of native cellulose. *T. reesei* cellulase has been found to contain two distinct domains: a catalytic domain and a cellulose binding domain. These domains are joined by an extended flexible "hinge" region. The cellulose binding domain contributes to maintaining a high concentration of cellulose on the solid cellulosic surface and is believed to interact preferentially with the crystalline region in cellulase. The catalytic domain, on the other hand, has a high affinity for the amorphous region. The flexible hinge enables the cellulose binding domain to attach the enzyme to the cellulose fiber with little restriction on the interaction of the catalytic domain with the cellulose [24]. The catalytic domain without the cellulose binding domain has a very limited overall action on cellulose.

The cellulolytic microorganisms produce a family of different cellulases with different specificities. Cellulases differ not only in the action mode, but also in the way they bind to the crystalline surface of the substrate. Because of the complexity of multiple and simultaneous actions of cellulase enzyme system, particularly those of fungal origins, the Commission of Biotechnology, IUPAC took the initiative to propose a Unified Assay System of the most important enzymes of the cellulase complex. Methods recommended in the Commission Report [25] form the basis for the comparison of cellulase activities reported by a large number of academic and industrial laboratories engaged in studies on cellulose saccharification. These are considered as global reference methods.

Cellulases are generally produced by submerged fermentation using the fungi *Trichoderma, Humicola, Aspergillus* and *Penicillium*. Cellulase synthesis in *Trichoderma* is governed by both induction as well as catabolite repression. While soluble sugars (cellobiose, sophorose, lactose) induce cellulase synthesis, the insoluble substrate cellulose has been found to be the best inducer to produce cellulases capable of degrading crystalline cellulose. Various approaches that have been suggested as a possible way of lowering the enzyme-associated costs include genetic improvement through mutagenesis, optimization of fermentation conditions, use of mixed fungal cultures and addition of surfactants and charged colloid materials. In the past thirty years, a more than 200-fold increase in cellulase productivity has been reported [26].

It is generally believed that the enzyme hydrolysis process does not currently provide adequate profit margins for suppliers at an enzyme cost that the selling price of ethanol can support. An assessment of cellulase enzyme availability for cellulose hydrolysis leads to the following conclusions: (i) a cellulase enzyme for cost effective bioethanol manufacture from cellulose is not available, (ii) with R & D efforts of enzyme manufacturers, ethanol manufacturers, (the ethanol program, and others) cost effective cellulase enzyme could be developed and (iii) industry does not appear likely to take the lead in developing a cost effective enzyme unless catalyzed to do so. The current cost of producing the enzyme at US $ 0.25–0.45 per gallon of ethanol is considered not economical [27].

The most desirable attributes of cellulases include the ability to produce a complete cellulase system with high specific activity, high catalytic activity against crystalline cellulose, thermal stability, decreased susceptibility to enzyme inhibition by cellobiose and glucose, selective adsorption on cellulose and ability to withstand shear forces. Suggested strategies to improve cellulases include discovering new enzymes through bioprospecting, creating new/better mixtures of enzymes and developing greatly improved expression systems through protein engineering [27, 28]. Creating a more effective cellulose binding domain in the enzyme molecule could be another approach to increase enzyme efficiency. There are reports on cloning and expression of cellulase genes from *T. reesei* and other cellulolytic organisms in *E. coli* and *S. cerevisiae* as well as other hosts. Protein engineering of cellulases may enable alteration of catalytic sites to produce hypercellulolytic mutants with greatly enhanced activities [29, 30].

Improvements in specific activity of the enzyme are being attempted using the following strategies: increased thermal stability, decrease in non-specific binding, decrease in feedback inhibition, increase in enzyme turn-over, and enzymatic decrystallization enhancement. Among these options, thermostability enhancement holds the greatest potential with 3 to 5 times improvement in specific activity. The Department of Energy has launched programmes for low-cost cellulase development by Genencor and Novozyme.

An important consideration in this regard is the possibility of producing transgenic plant cellulases [31]. The expression of *Acidothermus cellulolyticus* endoglucase and *T. reesei* cellobiohydrolase gene in transgenic tobacco and potato plants has been examined. The plant-based recombinant enzymes were found to be similar to those from microbial sources and the transgenic plants exhibited normal developmental characteristics. Preliminary economic analysis indicated endoglucanase production cost to be $ 1.40 per kg enzyme, which compares very favorably with the $ 5 per kg enzyme cost through the microbial route. One report indicates that by selling the potato tubers for food and vines for enzymes, farmers could increase their profits by as much as $ 100–200 per acre [32].

5 Ethanol Producing Organisms

The yeast *Saccharomyces cerevisiae*, due to its large size, thick cell wall, resistance to bacterial and viral infection and its ability to produce a high ethanol yield, has found greater acceptance for and use in industrial ethanol production. The bacteria *Zymomonas mobilis* produces ethanol at a much faster rate than the yeast and has a higher osmotolerance and alcohol tolerance.

Both *Saccharomyces cerevisiae* and *Zymomonas mobilis* can ferment only glucose, fructose and sucrose. On the other hand, *E. coli* utilizes all the sugars (glucose, mannose, xylose, arabinose and galactose), but lacks two key enzymes, namely, pyruvate decarboxylase and alcohol dehydrogenase, required for ethanol production. It is economically important that all sugars, hexoses and pentoses, obtained from biomass are used as substrate for ethanol production. Microorganisms that ferment both hexoses and pentoses have not been found in nature. Efforts are therefore being made by researchers to construct organisms that can meet these requirements.

The two main approaches to genetically engineer a microorganism for ethanol production are (i) insertion of genes into a potent ethanol producing organism, such as *Saccharomyces* and *Zymomonas*, to enlarge its substrate utilization range and (ii) insertion of ethanologenic traits into an organism capable of multiple substrate utilization, e.g., *E. coli*.

A number of yeasts (e.g., *Pichia stipitis*, *Pachysolen tannophilus*, *Candida shehate*) ferment xylose to ethanol. *P. stipitis* is known to be the best naturally occurring xylose fermenting organism. Specific ethanol productivities of xylose fermenting organisms are relatively much lower than those obtained with *Saccharomyces* and *Zymomonas* with glucose. Xylose fermenting yeasts also have low ethanol tolerance and xylose conversion is affected in the presence of glucose due to catabolite repression.

S. cerevisiae and *Z. mobilis* can utilize xylulose but not xylose. *Saccharomyces* and *Zymomonas*, however, follow different pathways to convert xylose to xylulose. *Saccharomyces* converts xylose to xylulose via xylitol through xylose reductase (Xr) and xylose dehydrogenase (Xd), whereas *Zymomonas* uses xylose isomerase (Xi) for the catabolism of xylose. The metabolism of xylulose to xylulose-5-P is catalyzed by xylulokinase (Xk). Xylulose-5-P is further metabolized to glyceraldehyde 3-P and fructose-6-P by the pentose phosphate (Pp) pathway. These intermediates are converted to pyruvate in the Embden-Meyerhoff-Parnas (EMP) or Entner-Duodoroff (ED) pathway. Pyruvate is converted to acetaldehyde by pyruvate decarboxylase (Pd) which is further reduced to ethanol by alcohol dehydrogenase (Ad) [33].

Improvement of the xylose utilizing ability of *S. cerevisiae* has been attempted by expressing heterologous genes to convert xylose to xylulose [34–36]. The cloning of Xr and Xd genes from *P. stipitis* to *S. cerevisiae* produced a recombinant strain which effectively cofermented both xylose and glucose [37]. The recombinant *Saccharomyces* gave an ethanol yield of 0.32 g of ethanol per g of xylose and a xylitol yield of 0.10 g of xylitol per g of xylose. Incorporation of the Xk gene directed the carbon flow towards the production of ethanol rather than xylitol [38]. Johnsson et al. [39] studied overexpression of Xk gene in *Saccharomyces* strains and observed that Xk overexpression increased ethanol yield by a factor of 2 and reduced xylitol yield by 70–100%. Xk overexpression, however, considerably reduced xylose consumption.

An ethanologen capable of fermentation at elevated temperatures (50°C or above) can greatly improve the efficiency of cellulose-based processes for ethanol production. Critical factors that increase heat and ethanol tolerance include high levels of esgosterol and unsaturated fatty acids in the plasma membrane, efficient plasma membrane H^+-ATPase activity, the capacity to stabilize cellular proteins with trehalose and heat shock proteins, and biochemical mechanisms that destroy oxidative radicals. Jeffries [40] approached this problem by cloning genes for thermotolerance, ethanol tolerance, arabinose utilization and cellulase secretion by expressing large blocks of genes from yeast artificial chromosomes (Yac) and then transferring the assembled Yac to an industrial yeast that had high fermentation activity.

Dilute acid hydrolysis generates a broad range of compounds, including phenolic compounds which inhibit the fermenting microorganism. To improve ethanol production from such an acid hydrolyzate, Larsson et al. [41] used recombinant *Saccharomyces cerevisiae*, which expressed laccase activity from the white rot laccase available from the fungus *Tremetes versicolor*.

Z. mobilis has also been genetically manipulated to enlarge its substrate utilization range. Zhang et al. [42] introduced Xi, Xk, transaldolase (Ta) and transketolase (Tk) genes into *Z. mobilis* thereby enabling the recombinant to ferment xylose to ethanol. On glucose alone, the recombinant had a slightly lower growth rate and ethanol yield. This reflected the metabolic burden on the recombinant *Zymomonas* due to high level expression of the introduced genes. Recombinant *Zymomonas* strains have also been constructed by cloning the genes necessary for pentose metabolism into the shuttle vector followed by introducing them into *Zymomonas* using tetracycline (Tc) as a selection marker.

In the absence of Tc the strain was not stable. The genetic stability of the recombinant in the absence of antibiotic selection was improved by integrating key genes encoding enzymes of the xylose assimilation, arabinose assimilation and Pp pathways into the *Zymomonas* genome [43]. To incorporate L-arabinose utilization capability of *Zymomonas*, genes encoding L-arabinose isomerase, L-ribulokinase, L-ribulose-5-phosphate-4-epimerase, transaldolase and transketolase isolated from *E. coli* were introduced into *Z. mobilis* under the control of constitutive promoters that permitted their expression in the presence of glucose. The recombinant *Zymomonas* metabolized arabinose almost exclusively to ethanol, but utilized glucose preferentially, reflecting the specificity of the indigenous facilitated diffusion transport system. Construction of a xylose and arabinose fermenting recombinant *Zymomonas* with facilitated entry of xylose into the cell was thus attempted by cloning the xylose transporter gene from *E. coli* and expressing it in *Z. mobilis* [44]. In order to further expand the industrial potential of *Zymomonas*, its acid-tolerant properties were augmented by introducing dps gene (which protects DNA from various assaults) from *E. coli* and a portion of small basic peptide (for supplementing the stress protection system) into *Z. mobilis* [45].

E. coli utilizes all the major sugars and is believed to be least burdened with genetic modification. This makes it a potential organism for ethanol production. Ingram's group [46] has done extensive work on the development of recombinant *E. coli* for ethanol production. *E. coli* genetically engineered to contain the PET operon (*Z. mobilis* PD and AD genes) produced high levels of ethanol. To overcome the requirement of antibiotics in the media and to improve genetic stability, Hespell et al. [47] constructed *E. coli* strains by transforming conditionally lethal *E. coli* with PET operon plasmid. The strains were capable of anaerobic growth and displayed no apparent plasmid losses after 60 generations. Ingram [48] identified the PD gene from *Z. mobilis* and the AD gene from *Bacillus stereothermophiles* to construct a thermophilic (60°C), Gram-positive ethanologen.

Ideally, for an ethanol fermenting organism the following characteristics are desirable: high ethanol yielding capacity and productivity, high ethanol tolerance, capability to ferment a broad range of sugars, resistance to inhibitory compounds (e.g., acetic acid, furfural, hydroxymethylfurfural, lignin degradation products etc. present in the pretreatment product stream), production of a low level of byproducts (e.g., acids and glycerol), ability to withstand high osmotic pressure (due to high sugar concentration), higher temperature and low pH tolerance, high cell viability for a repeated cell recycling and appropriate flocculation and sedimentation characteristics to facilitate cell recycle [49]. Serious research efforts are needed in India to design genetically engineered ethanologens for their use in conventional fermentations as well as for future application in cellulose-based bioethanol technologies.

6 Ethanol Recovery

Conventionally distillation is used for separating water from ethanol present in the fermentation broth. In order to obtain anhydrous ethanol additional energy is required to break the azeotrope. Various distillation configurations such as vapor recompression, low pressure distillation etc. have been used to make substantial savings in steam consumption for ethanol separation. Azeotropic distillation (with pentane, benzene and diethyl ether), molecular sieves and pervaporation have been used to obtain anhydrous product. Even though, in terms of degree of separation, distillation is very effective, new techniques have been proposed to further bring down the energy costs. Some non-distillation approaches such as selective sorption of ethanol, preferential adsorption of water, and membrane separation have shown promise.

Selective sorption of ethanol from fermentation broth by sorbents (with higher selectivity for ethanol than water), followed by stripping of the sorbed ethanol, represents a low energy option for ethanol separation [50, 51]. Based on the laboratory data using a polymeric hydrophilic sorbent with weak acidic groups and activated carbon, the energy requirement to concentrate ethanol from 8 to 94% was estimated to be 3.6 MJ per liter of ethanol which compared favorably with the energy requirement of more than 6 MJ/L by conventional distillation [50].

Potential methods to obtain anhydrous ethanol include the adsorption of water on some inorganic materials (e.g., calcium chloride, lime, barium oxide), starches (corn), and biomass material (e.g., rice straw, bagasse). Ethanol adsorbs at a slower rate and to a smaller extent than water at 35°C. At 80 to 100°C ethanol adsorption is minimal relative to water [52]. Further, considering that the energy required for ethanol separation increases disproportionally above 85% ethanol due to the shape of the equilibrium curve, it would be appropriate to consider an ethanol concentration up to 85% by distillation followed by an appropriate adsorption technique to obtain the anhydrous product. Such an approach was investigated by using rice straw and bagasse for selective adsorption of water [53]. The energy requirements for the concentration of ethanol from 8 to 85% by distillation and 85 to 98% by adsorption were estimated to be 3.8 MJ/L compared to 7.0 MJ/L by a conventional azeotropic distillation approach.

Ethanol separation by reverse osmosis is one of the many possible alternatives. Use of various polymeric membranes has been reported for the separation of ethanol. Cellulose acetate membranes are well known in reverse osmosis. However, for the separation of polar organic compounds, these membranes are found to be unsuitable due to their low separation properties and limited operation range [54, 55]. Moreover, ethanol impairs the stability of these membranes. Cellulose acetate membranes, however, can be suitably modified to overcome these problems. One such modified membrane was prepared by grafting styrene on cellulose acetate by irradiation [56, 57]. The styrene-grafted cellulose acetate membrane was tested for the recovery of ethanol from molasses-based fermentation broth. The performance of the membrane indicated a reasonably good separation efficiency(>92%), but moderate permeate flux (2.4 $L/m^2/h$). Prolonged exposure of the membrane to 20% ethanol concentration did not show any variation in either permeate flux or separation efficiency, indicating the stability of the membrane. A further improvement in permeate flux of the modified membrane will be necessary to make it commercially viable for the concentration of dilute fermentation broth such as those based on cellulose conversion technology where the ethanol concentration does not normally exceed 5%.

Distillation, although energy intensive, is still the most reliable method of ethanol recovery. Other approaches need further scale-up studies and collection of engineering data to realistically assess their potential. The future bioethanol technologies will use distillation for ethanol recovery, although with much improved energy recycling systems.

7
Integrated Lignocellulose Bioethanol Processes

The integrated lignocellulose bioethanol technology comprises several subtechnologies: pretreatment, enzyme production, cellulose hydrolysis, ethanol fermentation, product and coproduct recovery. Several groups are working on the commercialization of these technologies. Thus, technologies are being developed with various feedstocks (agricultural residues, hardwoods, soft-

Table 6. Lignocellulose bioethanol process parameters

Parameters	IIT, Delhi (1987)	NREL (1999)
Feedstock	Rice straw	Corn stover
Pretreatment	Autohydrolysis followed by solvent treatment	Acid prehydrolysis
Enzyme source	Mixed enzyme (*T. reesei* cellulase supplemented with *A. wentii* β-glucosidase)	*Trichoderma cellulase*
Enzyme requirement (FPU/g cellulose)	20	15
Fermenting organism	*Candida acidothermophilum*	Recombinant *Zymomonas*
Fermentation	SSF with vacuum cycling and step feeding	SSCF
Ethanol yield (m^3/tonne LCB)	0.23	0.27
Byproducts	Lignin, animal feed	Electricity

woods, and energy crops), using different pretreatment approaches (steam explosion, acid prehydrolysis, solvent delignification), cellulose hydrolysis systems (dilute and concentrated acid and enzymatic), involving high specific activity cellulases and ethanologens (recombinant *Saccharomyces cerevisiae*, *Zymomonas mobilis* and *E. coli*), generating coproducts (lignin as a source of process heat as well as starting material for other high value products) etc. The processes as proposed by Indian Institute of Technology (IIT), Delhi and National Renewable Energy Laboratory, Colorado, USA are discussed in detail here. The salient features of these two approaches are summarized in Table 6.

8 IIT Delhi Process

The interest in the bioconversion of lignocellulosic substances in India began in 1974 with research studies initiated at the newly established Biochemical Engineering Research Centre (BERC) at the IIT, Delhi, mostly as academic studies based on analyses of the logic of a two-step bioprocess, namely saccharification and ethanol production. In early 1980s analyses of a substantial amount of research data collected from a number of experimental approaches showed the potential of a novel and potential technology for the production of bioethanol and coproducts from lignocellulosic residues, such as rice straw and baggase (cane). In view of the research concepts and a large volume of laboratory data already generated, a demonstration facility for the integrated bioprocess to produce 50 liters of ethanol (95% v/v) per day was designed, fabricated and installed at IIT, Delhi which was approved by the DST and supported by the Department of Non-Conventional Energy Sources, Government of India. It is important to mention here that India's proposal for such a demonstration facility was elaborately considered by the UN Biomass Committee and later adopted by the UN

Conference of Non Conventional Energy held at Nairobi in 1983. It was the first global recognition of India's pioneering contributions towards establishing a novel technology of rendering lignocellulosic biomass into bioethanol.

Conversion of cellulose to ethanol is composed of saccharification followed by fermentation. In the enzymatic saccharification of cellulose, cellulases are inhibited by the sugars glucose and cellobiose. A method for production of alcohol directly from cellulose using cellulase enzyme and yeast was first reported by the group working at Berkeley [58]. In another report dealing with the effects of accumulated sugar and the inhibition exercised by the same on the ethanol fermentation appeared soon after [59]. However a rigorous analysis of all the inhibitions exercised by the various interacting components in simultaneous saccharification and fermentation was conclusively established [60] and reported in great detail [3]. Considerable amounts of experimental data have also been reported on the subject [61]. Given below is the scheme that emerged from these studies identifying the problems and solutions which are likely to cut back these problems to establish a duly analyzed process to provide a new technology (Fig. 1).

Although the extent of inhibition exercised by ethanol on cellulases is much less compared to sugars [62], the removal of ethanol from the system should result in an increase in the bioconversion of cellulose to ethanol. IIT, Delhi used this concept for the bioconversion of rice straw to ethanol (Fig. 2). The process consisted of a two-step pretreatment comprising autohydrolysis and solvent delignification to separate major components of rice straw for further processing. In the autohydrolysis step nearly 70% of the hemicellulose present in the straw was removed as soluble mixed sugars, mainly xylose. The pentose sugars present in the water extract from the autohydrolysis reactor were used to grow *Candida utilis* for its subsequent use as animal feed. The autohydrolyzed residue (cellulose 56.5%, hemicellulose 13%, lignin 15%) was further delignified with 50% (v/v) aqueous ethanol in the presence of a catalyst at 170°C for 30 minutes.

The treated straw (cellulose 76%) is sent to the SSF reactor for its conversion to bioethanol. The conversion process in the SSF reactor is initiated by treated

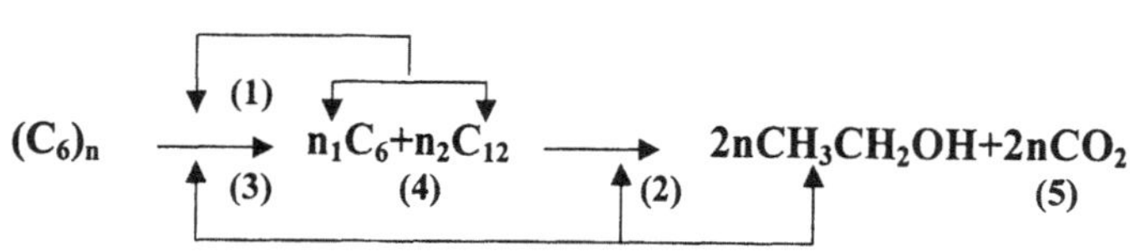

Problems:
1) Sugars (product I) inhibiting cellulase enzymes (saccharification),
2) Ethanol (product II) inhibiting yeast fermentation,
3) Ethanol (product II) inhibiting cellulase systems (saccharification),
4) Accumulation of C12 sugar causing less production of fermentable sugars, and
5) Product II (ethanol) separation is inconveniences by the presence of equimolar CO_2 produced.

Solutions:
(a) Simultaneous elimination of the three inhibitions,
(b) Availability of increased β glucosidase active protein and
(c) Removal of product II without associating CO_2 in the process of removal.

Fig. 1. Enzymatic saccharification of cellulose – inhibition and their elimination [62]

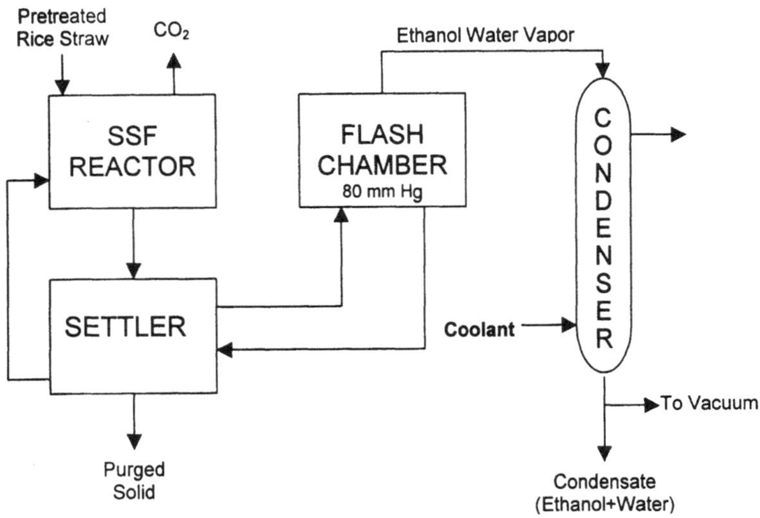

Fig. 2. Block diagram of SSF with vacuum cycling [63]

straw, mixed enzymes (cellulases from *T. reesei* supplemented with *β*-glucosidase from *A. wentii*) and yeast *Candida acidothermophilum*. The SSF operation is carried under programmed vacuum cycling coupled with intermittent substrate feeding [61–66]. The SSF reactor is coupled to a settler and a flash vessel operated in conjunction with a vapor recompression system. The system is basically a separation and purging unit to maintain a constant level of build-up of non-reactive components and transfer of the reacting fluid into a flash chamber to remove ethanol produced in the SSF reactor. The flash unit was connected with a thermocompression system operating at 40 °C and 80 mm Hg pressure. The operation was programmed to work between two discrete steps, namely, (i) vacuum to remove the bulk of the ethanol produced and (ii) feeding of fresh lots of cellulose equivalent to the ethanol removed. The flash vessel was connected with a condenser. During the repeated vacuum cycling operation, the average concentration of ethanol as observed in the receiver was 12.4 wt %. This was sent for ethanol recovery. The system was operated with 14 feedings of cellulose and 30 vacuum cycle operations covering a total period of 220 h corresponding to an average ethanol productivity of 4.4 g/L h. This is very high yield when compared to the prevailing conversion efficiency known in the bioethanol industry in India. The product yields were 230 liters ethanol, 120 kg lignin and 80 kg animal feed per tonne of straw processed. The integrated process (Fig. 3) shows the total mass balance of the system based on 1000 kg rice straw.

Based on the laboratory and pilot scale data, an order of magnitude economic analysis for a mid-sized rice straw-based plant producing 15 million liters of 95% (v/v) ethanol per year was made. The total capital investment was estimated to be US $ 18.3 million [64]. Considering the availability of rice straw at $ 10 per tonne, the production cost was 54.4 cents per liter ethanol (Table 7). Utilities accounted for 45% of the product cost. This was mainly due to the high

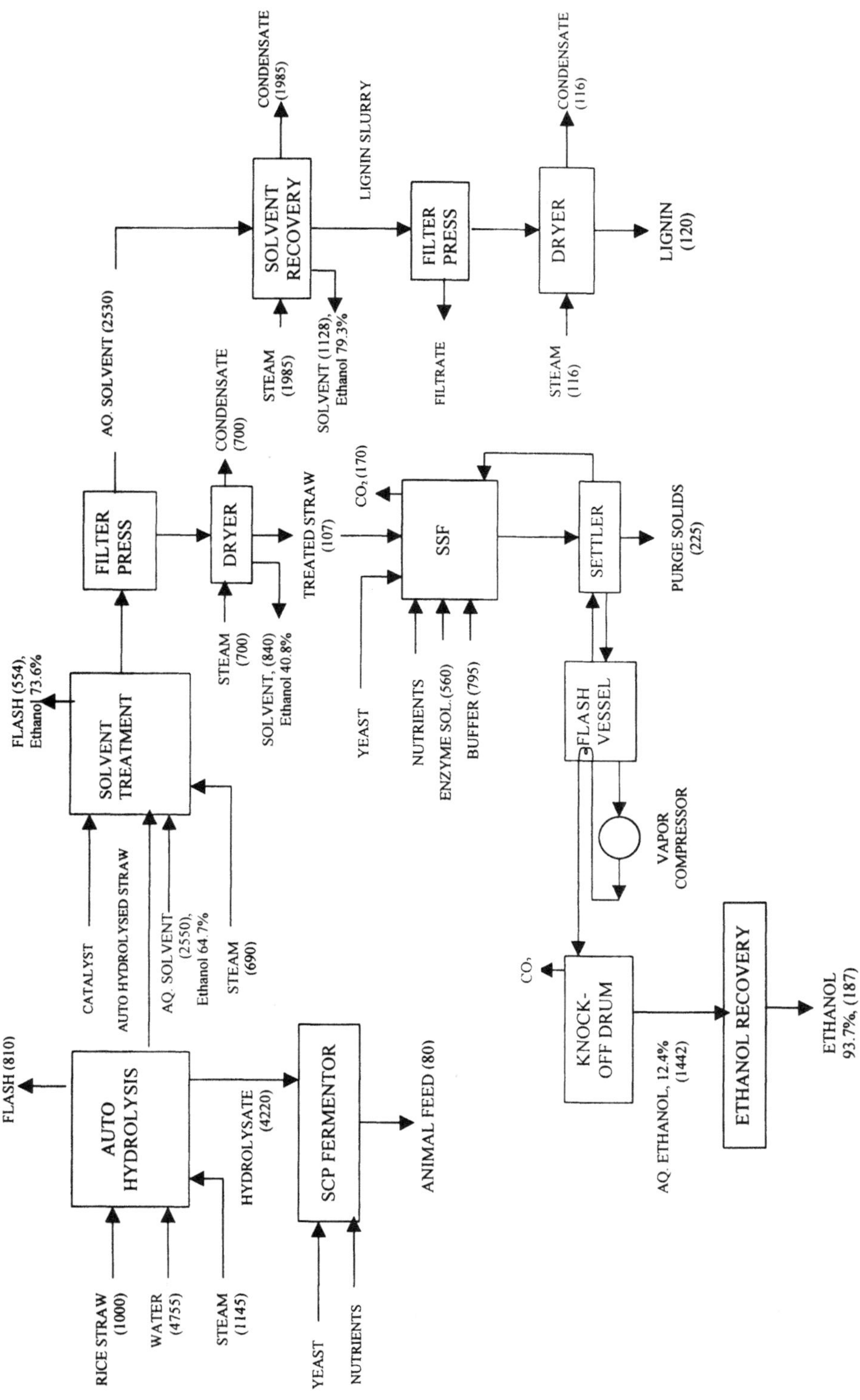

Fig. 3. Bioconversion of rice straw to ethanol and coproducts [64] (total mass balance and process efficiency)

Table 7. Summary of bioethanol production costs, IIT, Delhi (1987) [64]

Item	Cost (cents/liter bioethanol)
Raw materials	10.6
Utilities	24.9
Fixed costs	25.0
Depreciation of capital	11.1
Credits (lignin, animal feed)	(–) 17.2
Product cost	54.4

energy requirement of the solvent pretreatment process. About 20% of the product cost was due to raw materials. This is low compared to the conventional molasses to ethanol process, where the contribution of raw materials is about 75% of the production cost.

Coproducts considered in the analysis were lignin and animal feed. The value of lignin depends on its quality. Lignin can be combusted to produce heat and/or power for the ethanol process. Increasing the value of lignin by converting it to a high-value fuel additive can significantly enhance the competitiveness of bioethanol technology. It is to be recognized that the potential economic benefit of lignin with high-molecular weight is much greater than its value as a fuel. Unaltered lignin is obtained in the solvent pretreatment process and its value at $ 200/tonne considered in the analysis appeared reasonable. Katzen et al. [67] used $ 200/tonne of lignin in their economic analysis of the alcohol pulping process. Battelle [68] used a lignin value of $ 330/tonne in the phenol pulping process, whereas Myerly et al. [69] used a value of $ 176/tonne in their economic analysis when its use as modified lignin was envisaged.

The analysis on the dependence of product costs on the scale of operation and feedstock cost reflected that the product cost could be about the same at straw cost of $ 10 and $ 40 per tonne provided the scale of operation was doubled. The impact of the adsorption-desorption approach of ethanol separation (in place of distillation) and alkali pretreatment (in place of solvent process) on the product cost were also analyzed. With the alternate ethanol separation approach, the product cost increased by about 50%. This was mainly due to the high capital investment and additional expenditure on sorbents which needed frequent replacement. These additional investments offset savings in the energy for ethanol separation. The alkali pretreatment process required low capital and utility, but despite these advantages the product cost increased due to lower byproduct credits (lignin as a fuel). This emphasizes the importance of byproducts in the total process economics and underlines the necessity of converting all the major components of biomass into valuable products.

9 NREL Process

NREL analyzed several process technologies and chose the technology which involves dilute acid pretreatment of lignocellulosic biomass (corn stover), followed by simultaneous enzymatic saccharification of the remaining cellulose and cofermentation of the resulting glucose and xylose to ethanol [70]. At 190°C in the continuous pretreatment reactor (residence time 10 minutes), 75% of hemicellulose present in the stover is converted to soluble sugars (xylose, mannose, arabinose and galactose) and some of the lignin is solubilized. Acetic acid is also liberated as a result of hemicellulose hydrolysis. Degradation products of pentose and hexose sugars (furfural, hydroxymethylfurfural) are also formed. Following the pretreatment reactor, the hydrolyzate liquid and solids are flash cooled, which vaporizes large amounts of water, much of the furfural and HMF and a portion of the acetic acid. The remaining acetic acid present in the prehydrolyzate is removed using continuous ion exchange. The slurry is overlimed followed by gypsum removal. The liquid hydrolyzate and solid residue are remixed and sent to a simultaneous saccharification and cofermentation (SSCF) reactor. *T. reesei* cellulase enzyme (15 FPU/g cellulose), recombinant *Zymomonas mobilis* and corn steep liquor are added to the SSCF reactor. SSCF is conducted at 30°C for 7 days. Distillation and molecular sieve adsorption are used to produce nearly 100% ethanol. The distillation bottom contains lignin, cell mass, and dissolved chemicals. The lignin/cell mass cake is combusted to produce steam. The steam produces electricity through a multistage turbine generator. The dissolved chemicals are concentrated in an evaporator producing clean water fit for recycle.

The cost of ethanol production based on this technology was estimated to be 39.5 cents per liter ethanol (Table 8). The total capital investment for a 94.6 million liters ethanol per year corn stover-based plant is estimated to be $ 136 million (1999 $ value). About 36% of the production cost is on account of capital recovery. In the process, the byproduct stream (lignin, unconverted cellulose and hemicellulose etc.) is burnt to generate steam and electricity. After meeting the energy requirements for the process, it produces excess electricity which is taken as process credit.

The techno-economic analysis considered various future technology pathways for cost reduction [71]. This included more efficient pretreatment, conducting SSCF at higher temperature (55°C instead of 30°C), constructing recombinant ethanologens capable of fermenting all the sugars (glucose, xylose, arabinose, galactose, mannose), improving specific activity and productivity of the enzyme (e.g., 3-fold increase in specific activity and 8-fold improvement in enzyme productivity), etc. These technological achievements are expected to improve ethanol yield as well as decrease capital requirements and as a conse-

Table 8. Summary of ethanol production costs, NREL (1999) [71]

Item	Cost cents/liter bioethanol
Raw materials	18.7
Fixed costs	9.4
Capital recovery	14.3
Credits (electricity)	(-) 2.9
Product cost	39.5

quence decrease the production cost. According to the estimates, incorporation of these process improvements would result in decreasing the cost of producing bioethanol from 39.5 to 24.8 cent per liter of anhydrous bioethanol.

10 Concluding Remarks

Economics, oil prices, trade deficits, environmental concerns and how they relate to national needs are the major drivers for the development of bioethanol technology. India has the potential to produce 2.5 billion liters of ethanol from molasses in the next 5 years. Considering 10% ethanol supplementation in petrol and diesel, its demand in the transportation sector is projected to be nearly 12 billion liters. India has resources to meet the increasing demand of ethanol for transportation fuel as well as an ethanol-based chemical industry. More than 1 billion liters of ethanol can be produced using less than 15 million tonnes of sugarcane. The issues are: should cane juice, besides molasses, be used as a feedstock for ethanol production? Is there surplus sugar for this purpose and does this provide enough opportunity for value addition? India's alcohol-based chemical industry is apprehensive that production of fuel ethanol would increase the price of molasses, thus raising manufacturing cost for their products as well. This concern may not be reasonable. Starch-based feedstocks such as sorghum and maize can be explored as additional resource bases for ethanol production. Although India is the second largest producer of sorghum in the world with an annual production of about 10 million tonnes, its yield (840 kg/ha) is about 40% lower than the world average [72]. The sorghum yield, however, is projected to increase to 4250 kg/ha by the year 2010. Sorghum (starch 63–68%) has potential to give good ethanol yields (380 liters bioethanol per tonne sorghum) provided an efficient sorghum-based bioethanol technology is developed. This could be used for potable purposes.

Lignocellulosic biomass is less expensive than sugar or starch-based feedstocks, but its conversion to ethanol at present is more costly. Biomass-based ethanol technologies are still evolving and the commercialization of this technology has to overcome various bottlenecks. These include feedstock availability, scale of operation, cheaper pretreatment strategies, efficient hydrolytic agents, availability of recombinant organisms capable of cofermenting the whole range of sugars at a temperature compatible to optimum saccharification, and better coproduct value.

Logistics of raw material availability (collection, storage and handling) to meet large future demands is a major issue. For example, rice straw-based ethanol production would require the location of the plant within a reasonable distance from the rice farms. Moreover, seasonal availability of the feedstock would need either large storage facilities or would need plants to operate on multiple feedstocks for their continued operation throughout the year. In India ethanol plants are comparatively small in capacity. This brings to the fore another related issue: scale of operation vis-à-vis feedstock availability. Relatively large investments will be required to install lignocellulose bioethanol plants. Keeping in view the logistics of feedstock procurement, a decision is needed if it is advisable to build very large plants as increased feedstock cost (due to collection and transport of large amounts of feedstock) may offset savings due to economies of scale [73].

Pretreatment of lignocellulose continues to be a major barrier to both enzymatic saccharification and fermentation. We have to ascertain what is more important for enzymatic hydrolysis – the extent of delignification which requires harsher conditions for complete lignin separation or will the loosening of cellulose-hemicellulose-lignin bonds under milder conditions suffice for our purpose. The benefits of lignin solubilization need to be weighed against the potential for fermentation inhibition by soluble lignin derivatives. Pretreatment is expensive, but it is a necessary step. Some of the pretreatment approaches are expensive due to their high energy or chemical recovery costs or because of undesirable component losses. Other approaches are not so effective, in terms of rate and extent of hydrolysis, due to their limited cellulose decrystallization and lignin removal ability. Major R & D efforts are thus needed to resolve these problems and consequently evolve new pretreatment technologies. Development of energy plants with traits such as increased cellulose and hemicellulose and less lignin not only has the potential to improve ethanol yields but also the application of much simpler pretreatment technologies.

The most critical element for the success of bioethanol technology is the availability of cellulases at a cost that will dictate the ethanol price to be paid by the consumer. Currently, the major market for cellulase enzymes is the textile industry, and the enzymes produced are tailored to meet the requirement of this industry. It is important to recognize that biomass application needs are significantly different from textile applications. Cellulase enzymes are too expensive for bioethanol. There is, however, a good possibility of producing effective cellulases at a much reduced cost. For the hydrolysis of pretreated biomass, extremely complex cellulases may not be required; simpler cellulase systems may serve the purpose and can be derived either from nature or obtained by mixing cellulases of desired characteristics. Protein engineering can be used to design cellulases capable of overcoming end-product inhibition. Heterologous systems can be developed to express cellulases which can function at temperature as high as say 80–90°C.

Recombinant *S. cerevisiae*, *Z. mobilis* and *E. coli* have been developed which are capable of converting a wide spectrum of sugars to ethanol. A powerful enzyme has remained the most important issue. Initially in late 1960s work was confined to determining the specificity of the most important components of the complex in respect of its substrate affinity. Now rDNA technology is taking on all that matters in the final release of sugars. Today nothing except for engineered cells either for hydrolysis or glycolysis is relevant.

The projected cost of ethanol production from cellulose has declined significantly in the last twenty years [74]. Further reduction in ethanol production cost is possible in the near future. According to NREL projections, in another 15 years time a ethanol production cost of 20 cents per liter will be possible by employing a cellulose and hemicellulose-rich and lignin-lean feedstock (ethanol yield increase from 68 to 99 gal/ton), and using a highly efficient cellulase producer (cellulase yield improvement from 200 to 2000 FPU/g cellulose and cellulase productivity improvement from 75 to 2000 FPU/L h) and SSCF process improvement (decrease in residence time from 7 to 2 days). The cost projections indicate improved conversion technology to be the largest contributor to reducing the cost of ethanol production; the impact of conversion-related cost reduction has been shown to be 3 times larger than that associated with scale and 10 times larger than that associated with less expensive feedstock [14]. In the corn-based ethanol industry, in the past twenty years, the manufacturing costs were reduced from 60 to 23 cent per liter ethanol. Factors responsible for this cost reduction include: process innovation, equipment upgrading, energy integration and highly effective new enzymes [75]. These cost projections are useful indicators to decide our research priorities for ethanol production from lignocellulosic biomass.

There are several other issues concerning the commercialization of bioethanol and they need the attention of researchers, entrepreneurs and more importantly, the policy makers. One such issue is who will be the major promoter of the policy – agricultural interests, sugar industry or petroleum industry? Another related issue is the necessity and cost of government support to make this technology viable. The subsidy provided by the government for the promotion and development of the technology needs to be weighed against potential benefits such as employment opportunities and strengthening of the rural economy. One also needs to consider the subsidies provided to the petroleum industry. According to the analysis of California Bioethanol Program [12], the benefits of a biomass-to-ethanol production industry for its economy are potentially greater than the cost of state support for such an industry. The analysis estimates economic benefits of $ 1 billion (due primarily to employment opportunities) over a 20 year period, assuming state government incentives of $ 500 million for a 200 million gallon per year ethanol industry. Some analysts [76] claim that if the petroleum industry involves itself to ethanol production, as they control the distribution of the fuel, the chances of early commercialization of bioethanol technology are higher. Under what circumstances or conditions refiners might consider participation in the ethanol industry will continue to remain a big question. Only if the return on investment for ethanol plants exceed that of petroleum processing facilities and the demand hopefully continues to expand, some major companies might invest in this new venture.

The Government of India has evolved a National Auto Fuel Policy. It takes into consideration the availability and logistics of fuel supplies, economics of producing fuels and the possibilities of multi-fuel use in different categories of vehicles. Like may other countries in the West, it encourages the usage of alternative fuels like CNG, LPG, ethanol and electricity. The Cabinet has approved the Committees recommendation; representatives from industry have supported it. Hopefully soon this technology will become a "player" instead of just a "spectator". It is imperative that even after nearly two decades of waiting this vitally important energy chemical should occupy its rightful place in the fuel policy of the country so long as desired.

11 References

1. Distillery industry of India. http://www.aidaindia. org/aida/about_distillery.htm
2. Increasing user base of alco-chemical industry. (2002) http://www.indiachem 2002.com/indiachem 2002/chem2.htm
3. Ghose TK (1986) Proc. IIChE Symp. on Fuels and Feedstock for Chemical Industry. IIT, Delhi, pp 1–27
4. Mishra S (2000) India ethanol coalition. http://www.cleantechindia.com/neweic/ India.htm
5. Mishra S (2000) Background paper: Seminar on Use of ethanol in motor gasoline: in search of a viable strategy. New Delhi, http://www.cleantechindia.com/neweic/ seminar.htm
6. Naik R (2001) Ethanol and petrol: a sweet blend. The Hindu Business Line, June 21
7. Berg C (2001) World ethanol production. http://www.fo-licht.com
8. Utilization of sugar industry by-products. Indian Sugar Mills Association, http://www.indiansugar.com/story/byproducts.htm
9. Ghose TK (1972) US Patent 3,642,580
10. Mandels, M, Weber, JA (1969) Adv Chem Ser (ACS) 95:391
11. Lynd L, Wyman C (1999) Biomass processing in the 21st Century: potential challenges and a vision of the future. Plenary presentation summary, IEA Bioenergy Workshop, Natal, South Africa, Aug 25
12. California Energy Commission Report (2001) Costs and benefits of a biomass-to-ethanol production industry in California. http://www.energy.ca.gov/reports/2001-04-03_5 00-01-002+002A.pdf
13. Mantanis G (1999) Worldwide availability of agriwaste, NARLIT Ltd., Greece
14. Lynd LR (1996) Ann Rev Energy Environ 21:403
15. Ghosh P, Singh A (1993) Advances in Applied Microbiology 39:295
16. Millet MA, Baker AJ, Satter LD (1976) Biotechnol Bioeng Symp 6:125
17. Tassinari TH, Macey CF, Spano LA, Ryuddy (1980) Biotechnol Bioeng 22:1689
18. Datta R (1980) Biotechnol Bioeng 23:2167
19. Bouchard J, Abatzoglon N, Chornet E, Overend RP (1989) Wood Sci Technol 23:343
20. Pannirselvam PV, Ghose TK (1980) Proc. Second International Symposium Bioconversion and Biochemical Engineering, Delhi
21. Dale BE, Moreira MJ (1982) Biotechnol Bioeng Symp 12:31
22. Ghose TK, Pannirselvam PV, Ghosh P (1983) Biotechnol Bioeng 25:2577
23. Tsao GT (1978) Process Biochem 13:12
24. Glazer NA, Nikaido H (1995) Ethanol. In: Microbial Technology. WH Freeman & Co., San Francisco, p 359
25. Ghose TK (1987) Measurement of cellulase Activities. Pure and Appl Chem 59(2), 257–268 (Report by Comm. on Biotechnology, IUPAC)
26. Tolan JS, Foody B (1999) Adv Biochem Eng/Biotechnol 65:42
27. Cellulase assessment for biomass hydrolysis. http://www.ceassist.com/assessment.htm
28. Sheehan J, Himmel ME (1999) Biotechnol Prog 15:817
29. Godbole S, Decker SR, Nieves RA, Adney WS, Vinzant TB, Baker JO, Thomas SR, Himmel ME (1999) Biotechnol Prog 15:828
30. Himmel ME, Ruth MF, Wyman CE (1999) Current Opinion in Biotechnology 10:358
31. Production of cellulases in tobacco and potato plant bioreactors. Project Summaries (1998–99) US Department of Energy, http://bioenergy. ornl.gov/99summaries/cellulase.html
32. Two-for-one special: industrial enzymes and food grown in one plant (1999) Pacific Northwest National Laboratory, US Department of Energy, Press Release, July 12
33. Chandrakant P, Bisaria VS (1998) Critical Reviews in Biotechnology 18:295
34. Kotter P, Ciriacy M (1993) Appl Microbiol Biotechnol 38:776

35. Walfridsson M, Anderlund M, Bao X, Hahn-Hagerdal B (1997) Appl Microbiol Biotechnol 48:218
36. Tantirungkij M, Nakashima N, Seki T, Yoshida TJ (1993) J Ferm Bioeng 75:83
37. Ho NWY, Chen Z, Brainard AP (1998) Appl Environ Microbiol 64:1852
38. Toon ST, Phillippidis GP, Ho NWY, Chen Zd, Brainard A, Lumpkin RE, Riely CY (1997) Appl Microbiol Biotechnol 63:243
39. Johansson B, Christensson C, Hobley T, Hahn-Hagerdal B (2001) Appl Env Microbiol 67:4249
40. Jeffries TW (2000) Development of second generation ethanologenic yeast. Project Summaries 1998–99, US Department of Energy, http://bioenergy.ornl.gov/99summaries/fermentation.html
41. Larsson S, Cassland P, Jonsson LJ (2000) Appl Env Microbiol 67:1163
42. Zhang M, Eddy C, Deanda K, Finkelstein M, Picataggio S (1995) Science 267:240
43. Zhang M (2002) Improved *Zymomonas* for xylose and arabinose fermentation. Project summaries 1998–1999, US Department of Energy, http://bioenergy.ornl.gov/99summaries/fermentation.html
44. Conway T (2000) Pentose sugar transport in *Zymomonas*. Project summaries 1998–99, US Department of Energy, http://bioenergy.ornl.gov/99summaries/fermentation.html
45. Kaspar CW (2000) Enhancement of acid tolerance in *Zymomonas mobilis*. Project summaries 1998–99, US Department of Energy, http://bioenergy.ornl.gov/99summaries/fermentation.html
46. Ingram LO, Conway J (1988) Appl Environ Microbiol 54:397
47. Hespell RB, Wyckoff H, Dien BS, Bothast RJ (1996) Appl Environ Microbiol 62:4594
48. Ingram L (2000) Development of portable ethanol-producing operons that can be expressed in Gram-positive bacteria. Project Summaries 1998–1999, US Department of Energy, http://bioenergy.ornl.gov/99summaries/fermentation.html
49. Picataggio SK, Zhang M (1996) Biocatalyst development for bioethanol production from hydrolysates. In: Wyman CE (ed), Handbook on Bioethanol: Production and Utilization. Taylor and Francis, Washington DC
50. Malik RK, Ghosh P, Ghose TK (1983) Biotechnol Bioeng 25:2277
51. Pitt WW, Haag GL, Lee DD (1983) Biotechnol Bioeng 25:123
52. Ladisch MR, Voloch M, Hong J, Blenkowski P, Tsao GT (1984) I & EC Process Design & Development 23:437
53. Rakshit SK, Ghosh P, Bisaria VS (1993) Bioprocess Engineering 8:279
54. Mehta GD (1982) J Memb Sci 12:1
55. Leeper SA (1986) Membrane separation in production of alcohol fuels by fermentation. In: McGregor WC (ed), Membrane Separation in Biotechnology. Marcel Dekker, New York
56. Choudhury JP, Ghosh P, Guha BK (1985) Biotechnol Bioeng 27:1081
57. Choudhury JP, Ghosh P, Guha BK (1988) J Memb Sci 35:301
58. Cysewski GR, Wilke CR (1976) Biotechnol Bioeng 18:1297
59. Takagi M, Abe S, Suzuki S, Emert GH, Yata N (1977) In: Ghose TK (ed), Proc Bioconversion Symposium, IIT, Delhi, p 551
60. Roychoudhury PK, Tyagi RD, Ghose TK (1980) In: Ghose TK (ed), Proc Bioconversion Symposium. IIT, Delhi, p 443
61. Ghose TK, Roychoudhury PK, Ghosh P (1983) Biotechnol Bioeng 26:377
62. Ghosh P, Pamment NB, Martin WRB (1982) Enz Microbiol Technol 4:425
63. Roychoudhury PK (1985) Ph.D. Thesis, IIT, Delhi
64. Ghose TK, Ghosh P (1987) Sensitivity analysis of an integrated process of lignocellulose conversion. Proc. Conference Frontiers of Bioprocess Engineering, Colorado, 28 June–7 July, p 32
65. Roychoudhury PK, Ghose TK, Ghosh P (1992) Enz Microbiol Technol 14:581
66. Roychoudhury PK, Ghose TK, Ghosh P, Chotani GK (1985) Biotechnol Bioeng 28:972
67. Katzen RR, Fredrickson R, Brush BF (1980) Chem Eng Prog 76:62
68. Bungay HR (1983) Environ Sci Technol 17:24
69. Myerly RC, Nicholson MD, Katzen R, Taylor JM (1981) Chemtech 11:186

70. McAloon A, Taylor F, Yee W (2000) Determining the cost of producing ethanol from corn starch and lignocellulosic feedstocks. NREL Report NREL/TP-580–28893
71. Wooley R, Ruth M, Sheehan J, Ibsen K (1999) Lignocellulosic biomass to ethanol process design and economics utilizing co-current dilute acid prehydrolysis and enzymatic hydrolysis: current and futuristic scenarios. NREL Report NREL/TP-580–26157
72. Sheorain V, Banka R, Chavan M (2000) In: Chandrashekhar AJ, Bandyopadhyay R, Hall AJ (eds) Technical and institutional options for sorghum grain mold management, Proceedings of an international consultation, May 18–19, Patancheru, India, p 228
73. DiPardo J (2000) Outlook for biomass ethanol production and demand. Energy Information Administration, http://www.eia,doe.gov/oiaf/analysispaper/pdf/biomass.pdf
74. Wyman CE (1999) Ann Rev Energy Environ 24:189
75. Knauf M, Pilgrim C (2002) Enzymatic hydrolysis of cellulosic biomass. 7th Annual National Ethanol Conference: Policy and Marketing, http://www.ethanolrfa.org/NECO2-Knauf.pdf
76. Downstream Alternatives Inc. (2000) The current fuel ethanol industry: Transportation, marketing distribution and technical considerations. http://www.ott.doe.gov/biofuels/publications.html

Received: June 2002

Adv Biochem Engin/Biotechnol (2003) 85: 29–42
DOI 10.1007/b11044

Commercialization of a Novel Fermentation Concept

Kiran Mazumdar-Shaw[1] · Shrikumar Suryanarayan[2]

[1] Biocon India Limited, 20th Km, Hosur Road, Hebbagodi 561229, Bangalore, India.
E-mail: Kiran.Mazumdar@bioconindia.com

[2] Biocon India Limited, 20th Km, Hosur Road, Hebbagodi 561229, Bangalore, India.
E-mail: Shri@bioconindia.com

Fermentation is the core of biotechnology where current methodologies span across technologies based on the use of either solid or liquid substrates. Traditionally, solid substrate fermentation technologies have been the widely practiced in the Far East to manufacture fermented foods such as soya sauce, sake etc. The Western World briefly used solid substrate fermentation for the manufacture of antibiotics and enzymes but rapidly replaced this technology with submerged fermentation which proved to be a superior technology in terms of automation, containment and large volume fermentation. Biocon India developed its enzyme technology based on solid substrate fermentation as a low-cost, low-energy option for the production of specialty enzymes. However, the limitations of applying solid substrate fermentation to more sophisticated biotechnology products as well as large volume fermentations were recognized by Biocon India as early as 1990 and the company embarked on a 8 year research and development program to develop a novel bioreactor capable of conducting solid substrate fermentation with comparable levels of automation and containment as those practiced by submerged fermentation. In addition, the novel technology enabled fed-batch fermentation, in situ extraction and other enabling features that will be discussed in this article. The novel bioreactor was christened the "PlaFractor" (pronounced play-fractor). The next level of research on this novel technology is now focused on addressing large volume fermentation. This article traces the evolution of Biocon India's original solid substrate fermentation to the PlaFractor technology and provides details of the scale-up and commercialization processes that were involved therein. What is also apparent in the article is Biocon India's commercially focused research programs and the perceived need to be globally competitive through low costs of innovation that address, at all times, processes and technologies that exhibit high degrees of conformance to the international standards of regulatory and good manufacturing practice.

Keywords. Solid state fermentation, Bioreactor, Process control, Pharmaceuticals, Enzymes

Abbreviations

SSF	Solid-state fermentation
SmF	Submerged fermentation
USFDA	United States Food and Drug Administration
GMP	Good manufacturing practice
PlaFractor	The novel contained bioreactor described in this article

1 Introduction

Biocon India was formed in 1978 as a joint venture between an Irish multinational, Biocon Biochemicals and an Indian entrepreneur, first author of this article. The original mandate for this joint venture was to manufacture a plant-derived enzyme, papain and to develop a market in India for a range of enzyme products manufactured by the Irish partner. Fours years into the business, the author was excited with the emerging opportunities in the new realm of biotechnology and decided to embark on an R & D mission to develop novel enzymes that would complement the Irish partner's product range.

The rationale was to apply the Indian scientific skill base to provide low cost innovation to the Biocon Group world-wide. The importance to avoid replication of R & D programs between the Indian and Irish efforts were well recognized as a key success factor. The Indian R & D efforts were also to be internally funded through accrued profits and through local borrowings largely on account of the Irish partner's inability to provide an additional budget for this research activity.

It was this limitation of resources that provided the sharp focus for the Indian researchers and a high degree of improvisation that provided an inherent degree of creative thinking. The R & D program at Biocon India accordingly commenced in 1984 with a team of 3 scientists, a food scientist with a degree from Massachusetts Institute of Technology, a microbiologist from Bombay University and a biochemical engineer from The Indian Institute of Technology who headed the team (the second author of this article).

The R & D strategy involved two integrated programs: one that would prospect for novel microorganisms in search of novel enzymes and a second that would develop a novel technology based on solid substrate fermentation technology (SSF) for the production of these novel enzymes.

2 Learning from Basics

The R & D team at Biocon India faced an uphill task fraught with a number of challenges: solid substrate fermentation technology was not a well documented technology. There was limited published information on laboratory scale and plant scale fermentation procedures. The primitive state of Indian infrastructure in terms of communication and information technology at the time made it all the more difficult to access contemporary information. What this therefore entailed was an extensive amount of knowledge gathering and "re-inventing the wheel" so to speak. Nevertheless, it was the passion to succeed that provided the real impetus for innovation. What this effort achieved was the creation of a strong knowledge base in fermentation which encompassed developing skills in strain selection, strain mutation and process development. Proprietary skills were further accentuated by developing all of this knowledge on the base of solid substrate fermentation. Process development, on the other hand, called for innovation based on biochemical engineering, a skill base that has developed into a core competence today. What this investigated was design criteria based on heat transfer, ease of handling, mechanization and fabrication. Two years of painstaking R & D enabled the project to take some promising shape. By 1986, the team had developed a mutant strain capable of producing commercially viable levels of pectinase and by 1987, the team had developed a working pilot model of the tray culture solid-state fermentation technology. The next two years were focused on scaling-up this technology as well as developing other enzymes based on this new technology. Thus began Biocon India's foray into fungal enzymes and solid substrate fermentation. The commitment of the entire management to this endeavor was absolute and resolute. It is this ethos that is fundamental to successful innovation.

3 Operational Issues on Large-Scale Solid-State Fermentation

The next barrage of challenges in the scaling-up of this home-grown tray culture technology was by way of regulatory and quality issues. The enzymes proposed to be manufactured using this technology were destined for international markets for food applications. This compelled the company to conform to certain prescribed international standards of quality and food safety. The Joint Expert Committee on Food Additives or JECFA along with the Food Codex Committee or FCC had laid down certain stringent guidelines which involved the certifying of the strain used as a "safe strain". This involved a complete dossier development detailing the strain selection, mutation, characterization and identification which was to be certified by an external agency. The dossier also required exhaustive toxicology data which entailed the establishment of a mycotoxin testing facility and an intensive training of quality control personnel in this specialized function. Setting up enzyme assays that were harmonized with international norms like IUPAC specified assays, as well as developing novel "application-based" enzymatic assays that reflected the performance of various enzymes in real life and generating comparative data on competitor products was another prerequisite of scale-up. In addition, a product application laboratory was also set up to evaluate product performance and garner knowledge on dosage optimization, usage conditions, as well as performance enhancement through enzyme blends. This involved plant trials, investigating pH and temperature profiles of enzymes and characterizing enzyme side activities. All this generated a large knowledge bank which formed the strong foundation for future innovation. The intangible value of all biotech companies is measured by their ability to take products from the laboratory to the market and Biocon India's track record in this respect is impressive. Biocon India has scaled-up in several commercial enzymes and fermentation-derived pharmaceutical molecules over a span of just one decade and has a rich pipeline of products to be scaled-up in the future. Some of the enzyme products commercialized are cellulases (3 varieties), pectinases (4 varieties), xylanases (3 varieties), amylases (2 varieties), proteases (2 varieties), and tannase. Some of the pharmaceutical molecules commercialized include, lovastatin, simvastain, compactin, cyclosporin and mycophenolic acid.

4
Further Extension of Solid Substrate Fermentation

In 1997, a brainstorming retreat of the management team resulted in a strategy to leverage the proprietary solid substrate fermentation to areas outside the enzyme realm. Pharmaceutical secondary metabolites were identified as a potential area and research commenced on lovastatin, a cholesterol-lowering secondary metabolite which exhibited promising results on solid substrate fermentation. It was this initiative that brought about the much needed focus to accelerate the pace of R & D on the further development of the SSF technology. Both containment and automation were perceived to be important given the need to conform to international standards of cGMP and safety issues based on the requirements of organic solvent-based down streaming. The lovastatin project was exciting and challenging. Yields were impressively high on solid substrate fermentation and the process developed using this technology was truly novel and provided the company with a number of patent opportunities. This combination of low-production cost coupled with a high innovative content allowed Biocon India immediate access to several international markets. Biocon India has since earned the distinction of being the first and only Indian company to be approved by USFDA for lovastatin and even more noteworthy is Biocon India's distinction of being the first and only company in the world to be approved by USFDA for the production of lovastatin by solid substrate fermentation.

5
Limitations of Solid Substrate Fermentation

The original solid substrate fermentation technology based on a tray culture model was designed and operated in an aseptic manner. However, it was well recognized that this was not a "contained" system and had various parts of the process exposed to operators and the environment. It was felt important to further develop the SSF technology that would address these deficiencies very effectively. In fact, certain cytotoxic products such as immunosuppressants or highly sporulating cultures such as those used in agricultural biocontrol would automatically exclude themselves from the semi-automated tray culture-based solid state fermentation technology that was being currently practiced. Genetically modified organisms, when introduced, would also demand high levels of containment and would not be permitted by regulation to be cultivated through conventional solid substrate fermentation.

There are many literature reports of the use of solid substrate fermentation to develop many products of commercial importance including antibiotics [1–3] and other cytotoxic or immunosuppressive pharmaceuticals [4, 5] but none of these technologies lend themselves to scale-up for reasons of lack of containment of substance which have been known to cause severe allergenic reactions when handled or accidentally inhaled. These limitations and others related to automation and other fermentation features such as fed-batch, off-gas analysis, temperature and pH control etc., have now been overcome in an improved SSF technology christened the PlaFractor. The PlaFractor technology is a very enabling technology that will become the norm for solid substrate technology in the future.

6
Description of the PlaFractor Technology

The early success of the semi-automated tray culture technology developed at Biocon India was demonstrated by the fact that the first plant that was built based upon this technology (in 1987) was very quickly followed by another plant (in 1991) with a 3-fold scale-up of the first plant scale operation. Meanwhile, market feed-back and comparative data conclusively proved that many of the enzymes being developed with solid state fermentation were extremely cost competitive with their submerged fermentation counterparts. Even more significant was the efficacy of these products in different applications.

In addition to scientific, regulatory and process considerations, the fact that solid state-produced enzymes were commercially successful gave Biocon the underlying impetus to invest and develop the PlaFractor technology as the next generation technology with features that overcame various limitations of the conventional technology. In 1992 a dedicated research team at Biocon India embarked on a challenging project to develop this next generation technology. The design brief factors that the team ascribed to themselves were:

- The bioreactor would incorporate a high degree of containment. The air entering the reactor would be filtered to remove microorganisms as would the air leaving the system. At no stage during the operations would the live microorganism from within the reactor have a chance to come into contact with the environment. At the same time, the reactor would be effectively sealed from any contaminating influence of the environment.

- The reactor would remove metabolic heat from the bed using conductive means. Since air is a bad conductor of heat with a low specific heat capacity, the heat removal would be done by cooling water.
- The support matrix with the nutrients would be sterilized in place in the reactor and cooled in place.
- The inoculum would be added and mixed in an aseptic manner post-sterilization.
- There would be a provision to send air into the bed uniformly for aeration. The requirement of air for aeration would be de-linked from the cooling requirement. Thus, any kind of atmosphere could be maintained in the reactor that was optimal for productivity. The problem of bed drying due to excessive evaporation would also be avoided.
- There would be a provision to add nutrients during the fermentation. If necessary the mixing arrangement could be used to mix these nutrients through to ensure uniform distribution.
- It would be possible to sample the reactor aseptically during the fermentation.
- It would be possible to extract the product from the matrix without opening the reactor. It would be possible to use both organic and inorganic solvents for extraction.
- It would be possible to recover any residual solvent from the extracted bed by evaporation.
- Post-extraction, the reactor could be sterilized to destroy any live microorganisms, before being cleaned for the next cycle.
- It would be possible to operate the reactor in an automated fashion.
- A plant based upon this reactor would occupy much less space than the automated tray culture method.

Based on the above design criteria, the PlaFractor project team started their research effort which went through several iterations and several prototypes. In 1998 a pilot-scale PlaFractor was developed using commercially scaleable engineering fabrication techniques and the underlying principle of the PlaFractor technology was proven. Based upon this, the first technology demonstrator plant was commissioned in February 2000 (Fig. 5).

The construction and the operation of the PlaFractor is described in detail in the patent that has been granted for this device [6]. However, the key aspects are described here with reference to the Figs. 1–5.

The semi-automated tray culture process for solid-state fermentation is shown in Fig. 1. This process is successfully being used to produce many industrial products like enzymes, natural colors etc. on a large scale. A selected high-yielding strain of microorganism is propagated in several steps using shake flasks and submerged fermentors to develop a suitable inoculum. This inoculum is then used to inoculate a sterilized solid matrix on which the production

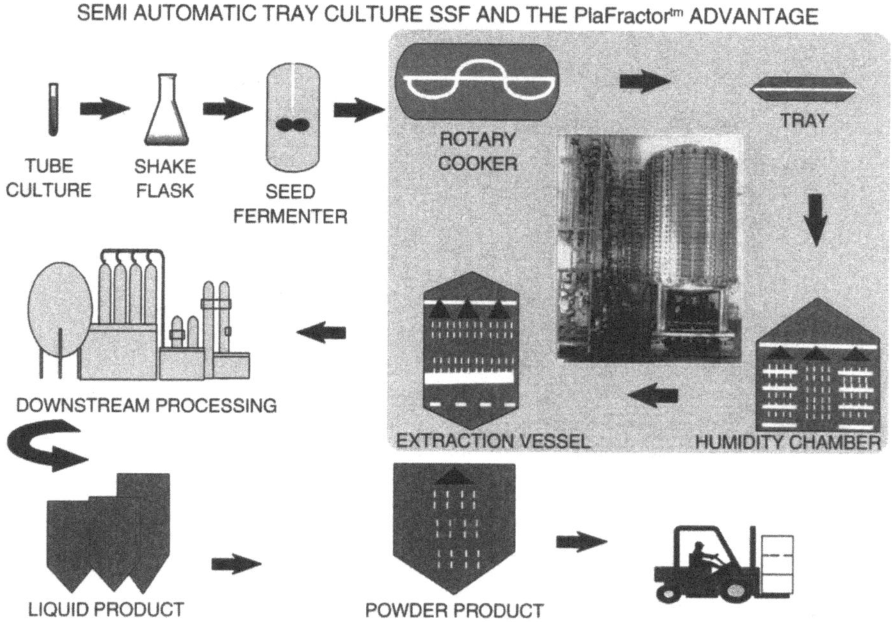

Fig. 1. This figure shows the process flowchart for the semi-automatic tray-culture solid-state fermentation process and highlights the advantage that the PlaFractor brings to the solid-state fermentation process. The gray boxed area in the figure – around the cooker, tray-layering operations, humidified incubation chamber and the extraction vessel – is replaced completely by one single equipment – the PlaFractor shown in the center of the gray box

actually occurs. The solid matrix, typically wheat bran, is sterilized in a large rotary cooker and then cooled down. The cooled matrix is then inoculated with the inoculum from the submerged seed fermentor and mixed well. The inoculated solid matrix is then automatically layered onto thousands of trays on a moving conveyor belt. The layered trays are then incubated in a large, temperature- and humidity-controlled incubation room for a period of time to allow the microorganism to colonize the solid matrix completely and produce the product. The solid matrix is then harvested from the trays into an extraction vessel. An extracting fluid is sent into the extraction vessel to extract the product, leaving behind the spent biomass which is then discarded. The extracted product is sent for down-stream processing and standardized for dispatch.

The PlaFractor is designed to replace the cooker, the trays, the incubation rooms and the extraction device in the semi-automated solid state fermentation tray culture process, with one compact equipment (Fig. 1). This results in a saving of space and a better control of the solid-state fermentation process.

The equipment consists of a set of circular tray modules (also known as plates), which are filled with the solid substrate or matrix prior to sterilization (item 3, Fig. 2). The modules are arranged upon each other to form a stack (Fig. 2). The modules have a special base containing channels in which cooling fluids can circulate to remove metabolic heat during the fermentation (the non-

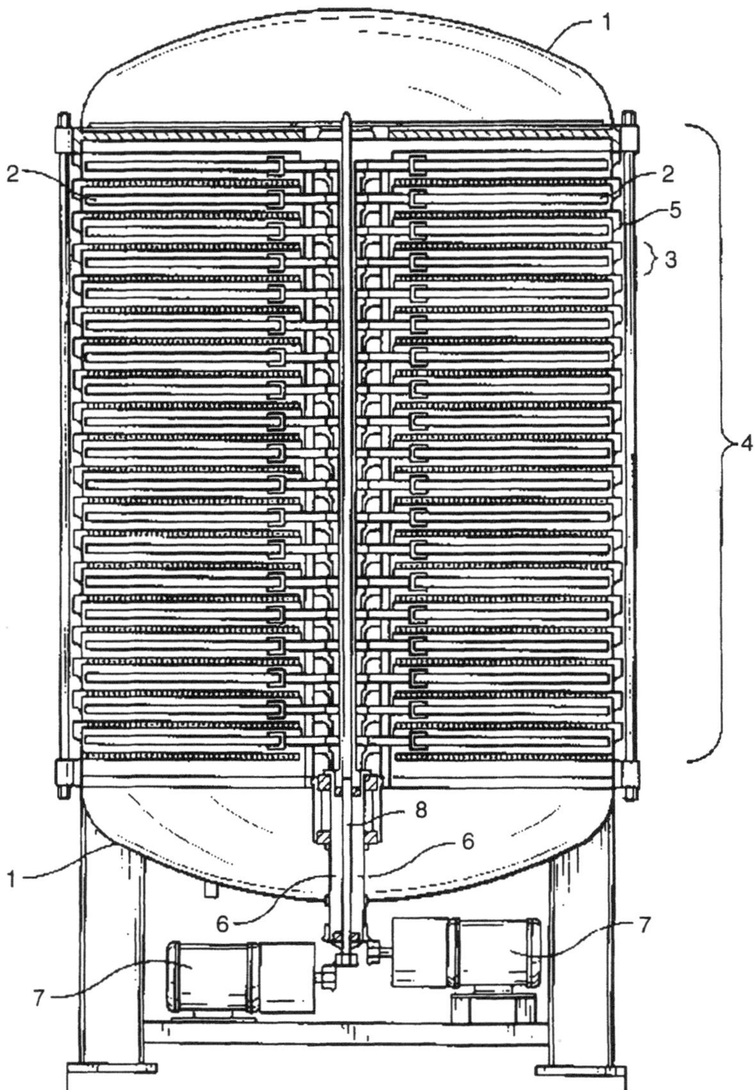

Fig. 2. A PlaFractor stack (*4*) is made of several tray modules (*3*) arranged upon each other and sealed by gaskets (*5*) in between. There is a top and bottom dish (*1*) above and below the stack of modules. Each tray module has a mixing arm (*2*) with blades, which rotate around the axis formed by the arms (*2*). The arms themselves revolve around the axis formed by the 2 concentric central shafts (*8* and *6*), which are independently driven by 2 motors (*7*). One motor provides the power for rotation for the blades of the mixing arms while the other provides the power for the revolving motion of the mixing arms around the central shaft

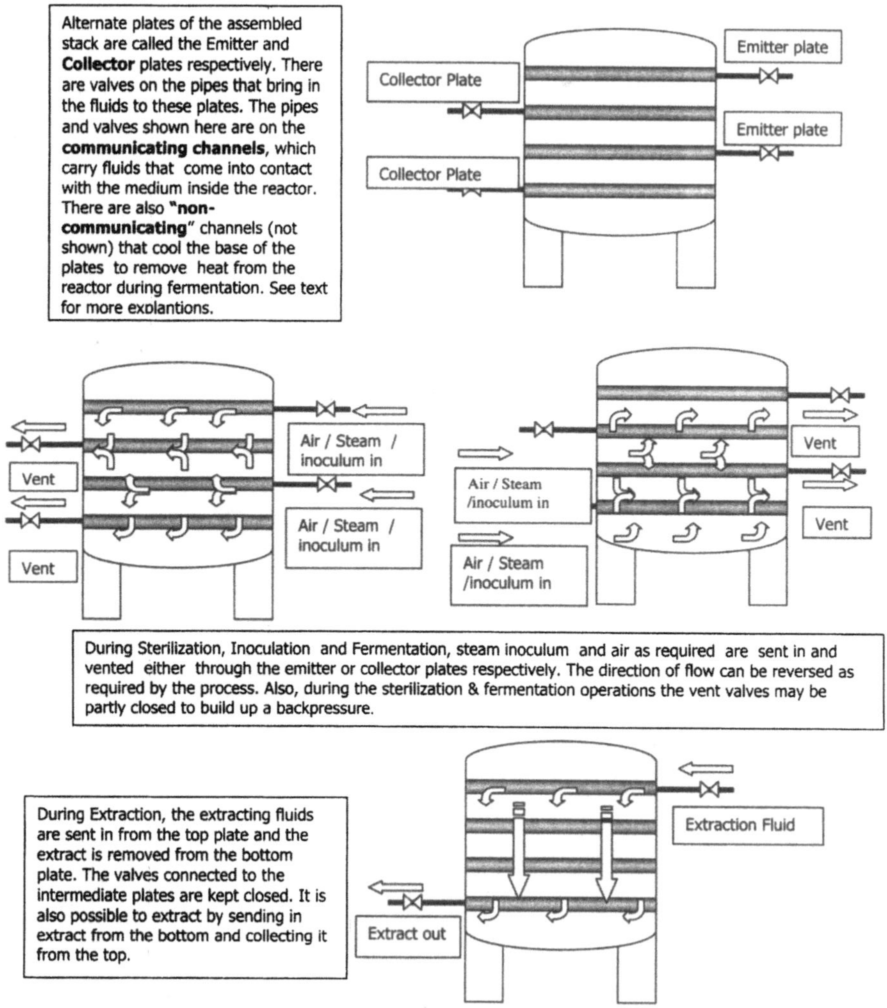

Fig. 3. PlaFractor – Modes of operation. The PlaFractor is connected to a piping arrangement (not shown) through a set of valves which can bring in fluids at various time to each of the trays; sterilization, cooling, inoculation or extraction can take place as desired

communicating channels). Another set of channels, called the communicating channels, also present in the base, can bring in sterile air into the modules, via perforations in the base, to provide aeration to the matrix during the fermentation [6]. Since the fluids that flow in these channels communicate with and come into contact with, the solid matrix, these channels are designated as the "communicating channels". During sterilization, the same channels are used to bring in steam to sterilize the matrix and during extraction, these channels double up to introduce the extracting solvent (see Fig. 3). As is clear from the above description, the plates are engineered in a way that prevents any cross-

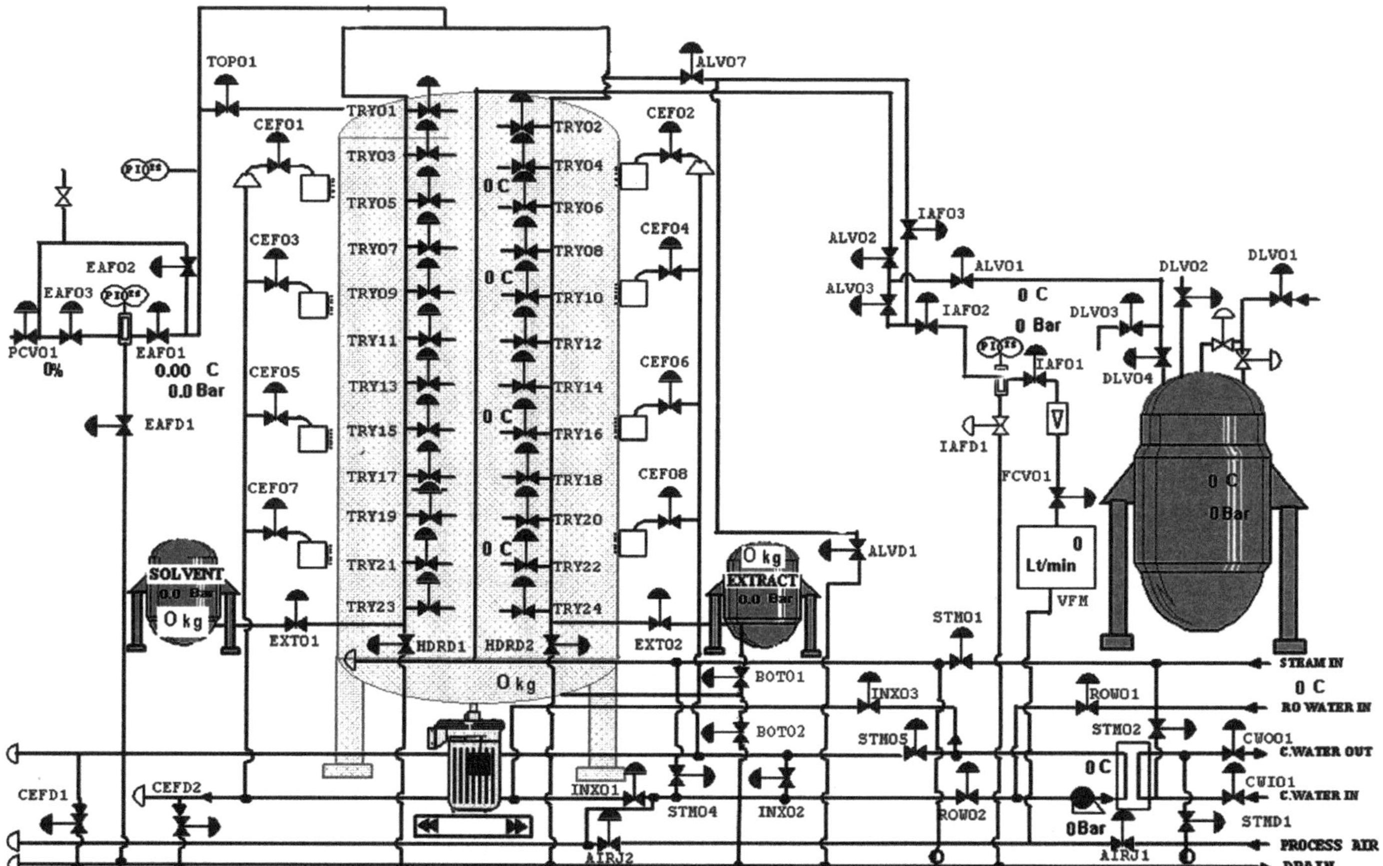

Fig. 4. The piping and instrumentation diagram that controls the PlaFractor The valves, shown in the picture are all controlled and sequenced by means of a programmable logic controller to allow the different modes of operation (sterilization, cooling, inoculation, fermentation and extraction etc.) to be achieved

Fig. 5. The PlaFractor arrangement along with the control pipe rack and interconnecting piping

contamination of fluids from the communicating channels to the non-communicating channels.

After assembling the plates into a stack, it is connected to a piping arrangement (Fig. 4) arranged as a pipe rack (Fig. 5) that distributes various fluids to the communicating and the non-communicating channels in each module via a set of valves. The opening and closing of the valves to bring in the correct fluid (air, water, steam, inoculum or extractant) is sequenced and controlled by means of a computer program.

The various operating modes of the PlaFractor are shown in Fig. 3. The PlaFractor is filled with the solid matrix and then connected to the pipe rack.

6.1 Sterilization and Cooling of the PlaFractor

Under the control of a computer program, a set of valves opens to bring in steam into the PlaFractor to sterilize it. There are temperature probes within the PlaFractor that can be used to monitor and control the temperature within the PlaFractor. Alternate tray modules (plates) in the PlaFractor stack are designated as the emitter and collector plates, respectively. The communicating channels of the emitter and collector plates are connected to independent headers. Thus, when steam is being sent through the communicating channels of the emitter plates into the PlaFractor, air from the PlaFractor can be vented through the header connected to the collector plates. When the sterilization is over, the computer switches the appropriate valves to bring in sterile air through a previously sterilized air filter, into the PlaFractor and hold it under positive pressure. An arrangement in the pipe rack (Fig. 4) allows the inlet air filter to be sterilized separately and cooled, independent of the PlaFractor. Simultaneously, cooling water is sent into the various modules of the PlaFractor, through the non-communicating channels in the base of each tray, to quickly cool down the matrix. The PlaFractor is cooled to the desired fermentation temperature and held there, under positive pressure until it is inoculated.

6.2 Inoculation and Fermentation Control

Well grown inoculum from a seed fermentor is sent into the PlaFractor using a pressure differential, through pre-sterilized transfer lines, and through the communicating channels of each tray and mixed well with the sterile matrix using a mixing arrangement that is present in each module of the stack. The mixing arrangement in each module is described in detail in reference [6]. Briefly, it is a biaxial mixing arrangement consisting of a set of blades in each module (see Fig. 2). The blades revolve in each circular module around the central axis of the PlaFractor and simultaneously rotate as they revolve thus providing efficient mixing in the vertical and horizontal planes. The driving force for this biaxial mixing arrangement is provided by modular concentric shafts in each tray module, which engage easily with each other when the PlaFractor is assembled and which are driven from the bottom by 2 electric motors.

The fermentation temperature is controlled throughout by sending heating and cooling fluids into the non-communicating channels of various modules depending on the temperature of the PlaFractor. The air flow rates can also be controlled based upon the carbon dioxide and oxygen concentrations of the exhaust gas. If necessary, a suitable back pressure can also be maintained during the fermentation.

All the fermentation parameters are automatically logged by the computer during the course of the fermentation.

Any nutrients that need to be fed during the course of the fermentation may be brought in into each of the trays, via the piping arrangement and mixed well into the growing matrix. Because of this arrangement the PlaFractor is a unique device to carry out fed-batch solid-state fermentation.

6.3 Recovery of Product

At the end of fermentation, the computer program switches the appropriate valves to bring in the extracting solvent, from a solvent holding tank, into the PlaFractor. The mixing arrangement is also turned on to mix the extracting solvent well with the solid matrix and improves the extraction efficiency. The PlaFractor is built to conform to explosion-proof standards and thus it possible to bring in flammable organic solvents also, to extract the solid matrix at the end of fermentation. Also, since it is not necessary to harvest the fermented material to a separate recovery device for product recovery, the whole recovery operation can be carried out in an extremely contained manner, thus allowing even cytotoxic products like some immunosuppressants to be handled very safely.

Once the extraction is over, any residual solvent that is trapped in the bed may be recovered by heating the PlaFractor and blowing air through it to evaporate the solvent. The solvent vapors are condensed to recover the solvent. The use of solvents and heat inactivates any live microorganisms that may be present in the PlaFractor thus allowing the spent biomass to be discharged safely. Depending on the product requirement any other appropriate denaturing conditions may be used prior to discharge of the spent biomass.

If the PlaFractor is extracted with aqueous solvents, which is the case when it is used to produce industrial enzymes, then it may be sterilized again, post-extraction to destroy any live microorganisms, before it is opened to discard the spent matrix. The spent matrix is discharged by closed pneumatic suction.

7 Commercial Application of the PlaFractor and Future Prospects

A production plant based upon the PlaFractor was commissioned in February 2000. Several different kinds of processes have been successfully demonstrated using the PlaFractor. These include the production of lovastatin (an antihypercholesteremic agent), mycophenolic acid (an immunosuppressant), *Trichoderma viridae* spores (for biocontrol purposes) as well as a fungal protease (enzyme for the food industry).

The PlaFractor is an enabling technology that opens up new possibilities for solid substrate fermentations. Using the PlaFractor, it is possible to carry out even fed-batch solid-state fermentations on a commercial scale. The whole operation of solid-state fermentation can be carried out aseptically and in a contained manner with a high degree of control and consistency. Parameters like temperature and air-flow rate can be varied during the fermentation and profiled automatically if necessary using the computer that controls the PlaFractor. The cost of producing many of the products mentioned above compares very favorably with the cost of producing them by submerged fermentation. Biocon will look to expand the applications of this technology more in the future by its own efforts and in collaboration with others.

8 Conclusion

Biotechnology is a challenging realm that demands a multiplicity of skill bases: biochemical engineering, microbiology, molecular biology, general biology, software development, chemistry including proteins, and medicine. The real challenge is to train and integrate these skills on a low-cost innovation platform. The R & D success at Biocon owes itself to the strong commercial focus of its programs and the perceived need to be globally competitive which results in setting ambitious targets and short time lines. Low cost of innovation is not the cost of scientific manpower but the cost of highly innovative manpower. This coupled with a high success of commercial scalability and the ability to access global markets through conformance to international standards is what has been Biocon India's recipe for success. Because of this, Biocon today enjoys in excess of a 60% market share of the Indian enzyme markets and a similar share of the markets for the bulk pharmaceutical ingredients that it produces. The bulk of the remaining part of the market is made up mainly by imports and a smaller percentage by other enzyme companies that produce a few selected enzymes locally, mostly using imported technologies.

9 References

1. Robinson T, Singh D, Nigam P (2001) Appl Microbiol Biotechnol 55:284
2. Barrios-Gonzalez J, Castillo TE, Mejia A (1993) Biotechnol Adv 11:525
3. Kota KP, Sridhar P (1999) Process-Biochem 34:325
4. Murthy MVR, Mohan EVS, Sadhukhan AK (1999) Process-Biochem 34:269
5. Sadhukhan AK, Murthy MVR, Kumar RA, Mohan EVS, Vandana G, Bhar C, Rao KV (1999) J Ind Microbiol Biotechnol 22:33
6. Suryanarayan S, Mazumdar K (2001) US Patent 6,197,573 B1

Received: May 2002

Adv Biochem Engin/Biotechnol (2003) 85: 43–93
DOI 10.1007/b11045

Bioprocessing of Therapeutic Proteins from the Inclusion Bodies of *Escherichia coli*

Amulya K. Panda

Product Development Cell, National Institute of Immunology, Aruna Asaf Ali Marg, New Delhi – 110067, India. *E-mail: amulya@nii.res.in, amulya_p@hotmail.com*

Escherichia coli has been most extensively used for the large-scale production of therapeutic proteins, which do not require complex glycosylation for bioactivity. In recent years tremendous progress has been made on the molecular biology, fermentation process development and protein refolding from inclusion bodies for efficient production of therapeutic proteins using *E. coli*. High cell density fermentation and high throughput purification of the recombinant protein from inclusion bodies of *E. coli* are the two major bottle necks for the cost effective production of therapeutic proteins. The aim of this review is to summarize the developments both in high cell density, high productive fermentation and inclusion body protein refolding processes using *E. coli* as an expression system. The first section deals with the problems of high cell density fermentation with an aim to high volumetric productivity of recombinant protein. Process engineering parameters during the expression of ovine growth hormone as inclusion body in *E. coli* were analyzed. Ovine growth hormone yield was improved from 60 mg L^{-1} to 3.2 g L^{-1} using fed-batch culture. Similar high volumetric yields were also achieved for human growth hormone and for recombinant bonnet monkey zona pellucida glycoprotein expressed as inclusion bodies in *E. coli*. The second section deals with purification and refolding of recombinant proteins from the inclusion bodies of *E. coli*. The nature of inclusion body protein, its characterization and isolation from *E. coli* has been discussed in detail. Different solubilization and refolding methods, which have been used to recover bioactive protein from inclusion bodies of *E. coli* have also been discussed. A novel inclusion body protein solubilization method, while retaining the existing native-like secondary structure of the protein and its subsequent refolding in to bioactive form, has been discussed. This inclusion body solubilization and refolding method has been applied to recover bioactive recombinant ovine growth hormone, recombinant human growth hormone and bonnet monkey zona pellucida glycoprotein from the inclusion bodies of *E. coli*.

Keywords. Fed-batch fermentation, Volumetric productivity, Recombinant protein, Inclusion body, Refolding, Purification

Symbols and Abbreviations

ATP	adenosine triphosphate
ATR-FTIR	attenuated total reflectance Fourier transform infrared
bGH	bovine growth hormone
BFGF	basic fibroblast growth factor
bmZPC	bonnet monkey zona pellucida glycoprotein C
βgal	beta-galactosidase
CD	circular dichroism
CDNA	complementary deoxyribonucleic acid
CER	carbon dioxide evolution rate
C/N	carbon to nitrogen ratio
DEAE	diethylaminoethyl
DO	dissolved oxygen
DTT	dithiotheritol
DMSO	dimethyl sulfoxide
EDTA	ethylenediaminetetraacetic acid
GSH	glutathione reduced
GSSH	glutathione oxidized
hGH	human growth hormone
IFN-γ	interferon-γ
IGF-1	insulin like growth factor-1
IL-1β	interleukin-1β
IL-2	interleukin-2
IPTG	isopropyl thio β-D galactopyranoside
kDa	kilodalton
LB	luria bertani
M	mass
$NADH_2$	nicotinamide adenine dinucleotide dihydrogen
Ni-NTA	nickel-nitrilotriacetic acid
OD	optical density at 600 nm
oGH	ovine growth hormone
PAGE	polyacrylamide gel electrophoresis
PDI	protein disulfide isomerase
PEG	polyethylene glycol
PHB	polyhydroxybutyrate
PMSF	phenylmethylsulfonyl fluoride
PO_2	oxygen partial pressure
PPI	prolyl-peptidyl isomerase
r-bmZPC	recombinant bonnet monkey zona pellucida C
rhaBAD	rahmnose BAD
r-hGH	recombinant human growth hormone

RIA	radioimmunoassay
RNA	ribonucleic acid
mRNA	messenger ribonucleic acid
rRNA	ribosomal ribonucleic acid
r-oGH	recombinant ovine growth hormone
SDS	sodium dodecyl sulfate
S-200 HR	Sephacryl 200 high resolution
TCA	tricarboxylic acid
tRNA	transfer ribonucleic acid
V	volume
ZP	zona pellucida
ZPA	zona pellucida A
ZPB	zona pellucida B
ZPC	zona pellucida C
μ	specific growth rate

1 Introduction

The ultimate goal of recombinant fermentation research is the cost effective production of desired protein by maximizing the volumetric productivity, i.e., to obtain the highest amount of protein in a given volume in the least amount of time. Such bioprocessing for recombinant protein using genetically modified organisms requires a stable high-yielding recombinant culture, a high productive fermentation process and cost effective recovery and purification procedures. *Escherichia coli* species have been most widely used as host for the expression of recombinant proteins [1, 2]. Advantages of using *E. coli* as expression system is the enormous amount of data available on its cell biology, fermentation process development and its ability to produce large quantities of recombinant proteins in an inexpensive way. The successful large-scale cost-effective production of insulin by Eli Lilly (USA) and bovine growth hormone by Monsanto Corporation (USA) attest to the versatility and economic potential of *E. coli*-based therapeutic protein production. Although *E. coli* cannot be used to produce complex glycoproteins or proteins having multiple disulfide bonds, in past 20 years recombinant DNA technologies have enabled us to produces huge quantities of therapeutic proteins that might otherwise have been difficult [3, 4].

Recombinant protein expression using *E. coli* as host is frequently associated with the formation of intracellular aggregates as an inclusion body [5]. The volumetric yield of the protein is thus is a function of both unit cell concentration and specific cellular protein yield. Optimization of high cell density fed-batch fermentation processes is thus one of the key steps for enhancing the volumetric yield of recombinant proteins [6, 7]. High level expression of protein in the form of an inclusion body facilitates the isolation of the protein of interest from the cytoplasm at the cost of its native structure. Renaturation of recombinant proteins from inclusion bodies into the bioactive form is cumbersome, results in low recovery of the final product and also accounts for the major cost in overall production of recombinant proteins [5, 8]. However, in the cases where a simple high-yielding protein refolding process is developed for the aggregated recombinant protein, high-level expression of protein as inclusion body provides a straightforward strategy for the cost-effective production of therapeutic protein. Thus, in spite of the problems associated with the inclusion body in *E. coli*, they have been extensively used for the commercial production of therapeutic protein. High cell density fermentation and improved refolding of the inclusion body proteins are thus the two major bioprocess engineering considerations for enhancing the overall yield of recombinant proteins from *E. coli*.

The objective of the present review is to emphasize the importance of high productive fermentation as well as high throughput purification of bioactive therapeutic protein from the inclusion bodies of *E. coli*. Understanding of basic biological aspect of the expression system at the molecular level and translating this information at process level is imperative for efficient and cost-effective production of therapeutic compounds. Parameters that influence the high cell density fed-batch aerobic growth of *E. coli* while maintaining a stable plasmid of interest have been analyzed. Novel ways of fed-batch fermentation process considering most of these factors have been discussed in detail to maximize the volumetric yield of recombinant ovine growth hormone expressed as inclusion body in *E. coli*. Solubilization and refolding of inclusion body protein to the bioactive conformation severely limits the overall efficiency of the therapeutic protein production from *E. coli*. Solubilization of the inclusion body protein without disturbing the existing native-like secondary structure while using a low concentration of a chaotropic salt, its refolding and purification into bioactive forms have been described. Finally the novelty of high cell density fermentation processes and improved refolding of inclusion body proteins have been applied to a few other proteins expressed in *E. coli* and process development strategies have been discussed. Apart from reviewing the recent trends in bioprocessing of recombinant protein from *E. coli*, the review discusses the fermentation and inclusion body protein refolding process developed at the National Institute of Immunology, New Delhi.

2 Parameters Influencing Productivity of Therapeutic Protein from *E. coli*

Numerous genetic and environmental factors influence the expression of cloned gene product in recombinant *E. coli* which is most frequently used prokaryotic expression system for the production of heterologous proteins [9]. At the molecular level, strength of transcriptional promoters, plasmid stability, copy number, mRNA stability, translational efficiency, localization, status and the stability of the expressed foreign protein in the host influences the expression levels. These factors influence the metabolic state of the host during gene expression, which in turn, can be controlled and manipulated during fermentation to maximize the yield of the expressed protein [10]. The expressed protein, depending on its localization, can be purified and recovered in the bioactive form. It is interesting to note that even though *E. coli* does not provide an oxidizing environment for disulfide bond formation leading to the aggregation during expression, such aggregated proteins have been successfully refolded in vitro to achieve the bioactive conformation. The successful production of insulin [11], bovine growth hormone [12] and tissue plasminogen activator [13] from inclusion bodies of *E. coli* indicates that complex proteins can be expressed and purified using an *E. coli*-based expression system.

With recent advances in gene cloning, development of better cell growth process and improvement in refolding yield of the inclusion body protein, the efficiency of protein production using *E. coli* has increased rapidly. It has been realized that most of the factors which influence the efficiency of the overall protein production process act in a very complex and interactive way at different stages. Putting the gene in front of the promoter is no longer considered as the end of a successful recombinant expression system. In fact, it is realized of late that the beginning of the problem for successful production of recombinant protein starts only after a stable recombinant construct is ready for expression at the shaker flask culture level. Innumerable overlapping factors need to be taken into consideration for optimization of a recombinant fermentation process. These factors can be divided primarily into two broad groups: the first cell and molecular biology considerations and the second process engineering considerations. Cell and molecular biology considerations deal mainly with the level of expression, destination, location and state of the protein produced using *E. coli* as expression system. Factors that influence these things in an expression system are host, vector/promoter system and the origin and the nature of the protein of interest. By contrast, process engineering considerations deal with the large-scale culture of the recombinant organism and the recovery of the expressed protein. It aims at high volumetric yield and high throughput recovery of the expressed protein in bioactive form. The interactions of these broad groups of factors also need careful analysis to optimize the level of protein production using recombinant *E. coli.*

2.1 Cell and Molecular Biology Considerations

Most of the cell biology parameters that influence the productivity of the recombinant protein from *E. coli* include the host organism: expression vector and promoter system, gene dosage or plasmid copy number, stability of plasmid, promoter strength, induction strategy, mRNA stability, ribosomal population, tRNA concentration, codon bias, amino acid concentration and finally the localization of the expressed protein [2, 9]. Extensive reviews highlighting the influence of above factors have been published from time to time [9, 14, 15]. What is more important is to consider the above cellular factors and see their implications during the large-scale processing of recombinant proteins, particularly during high cell density fermentation and high throughput purification of the protein. In spite of extensive knowledge of the genetics and molecular biology of *E. coli,* it is sometimes difficult to express a gene efficiently in this organism. This may be due to unique structural features of the gene sequence, initiation of translation, stability and translational efficiency of mRNA, major differences in codon usage between the foreign gene and native *E. coli*, toxicity of the protein and degradation of expressed protein by host cell proteases. Varieties of promoters are now available for the efficient expression of a therapeutic protein in *E. coli* [16, 17]. A chemical inducer like IPTG or heat induction provides the best available promoter for gene expression having high strength and regulatory capability and have been mostly used for the production of therapeutic proteins. The recently developed arabinose promoter is probably the closest ideal promoter in terms of strength, regulation and controlled expression [18, 19]. The rhamnose inducible promoter (rhaBAD) has also been used for expression of L-N carbamoylase using *E. coli* as the host [20]. The use of a dual promoter has also been reported to have helped in high level expression of recombinant protein in *E. coli* [21]. The most important cellular parameter for controlling and optimizing gene expression is the translation process [22, 23]. Stability of mRNA, secondary structure of the mRNA [24, 25] and ribosomal population affect the overall yield of the recombinant protein [9, 26]. It has been widely documented that the translation capacity of the cell remains the most crucial factor for the efficient expression of a recombinant protein [9]. Efficient design of the expression vector considering the end effects of translational initiator, enhancers, terminator and m-RNA stabilizer need careful assessment for high-level expression of the foreign protein using *E. coli* as host.

The major draw back of the *E. coli* as expression system is the inability to do many post-translational modifications found in eukaryotic protein, limited ability to facilitate disulfide bond formation due to the reducing nature of cytoplasm and lack of secretion mechanism for the efficient release of the expressed protein into the culture medium. The decision to target the expressed protein into the cytoplasm, periplasm or the culture medium depends on balancing the advantages and disadvantages offered [15]. Real secretion of protein in to the extracellular medium is rare and periplasmic expression most of the time results in low level of recombinant protein expression. Exceptions are the secretion of leptin [27] and the high-level accumulation of IGF using a dual promoter system [21]. High-level expression of protein into cytoplasm leads to accumulation of denatured protein in the form of inclusion bodies [28, 29]. However, with recent understanding of the structure function of the inclusion body protein, recoveries of bioactive protein with reasonable yield have been achieved for many proteins [30]. The high initial level of expression compensates loss during recovery of protein from inclusion bodies. In fact, most of the commercially available proteins from *E. coli* are expressed as inclusion bodies and then suitably refolded into the bioactive form. Thus for high level expression of recombinant protein which does not require post-translational modification for bioactivity, expression in the form of inclusion body and its subsequent purification and refolding becomes the most cost effective way of therapeutic protein production.

Another molecular biology consideration, which needs attention, is the improvement of host cell metabolism for improved expression. This includes co-expression of chaperone [9, 14], use of tag or fusion protein for expression in soluble form or efficient purification [15, 31]. Using a dual promoter system and by delineating the cell growth phase from that of expression phase, very high levels of recombinant IGF1 have been successfully expressed in *E. coli* [21].

Metabolic engineering of host cells for low acetate secretion [32, 33] and incorporation of gene for improving oxygenation of the large-scale culture [34] have also been used for maximizing the level of foreign gene expression in *E. coli*.

During large-scale processing of the recombinant *E. coli* the above cellular parameters influence the metabolic status of the cells which, in turn, necessitates special attention both during cell growth and protein recovery to optimize the overall protein production process. More important parameters like plasmid stability, inducer concentration and time of induction, kinetics of growth product relations, harvest time and efficient protein recovery processes need special attention to scale-up the shaker flask result successfully to high cell density fed-batch fermentation. Interrelations and effects of many such factors influence the overall yield of the finished products and thus need to be analyzed in the context of the overall process performance rather than the single unit operation stages.

2.2 Process Engineering Considerations

In most of the cases, *E. coli* expressed proteins are associated with intracellular accumulation as inclusion bodies. The volumetric yield of a recombinant protein will depend on both the biomass concentration as well as specific cellular protein yield [6, 7]. Thus, after successful construction of the recombinant culture for optimal expression of the protein, high productive fermentation and efficient protein recovery process development are the two major constraints for efficient cost effective production of therapeutic protein. In recent years, tremendous progress has been achieved both in terms of development of high productive fermentation and high throughput protein purification from inclusion bodies. These two factors are discussed in detail in this review with an aim to increase the volumetric production of therapeutic protein using *E. coli*.

High volumetric yield of the protein can be achieved by increasing the cell concentration in the reactor volume. High cell density fermentations using fed-batch culture techniques are routinely used to maximize the yield of recombinant proteins in *E. coli* [35, 36]. Most of the high cell density fermentation processes have been developed using recombinant *E. coli*. The highest *E. coli* cell concentration around 200 g dry cell weight per liter of fermentation broth has been achieved for polyhydroxybutyrate expression [37]. Recombinant *E. coli* expressing polyhydroxybutyrate in high cell density fermentation produces as high as 2.8 g L^{-1} d^{-1} of product, emphasizing the enormous capacity of the *E. coli* system to produce heterologous protein. It is expected that with proper bioprocessing the yield of product can be matched to the chemical synthesis process of polymer production [38]. Assuming a 20% expression level of any recombinant therapeutic protein at such a high cell concentration, it is expected in the near future that an expression level of 10–20 g L^{-1} of the recombinant protein can be easily produced through high cell density fed-batch fermentation of recombinant *E. coli*.

Efficient recovery of bioactive protein from the inclusion bodies is the major constraint for the successful production of therapeutic protein from *E. coli*. In general, the yield of bioactive recombinant protein from inclusion bodies of *E. coli* is around 10–20% of the total expressed protein. The solubilization of protein from the inclusion body by high concentrations of chaotropic reagents results in the loss of its secondary structure, leading to the random coil formation of the protein structure [39, 40]. Loss of secondary structure during solubilization and the interaction among the denatured protein molecules leading to their aggregation are considered to be main reasons for the poor recovery of bioactive proteins from the inclusion bodies [41]. Refolding at low protein concentration and use of urea for solubilization have been reported to be the most costly factors in the production of recombinant insulin from *E. coli* [8].

Apart from the above-mentioned problems mainly encountered during the bioprocessing of therapeutic proteins, other factors that are inherent to the fermentation process need careful consideration. Recovery to fermentation cost for bioactive protein from *E. coli* is around 3 to 5, whereas for antibiotics like penicillin it is around 1, indicating the importance of downstream operation for cost effective production of therapeutic proteins [42]. Like many fermentation processes, therapeutic protein production involves lots of water for different processing steps. More importantly, protein folding and polishing need high quality water, which has to be treated specially before discharge into the environment. Huge water requirements during therapeutic protein production will adversely affect the process efficiency and power consumption and increase many-fold the overall cost of the product [8, 13]. Thus it is also essential to develop low water requiring, environmental friendly technology for the production of therapeutic proteins, particularly during refolding of inclusion body proteins.

3 High Cell Density Fermentation

Essentially, high cell density growth involves the culturing of a recombinant organism to a very high cell concentration (>20 g L^{-1} dry cell weight) by employing fed-batch fermentation methods [6, 36, 43, 44]. Operation of fed-batch fermentation helps in increasing the unit cell concentration in the reactor and thus improves the volumetric yield of the protein. As recombinant protein expression using *E. coli* results mostly in intracellular accumulation, the volumetric yield depends on both the final cell concentration as well as the specific cellular protein yield [6]. In high cell density fermentation, maximizing cell concentration helps in increasing the volumetric productivity of recombinant proteins. High cell density culture of *E. coli*, apart from improving the volumetric productivity, also provide advantages such as reduced culture volume, enhanced downstream processing, reduced waste water, lower production cost and reduced investment on equipment [7]. Apart from this, it is essential that cell growth be achieved in an optimal time period to improve the overall productivity of the recombinant protein. Toxicity of acetate, slow growth rate, instability of plasmid, depletion of amino acid pools to sustain a high rate of protein synthesis affect the specific cellular yield of recombinant protein at high cell concentrations. It is expected that an analysis of all these parameters during high cell density fed-batch growth of *E. coli* will lead to high volumetric productivity of the desired protein. Such approaches have been used extensively to increase the volumetric yield of recombinant products in prokaryotic systems, particularly those employing *Escherichia coli* as host [36].

3.1
Parameters Affecting High Cell Growth of *E. coli*

Innumerable overlapping factors must be taken into account while optimizing high cell density *E. coli* growth for protein expression [35, 36]. Composition of medium, physical parameters during growth and operating conditions are the most important factors that influence the cell growth. Limitation and/or inhibition of substrates, limited capacity of oxygen supply, formation of metabolic byproducts and instability of plasmid during long hours of cultivation are the major problems encountered during high cell density growth of *E. coli*. Most of the time these depend on host strain, vector and strength of promoter. Dense culture requires large amounts of O_2 to support good growth and thus necessitates an unconventional aeration strategy to maintain the dissolved oxygen concentration at a suitable level throughout the growth period. In most cases with *E. coli* used as a host for recombinant protein, the production phases start after induction with a suitable inducer. Thus, in principle, the growth phase and the production phase can be delineated in the same vessel for a high volumetric yield of the recombinant protein. However, in many cases, the operation of the reactor during cell growth influences the specific yield of the recombinant proteins. Thus, while developing fed-batch operation to increase unit cell growth in the reactor, it is equally essential to take care of the factors which affect the specific yield of the recombinant protein.

3.1.1
Nutrient Formulation

One of the essential requirements during fed-batch operation is to supply nutrients to promote cell growth [45, 46]. To limit their toxicity to the growing cells nutrients such as glucose, ammonia, salt are fed in approximation of their requirement. The accumulation of nutrients at high concentration inhibits growth and recombinant protein expression [6]. High glucose causes the Crabtree effect, leading to the accumulation of acetate which is inhibitory to cell growth [47]. Ammonia inhibits gene expression [48] and the absence of a metal salt may hamper the enzymatic activities of many enzymes vital for cell metabolism [45]. In general, most of the media used for high cell growth of *E. coli* have mostly glucose as carbon source, with major salts like phosphate, sodium potassium, magnesium, ammonia, sulfate, iron, minor trace elements and complex nitrogenous materials. High-density growth in general is initiated with a low concentration of the most required substrate and the nutrients are added later in the growth period [49]. Typically, glucose >50 g L^{-1}, ammonia>3 g L^{-1}, P>10 g L^{-1}, Mg>8. g L^{-1}, Fe>1.2 g L^{-1}, Mo>0.8 g L^{-1} B>44 mg L^{-1}, Cu>4 mg L^{-1}, Mn>68 mg L^{-1}, Co>0.5 mg L^{-1}, Zn >38 mg L^{-1} inhibit *E. coli* growth [35]. Ideally,

Table 1. Examples of high cell density growth of *E. coli*

Host strain (*E. coli*)	Culture method/feeding	Cell density ($g\ L^{-1}$)	Year	Reference
B	DO stat	125	1979	[49]
HB101	DO control with cell recycle	145	1990	[51]
TG1	Specific growth rate controlled	110	1991	[52]
KA197	Exponential feeding	77	1991	[43]
X90	Exponential feeding	95	1993	[44]
W3110	Exponential feeding	45	1994	[53]
TG1	Exponential feeding	148	1995	[54]
K12	Dialysis culture	190	1997	[55]
BL21	Exponential feeding	80	1997	[56]
N48301	Specific growth rate	50	1998	[57]
XLi-Blue	pH-stat	201	1999	[37]
CGSC4401	pH stat	119	2000	[58]
K12	Dialysis culture	190	2002	[59]

the components should be added to the fermenter at the same rate at which they are consumed so as to prevent nutrient accumulation up to toxic levels while still promoting good growth. Another factor which needs attention during medium formulation is the solubility of many components, particularly while making concentrated solutions for fed-batch addition. High concentrations of glucose, yeast extract and trace elements need careful composition to avoid precipitation. The osmolarity of the medium can affect nutrient yield, cell growth and specific recombinant protein yield [6, 50]. Hence it is also essential to take care of medium's osmolarity so that its detrimental effect is minimum during high cell density growth. Table 1 summarizes examples of high cell concentrations of *E. coli* at a range of 50 to 100 $g\ L^{-1}$ achieved using fed-batch fermentation.

Apart from medium formulation, the operating conditions such as pH, temperature and, more importantly, O_2 supply are very very essential for supporting high cell growth. The solubility of oxygen in the medium is very low and, with increases in cell concentration during fed-batch growth, the solubility is reduced. At very high cell concentrations use of air is not sufficient for the respiratory demand of the rapidly growing *E. coli* cells. Increasing aeration rate, feeding O_2-rich air, decreasing temperature, increasing partial pressure of the culture vessel are some of the methods employed to maintain aerobic conditions during cell growth. It has been widely documented that oxygen not only influences the cell growth but also has an affect on gene expression by influencing the oxidative status of many enzymes [60]. Hence, its essential that, with proper feeding of nutrients, the oxygen supply should be at an optimal level to support good growth and provide an oxidizing environment for quality protein synthesis [61].

3.1.2
Acetate Inhibition

Escherichia coli excretes 10–30% of the carbon flux from glucose to acetate during its aerobic growth on glucose [33]. Glucose-mediated aerobic acidogenesis known as the "Crabtree effect" which is most readily observed when *E. coli* cells are grown at a high growth rate [47]. To meet the energy requirements at high growth rate, *E. coli* uses the pathway of acetate production from acetyl CoA since it generates the second largest amount of ATP and $NADH_2$ [62]. Limitation in anaplerotic fluxes during aerobic growth of *E. coli* on glucose has also been suggested to be partly responsible for the secretion of acetic acid [36]. Acetic acid secretion during aerobic growth of *E. coli* is a major constraint of the high productivity recombinant fermentation process [50, 63]. Acetic acid, being lipophilic, is harmful to cell growth and acts as a typical uncoupler [64]. It directly dissipates the ΔpH contribution to proton motive force and thus inhibits the bioenergetic work. Consequently, acetate decreases both biomass and expected cell densities [65]. The detrimental effect of acetate is further augmented by salt that accumulates in the reactor as a result of the acid or base used for pH control. Inhibition of acetate is more significant for cells cultured in a defined medium than in a complex medium [66]. Inhibitory effects of acetic acid on the growth of *E. coli* cells and affecting the volumetric productivity of recombinant proteins during high cell density fermentation have been widely documented [6, 36, 50, 55].

Acetate formation during the growth *E. coli* is a function of oxygen supply, glucose concentration and anabolic requirements of the cells. It also depends on the host strain, the medium used and the specific growth rate of the culture. Acetate formation at different specific growth rates for complex and synthetic media have been reported with out any clear-cut information. However the critical specific growth rate that leads to acetate formation varies among strains, medium and cultivation conditions used. Reducing the anabolic requirements while supplying nitrogenous nutrients has been reported to reduce the formation of acetate [65, 67].

A number of strategies have been adopted to reduce the acetate accumulation during high cell density fed-batch cultivation of *E. coil*. High cell density *E. coli* fermentation runs for extended periods of time, thus the level of acetate formation is always higher than the batch culture. Most of the strategies have been based on reducing the acetate accumulation by decreasing the anabolic requirements in terms of reducing glucose uptake or increasing the oxidative metabolic capacity [65, 67]. Controlling the specific growth rate by manipulation of nutrients or nitrogen source has also been attempted to reduce acetate formation during high cell density growth of *E. coli* [68]. Alternatively, using a dialysis reactor can reduce the inhibitory effect of acetate, which facilitates the removal of acetate from the culture broth [55].

Genetic engineering techniques to reduce the secretion of acetic acid [69], use of fructose as an alternate carbon source [70] and selection of low acetate producing strains [71] have been successfully tried to lower the secretion of acetic acid during aerobic growth of *E. coli*. Activation of glyoxalate shunt pathways has also been proposed to reduce the accumulation of acetic acid during the growth of *E. coli* on glucose [72]. The accumulation of acetate during *E. coli* fermentation is net result of its formation and uptake by the cells. It is thus essential to control the formation of acetate and enhance its uptake to maintain a low level of acetate during fed-batch growth of *E. coli*. As acetate accumulation affects the substrate utilization, inhibits cell growth and recombinant protein production, it needs special attention during the development of fed-batch fermentation processes.

3.1.3
Nutrient Feeding Strategy for High Cell Growth

One of the essential requirements to obtained good cell growth during cultivation is to supply nutrients in a manner that is desirable. Ideally, by providing proper nutrient and operating conditions, exponential growth of *E. coli* can be maintained so that a high cell concentration is achieved in less time. However, as the cell concentration increases beyond 10 g L^{-1}, maintenance of exponential growth becomes very difficult due to various operational reasons. Oxygen supply, saturation of the oxidative capacity of cells at high glucose concentration, build-up of acetate to toxic levels, plasmid instability and low productivity associated with cell growth at a high specific growth rate have led to the development of different feeding strategies to achieve high cell growth at a reasonable time [6, 36].

The simplest way to achieve high cell growth is to grow the cells at a very low growth rate to take care of acetate accumulation, however, this ends up in extending the batch time and thus affecting the overall volumetric productivity. High cell density growth is normally carried out under nutrient (carbon) limiting conditions. Two principal methods have been used to achieve high growth rate during the fed-batch culture of *E. coli* cells. These are presented in Table 2 and have their own pros and cons.

Table 2. Feeding methods used for high cell density culture of *E. coli*

A. **Without feedback control**
1. **Constant feeding:** Feeding nutrient at a predetermined (constant) rate. The specific growth rate continuously decreases, longer batch duration for high cell density growth
2. **Increased feeding:** Feeding nutrient at an increasing (gradual, stepwise, or linear) rate. The decrease in specific growth rate can be compensated. High cell growth achieved but may result in toxic metabolite build-up
3. **Exponential feeding:** Feeding nutrient at an exponential rate. Constant specific growth rate can be achieved. Helps in high cell growth with low level of toxic metabolite accumulation at less time

B. **With feedback control**
1. **DO-stat:** Feeding nutrient when there is a rise in the concentration of dissolved oxygen (DO), which results from depletion of the substrate
2. **pH-stat:** Feeding nutrient when the pH rises as a result of depletion of the principal carbon source
3. **Carbon dioxide evolution rate (CER):** This is estimated on-line using a mass spectrometer, and is used to control nutrient feeding. The CER is roughly proportional to the rate of consumption of the carbon source. This method is most frequently used to control the specific growth rate
4. **Cell concentration:** The nutrient feeding rate is determined from the cell concentration, which is measured on-line using a laser turbidimeter

One is using predetermined feeding strategy and the other is using feedback control strategies [36]. Feeding strategies such as constant/linear feeding [4, 73], step-feeding [74, 75], exponential feeding [37, 56, 76] have been used to achieve high cell growth in fed-batch culture. Feedback control strategies such as pH stat [37, 58], DO stat [32, 49], acetate [77] and culture fluorescence monitoring using green fluorescence protein [78] have also been used for achieving high cell growth of *E. coli* cells.

3.1.4
Maximum Achievable Cell Concentration

For achieving a high cell concentration of *E. coli*, it is of interest to calculate what maximum cell concentration can be achieved in a reactor. The theoretical maximum cell number per liter can be calculated assuming that the cells are tightly packed inside the reactor and there is no space in the reactor. Using the concept of average cell mass ($M=10^9$ pg/cell) and mean cell volume ($V=0.4\times2^{\mu}$) and taking an average growth of cells with a doubling time of one hour, the maximum cell concentration achievable can be around 400 g/L [35]. Markl et al. in 1993 [79] calculated that with culture cells having 3 μm length and 1 μm diameter, around 75% of the reactor volume could be filled with cells. Considering that the dry weight of the cells is 20–25% of the wet weight, the maximum cell concentration achieved is around 160–200 g/L [79]. A maximum *E. coli* cell concentration of ~200 g L^{-1} appears reasonable. In a stirred tank reactor, the maximum cell density is lower than the predicted because the measurement includes nutrient solutions along with cells. More importantly, it is impossible to have cells totally packed in the reactor volume because mass transfer limitations will be very high to providing nutrients uniformly at such a very high cell concentration. Maximum *E. coli* cell concentrations of 190 g L^{-1} [59] and 175 to 203 g L^{-1} [37] reported to date approach that estimated by theoretical calculations.

3.1.5
Plasmid Stability

One of the most important factors that affects the expression of recombinant protein in from *E. coli* is the maintenance of the plasmid within the host cells [80]. High cell density growth of *E. coli* in fed-batch mode needs a longer time and a higher generation number for cultivation in comparison to normal shaker flask culture. Because of this, plasmid instability problems are more serious in large-scale culture [88]. There are primarily three different ways for plasmid instabilities to occur during the growth of recombinant *E. coli* cells. Segregational instability occurs due to random or defective partitioning of the plasmid during cell division whereas structural instabilities occurs due to undesirable plasmid modification resulting from insertion, deletion or rearrangement of DNA. Host cell regulatory mutation also leads to plasmid instabilities resulting in a lowering in the productivity of the expressed protein [10, 81]. All these factors lead to higher growth rate advantages of the plasmid-free cells compared to plasmid-bearing cells, resulting in the accumulation of unproductive plasmid-less cells in the reactor. It is thus essential to have a host/vector system as well as cultivation conditions that promote better plasmid stability during high cell density cultivation. Even though in many cases antibiotic pressure is used to overcome plasmid instability, the stability of the antibiotic solution during a long cultivation period should be taken into consideration. Most of the antibiotic has very a short half-life and it is thus essential to supply it throughout the cultivation period so the high plasmid stability is maintained during the fed-batch cultivation period [44, 73]. Apart from this, novel vectors such as runaway replication vector, vectors having self-destroying properties of killing plasmid-less cells such as using hok/sok sequences can be used to overcome the plasmid instability problems [10]. Use of nutrient addition and bioreactor configurations have also been used to improve plasmid stability during high cell density fed-batch fermentation [2, 10, 81].

3.2
Induction Strategy for High Level Protein Expression

In high cell density fermentation, maximizing the cell concentration helps in increasing the volumetric productivity of recombinant proteins usually at the cost of a lower specific cellular protein yield [7]. Lower specific cellular protein yields during high cell density fed-batch fermentation in comparison to batch fermentation have been reported for recombinant porcine growth hormone [82], recombinant β-gal fusion protein [43] and recombinant insecticidal protein expressed in *E. coli* [83]. This is primarily because most of the investigations related to high cell density fermentation have been focused on increasing the unit cell concentration with very little attention to the specific cellular yield of the recombinant proteins. Thus it is essential to maintain the specific cellular yield during high cell density growth.

Even though in principle high cell growth leads to high protein expression per unit reactor volume, the induction strategy needs optimization to maintain the specific cellular protein yield during high cell density fed-batch fermentation. The level of gene expression, localization of the expressed protein, its toxic effect either due to the gene product or due to a high level accumulation of the foreign protein and the product degradation characteristics decide the induction strategy. For chemical inducers, the concentration of inducer is also important to completely de-repress the promoter in order to achieve maximum protein expression during high cell density fermentation [56, 83]. Final synthesis of the protein takes place from the amino acids, thus its also important to see that the amino acid pools of the host cells meet the demand during the high-level synthesis of the foreign protein. For positive feedback inducers like tryptophan, arabinose and rhamnose, it is also essential to add the inducer at the proper time for the maximum synthesis of the protein. With temperature-inducible promoters for the expression of recombinant protein, its also important to evaluate the overproduction of heat shock protein associated with gene expression. All these parameters which affect the induction capacity are discussed in detail to optimize the volumetric yield of the expressed protein.

3.2.1
Effect of Inducer Concentration and Time of Induction

To maximize the volumetric yield it is always desirable to achieve high growth and then to induce with the suitable inducers for maximizing the volumetric yield. In most of the high cell density fermentations of recombinant organisms involving inducible promoters, optimal productivity is achieved when the growth and production phases are separated [6]. In fed-batch culture separation of the two phases can be achieved by delaying induction of the culture until the culture has completed its growth and attained the required high densities. Alternatively, if the plasmid is stable and the inducer is non-toxic to the culture, a repeated fed-batch system may be used to increase the productivity. If the inducer or the product is highly toxic to the cell, the induction phase needs to be separated physically from growth. Considering these aspects, it is essential to decide the induction time so that both cell growth and specific yield are maximized which will result in a high volumetric yield of the protein. It has been widely documented that gene expression places a metabolic burden on growing cells [84, 85]. Depending on the host strain and the level of expression, the metabolic burden varies and most of the time it results in complete cessation of cell growth. Thus, unless the time of induction is optimized, the batch productivity may be compromised due to low cell growth. In most of the gene expression cases using a strong promoter, it has been observed that gene expression kinetics plateaued at 3–4 hours post-induction. Thus depending on the cell growth achieved and the metabolic burden it exerts on cell growth, the induction time should be optimized. Apart from the time of induction, depending on the promoter used, it is also essential that the inducer concentration be optimized to de-repress the gene completely. The result of the shake flasks should optimally be scaled to provide the required concentration of the inducer at high cell concentrations. This is more important for chemical inducers whose concentration on a molar basis should be used at high cell concentrations to fully de-repress the promoter [56].

3.2.2
Maintenance of Specific Cellular Protein Yield

Two major constraints of high productive recombinant protein expression during aerobic growth of *E. coli* are the secretion of acetic acid and lowering of the specific cellular protein yield. To avoid secretion of acetic acid, cells are generally grown at lower specific growth rates to achieve high cell concentrations. Operation of fed-batch fermentation at lower specific growth rates of *E. coli* extends the duration of the batch time and hence affects the volumetric productivity of recombinant protein. More importantly, growth at lower specific growth rate also affects the specific cellular protein yield due to the lower biosynthetic capacity of the *E. coil* cells [84]. This is because cells growing at lower specific growth rate are already under nutrient limitation and correspondingly when induced for protein synthesis lack sufficient cellular resources for foreign protein synthesis, resulting in a lowering of the specific cellular protein yield. Induction of foreign protein synthesis exerts a metabolic burden on the cell physiology, which is associated with a reduction in the specific growth rate of the growing cells [85]. If nutrient feeding during protein expression phase is not adjusted according to the reduction of the specific growth rate of the cultures, accumulation of toxic by-products, particularly acetic acid, and build-up of high concentrations of nutrients may affect cell the physiology, resulting in a lowering of the specific cellular protein yield [86]. Furthermore, the specific cellular protein yield also depends on the nutrient concentration, which favors not only growth but also provides essential nitrogenous sources, i.e., amino acids for protein synthesis or precursors for macromolecular synthesis. This is particularly true during the induction of gene expression at a very high cell concentration, where the nutrient demand increases suddenly for the synthesis of plasmid-derived proteins, resulting in a drainage of the amino acid pool of the cell [87]. Thus, if high cell density fermentation is carried out with an optimal complex nitrogenous source, it will help in reducing the anabolic requirements of the cell, will control the acetic acid formation, and will also provide precursors to meet the demand for high-level synthesis of the expressed protein at a high cell concentration. This will help in maintaining the specific cellular yield of recombinant proteins during high cell density fermentation. Such a high level supply of optimized nitrogenous nutrients along with glucose has helped in achieving a high volumetric yield of pro-mini insulin [56].

3.2.3
Metabolic Burden due to Gene Expression

Large-scale production of recombinant protein from *E. coli* most of the time uses a strong promoter for the high level synthesis of recombinant protein. Thus, after induction, plasmid-directed synthesis of the recombinant protein exerts a metabolic burden on the host cells [88]. This is manifested on the host cell growth as well as on its metabolism, depending on the extent of gene expression [89]. If the expressed product is toxic, it may lead to cell death. Otherwise in most of the cases, recombinant gene expression is always associated with a reduction in the specific growth rate of the culture. The reduction in cell growth due to gene expression may be either due to degradation of rRNA or ribosomal population, which affects the host cell growth. Depending on the expression level and its toxicity, the growth rate reaches a plateau in a programmed manner over a period of time [90]. The metabolic burden due to gene expression depends on the level of expression, the higher the amount of gene expressed the more severe is the reduction in the specific growth rate of the cells during the post-induction period [85]. This needs careful attention as it influences the harvest time, thus helping in optimization of batch duration. Extension of batch time after the completion of gene expression may result in cell death and protein degradation which will complicate the downstream operation. More importantly, the reduction in growth rate of the induced cells necessitates that the feed be supplied as it is required during the post-induction phase. It has been reported that post-induction cell growth is independent of the nutrient feeding rate [91]. Continuation of similar nutrient feeding after induction will lead to the accumulation of nutrients, which may be toxic to the growth, and the physiology of the cells. Thus, the development of a feeding strategy in tune with the reduced growth rate of the cells during the post-induction phase needs a similar optimization to achieve optimal results [90]. Hence, its essential to design the feeding strategy according to the reduction in the growth rate of the cells, particularly during the post-induction stage in order to maximize the substrate utilization, optimize the harvest time and reduce the degradation of the product.

3.2.4
Effect of Oxygen on Gene Expression

In recent times the secondary role of oxygen on the maintenance of cell physiology and quality of the recombinant protein has been a major concern for the high volumetric yield of recombinant protein from *E. coli* [61]. Fluctuations in oxygen contents during high productive fermentation processes can cause oxidative stress within the cells leading to limitations in amino acid production, plasmid instability and, more importantly, oxidation of proteins. These effects all together may affect the quality of the final product. Hence, it is essential to evaluate the secondary effect of oxygenation during the high productive fermentation of recombinant protein from *E. coli*. This is particularly significant at high cell density fermentation where mixing limitations and low solubility of oxygen affect its primary use as nutrient and its secondary effect on metabolism. Anaerobesis, build up of acetate, oxidation of protein, oxidation of DNA, and plasmid replication are affected by fluctuations of oxygen in the broth and affect the overall protein yield in terms of both quantity and quality [92]. Amino acid synthesis is influenced by the aerobic conditions during growth due to the involvement of both glycolytic and TCA cycle enzymes [60]. There are reports that reduced oxygen tension has a detrimental effect on plasmid replication and stability [93]. *E. coli* has its own method to deal with oxidative stress conditions [94]. This can be achieved either by clearance, repair and degradation of oxidative species. There are at least five regulons that serve to deal with oxidative stress conditions [61]. Protein oxidation which causes damage to the quality of protein can be due to metal catalyzed oxidation, improper disulfide formation, methionine sulfoxide formation, oxidation of iron-sulfur centers and cross linking of proteins to sugars or fatty acids [95]. The detrimental effect of oxygen on protein expression is more profound at high cell concentrations and at higher scales of operation. For human growth hormone, it has been recently reported that reduced oxygen concentration, apart from promoting build up of growth promoting acids like formic acid, results in high degradation of the expressed protein [96]. Significant efforts have been made to reduce the effect of oxygen tension by expressing hemoglobin in *E. coli* [34]. Improvement in batch productivity has been achieved by regulating the oxygen tension using hemoglobin expression for the production of recombinant amylase. Hence, while expressing huge amount of recombinant protein, it is essential to evaluate the detrimental effect of oxygen tension, particularly on the quality of the recombinant protein as it will have not only adverse effects on the final yield of protein but may also lead to the accumulation of truncated or aggregated species which will be highly immunogenic and thus will be undesirable.

3.2.5
Amino Acid Mis-Incorporation During Protein Expression

Mis-sense amino acid substitution and processivity errors occur mostly during translation of heterologous gene and result in the production of a therapeutic protein having a different primary structure and may lead to the synthesis of a truncated or immunogenic version of the protein [97]. These amino acid substitutions are very difficult to detect due to their rare occurrence and impact on the overall protein structure, but may have undesirable side effects for human consumption [98]. Frame shifting, premature termination, hopping and hungry codon syndrome [98, 99, 100] account for most of the mistranslated protein accumulation during high level expression of the gene from a strong promoter. Mis-incorporation of amino acids has been widely reported during expression of recombinant protein in *E. coli*, the most prominent one being the incorporation of norleucine in the place of methionine [101]. Histidine in place of glutamine [102] and lysine for arginine [103] have also been detected during the high level synthesis of foreign protein in *E. coli*. Amino acid mis-incorporation which even the mRNA does not code for has been detected for mouse epidermal factor fusion protein where significant amounts of phenylalanine were detected in the expressed protein [104]. Cloned genes in *E. coli* often contain codons that are normally underutilized by their host cell. Otherwise the high level expression of a gene may place demands on host protein's synthetic apparatus that are not matched to its normal tRNA population. In both the cases amino acid mis-incorporation may lead to heterogeneity in the quality of the recombinant protein produced using *E. coli*. Use of proper codons favorable to the host [105], amino acid incorporation in the medium [106] and simultaneous expression of cloned gene for the tRNA that reads the codon [107] have been used for reducing the extent of amino acid mis-incorporation during high level expression of foreign genes.

With the analysis of most of the factors associated with high cell density growth and protein expression, the levels of protein yield from *E. coli* have increased tremendously. Table 3 provides a list of therapeutic proteins expressed at high concentrations using the high cell density fed-batch fermentation process. It is worth noting that yields of around a few g L^{-1} have been a common feature of *E. coli* base expression in recent days. With the capacity of *E. coli* for expressing as high as 80% of the total protein as foreign protein as in case of polyhydroxybutyrate [58], it is expected that combined with high cell density fermentation technology a protein yield of 10–20 g L^{-1} as inclusion bodies will be a regular feature in years to come.

3.3
Developments of High Cell Density Fermentation Process for Ovine Growth Hormone

Ovine growth hormone (oGH) is a single chain polypeptide consisting of 191 amino acid residues with two disulfide bridges formed by residues 53–164 and

Table 3. Recombinant proteins expressed in *E. coli* using high cell density fermentation process

Protein	Cell conc ($g L^{-1}$)	Ferment time (h)	Protein conc ($g L^{-1}$)	Reference
IL-1β	55	12	2.2	[73]
hGH	50	35	1.9	[50]
β-Gal fusion	77	22	19.2	[43]
bGH	40	26	2.9	[53]
BFGF	61	35	4.9	[76]
pro-Insulin	75	50	7	[56]
hGH fusion	90	50	9	[108]
IFN-γ	45	24	7.4	[57]
Insulin	100	45	4.5	[109]
Bioadhesive protein	48	10	5.3	[91]
IGF-1	50	40	8.5	[21]
oGH	48	16	3.2	[90]
PHB	203	50	158	[37]
Leptin	60	24	9.7	[110]
PHB	119	37	96	[58]

181–189 [111]. It is synthesized by the anterior pituitary and is required for normal growth and lactation in ovine species. Recently, studies have shown that administration of exogenous growth hormone enhances milk yield and normal growth in diary animals. Ovine growth hormone is also used for superovulation during embryo transfer technology in farm animals. Thus, to meet the demand of oGH for its large-scale application in diary animals, recombinant technology using *E. coli* as host was used for its large-scale production. Most of the bioprocessing parameters that affect the cell growth and protein expression were studied in detail. The aim was to develop a fed-batch fermentation process for high volumetric yield of recombinant oGH expressed as inclusion bodies in *E. coli*. The expressed protein in the form of inclusion bodies was purified and refolded into the bioactive conformation.

3.3.1
Expression of Ovine Growth Hormone

A c-DNA fragment coding for oGH was cloned in the pQE-30 expression vector (Qiagen, USA) containing a histidine tag at the N-terminal end [112]. Briefly, *E. coli* M15 transformed cells containing the recombinant expression plasmid (pQE 30-oGH) were grown in Luria Bertani (LB) or complex medium in the presence of kanamycin (25 $\mu g\ mL^{-1)}$ and ampicillin (50 $\mu g\ mL^{-1)}$. The grown cultures, when induced with 1 mM IPTG, expresses the recombinant oGH as inclusion bodies. Recombinant oGH was expressed as a 22 kDa fusion protein with a histidine tag and was purified using Ni-NTA chromatography [112]. In shaker flask culture, around 60 mg L^{-1} of oGH were expressed as inclusion bodies. The cell OD at 600 nm in shaker flask culture was around 2, thus the specific r-oGH yield was around 26 mg L^{-1} OD^{-1} of the culture. Recombinant *E. coli* expressing oGH was used for both development of high cell density fermentation and improvements in refolding of the expressed protein from the inclusion bodies.

3.3.2
Effect of Acetate on Cell Growth and r-oGH Expression

As acetate formation is a major constraint for the development of high cell density fermentation process, its inhibitory effect on cell growth and gene expression was evaluated before the development of a fed-batch fermentation process. To delineate the inhibitory role of acetate on cell growth and recombinant protein expression, *E. coli* cells expressing ovine growth hormone were grown in LB medium containing different concentrations of sodium acetate. It was observed that with increases in the concentration of acetate, the cell growth decreased (Table 4A). In the presence of 15 g L^{-1} sodium acetate (10.9 g L^{-1} acetate), the cell concentration in terms of OD at 600 nm decreased to 40% of the control culture. The effect of acetate on the expression of r-oGH was analyzed and the results are presented in Table 4A. It was observed that even though addition of acetate reduced the concentration of r-oGH, the specific yield of r-oGH was ~20 mg L^{-1} OD^{-1} of cells in all the cases. With increases in the concentration of acetate in the medium, the expression of r-oGH decreased concomitantly with the reduction in cell growth. However, the specific cellular yield of r-oGH was constant in all cases of acetate addition. In the absence of glucose, acetic acid was utilized preferably as a carbon source. A similar inhibitory effect of acetate was observed in a glucose-containing medium. The lower production of r-oGH in the presence of acetate was concomitant with the low growth of *E. coli* cells. The specific cellular yield of r-oGH was unaffected by increasing concentrations of acetate in the medium, thus confirming the inhibitory role of acetate on cell

Table 4. (**A**) Effect of acetate on expression r-oGH (**B**) Acetate utilization during *E. coli* growth

A Sodium acetate (g L^{-1})	Cell OD at harvest	r-oGH concentration (mg L^{-1})	r-oGH yield (mg L^{-1} OD^{-1})
0	1.8	35	19.4
2.5	1.7	34	20.00
5	1.21	23	19.08
10	1.02	18.9	19.5
15	0.72	14.00	20.4

B Sodium acetate (g L^{-1})	Initial acetate (g L^{-1})	Acetate before induction (g L^{-1})	Acetate at harvest (g L^{-1})
Control – 0	–	0.25	–
2.5	1.8	0.835	–
5	3.6	1.79	–
10	7.3	3.45	1.32
15	10.9	6.27	6.06

growth during the expression of recombinant protein. This indicated that the inhibition of growth by acetate is the main reason for the lower volumetric productivity of the expressed protein in *E. coli*. An acetate concentration around 2.5 g L^{-1} has very little effect on cell growth during the expression of oGH in *E. coli*.

Acetate is also used as a carbon source through the glyoxalate shunt pathway during the growth of *E. coli* [72]. Thus, the total acetate concentration in the growth medium is the result of acetate production as well its consumption by the organism. This was also observed during the expression of r-oGH, indicating the utilization of the exogenously added acetate through the glyoxalate shunt (Table 4B). In the case of a medium containing glucose, acetate uptake was low and growth inhibition by acetate was more due to the lower uptake of glucose. Acetate assimilation by *E. coli* has been reported to be favored in the presence of low levels of glucose [63] as well as in the presence of complex organic nitrogenous nutrients [67]. Under carbon limitation, assimilated acetate is also used for the biosynthesis of cell mass thus resulting in a lower accumulation of acetic acid [67, 83]. The assimilation of acetic acid by *E. coli* cells, particularly at low concentration of glucose is an important observation for the development of glucose limited fed-batch culture. This will not only reduce the formation of acetic acid but also help in its utilization. Considering that *E. coli* can utilize acetate through the glyoxalate shunt and that uptake of acetate is better in the presence of low levels of glucose along with a complex nitrogenous source, it will be ideal to develop a feeding strategy such that, at a given time, the glucose concentration is so low that it allows only utilization of the acetic acid. In conclusion, it was observed that during expression of a gene from a strong promoter, as is the case of r-oGH expression, the presence of external acetate has very little influence on the specific cellular yield of the expressed protein. A high acetate concentration only inhibits cell growth, which affects the volumetric productivity of the expressed protein. As the inhibitory role of acetate is limited to cell growth, it will be more preferable to suppress its formation rather than to secure its removal from the medium during recombinant *E. coli* fermentation. As an acetate concentration around 2.5 g L^{-1} has a very little toxic effect on cell growth during the expression of oGH, a high cell density fermentation process was developed keeping in mind that, at any stage of nutrient feeding or fermenter operation, the level of acetate was below 2.5 g L^{-1}.

3.3.3
Kinetics of Inclusion Body Production During Batch Fermentation

Batch fermentation of *E. coli* expressing r-oGH was carried out in a 2 L fermenter under a controlled environment [90]. To understand the detailed kinetics of recombinant protein expression as inclusion bodies and its effect on the specific growth rate of *E. coli*, batch fermentation data of the culture were analyzed. Cultures were induced with 1 mM IPTG at the mid-log phase (OD of 4) and samples were taken every half an hour to analyze cell growth, acetic acid and recombinant oGH. It was observed that induction with IPTG was associated with a reduction in cell growth (Fig. 1). Maximum cell concentrations of 15 OD

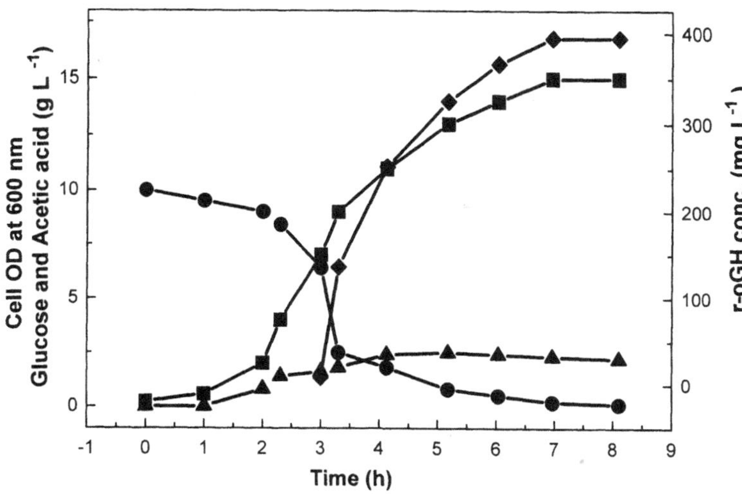

Fig. 1. Batch fermentation kinetics of *E. coli* expressing r-oGH. Cells were induced with 1 mM IPTG at cell OD of 4. Samples were taken at regular interval to monitor different parameters (*black square*) cell concentration, (*black diamond*) r-oGH concentration, (*black circle*) residual glucose concentration and (*black triangle*) acetate concentration

were achieved at 4–5 hours post-induction. Cell growth stopped almost 4 hours after IPTG induction. A maximum of 400 mg L^{-1} of recombinant oGH was obtained in four hours of IPTG induction after which the cell growth and protein expression plateaued. Acetic acid formation was around 2 g L^{-1}, and its secretion was more during the protein expression phase.

The kinetics of inclusion body formation and its effects on the specific growth rate of the culture were analyzed during the induction period and are presented in Fig. 2. Immediately after the addition of IPTG, cells grew with the same specific growth rate until the inclusion bodies were detected in the cells. At one hour post-induction, the reduction in specific growth rate was very low (around 5% of the original value). The maximum reduction in specific growth rate of the cultures was observed during 2–3 hours of post-induction period where it decreased from 0.55 h^{-1} to 0.1 h^{-1}. At four hours post-induction, the culture almost stopped growing and protein expression also plateaued. In terms of doubling time of the organism, it was observed that cells upon induction with IPTG grew with almost the same specific growth rate for one cycle after which the effect of IPTG on specific growth rate was observed. The specific growth rate declined by 40% of its original value in two hours and by 80% at three hours post-induction, complete cessation of growth occurred four hours after IPTG induction. However, irrespective of the specific growth rate at which the culture was growing, the specific growth rate decreased almost in the same manner with time upon induction with IPTG: 5% reduction in first hour, 40% reduction in 2nd hour, 80% reduction in 3rd hour and complete cessation of growth after 4 hours of induction. The specific cellular yield of the r-oGH also increased with time after induction with IPTG and plateaued in four hours. The increase in the percent expression of the recombinant protein was associated

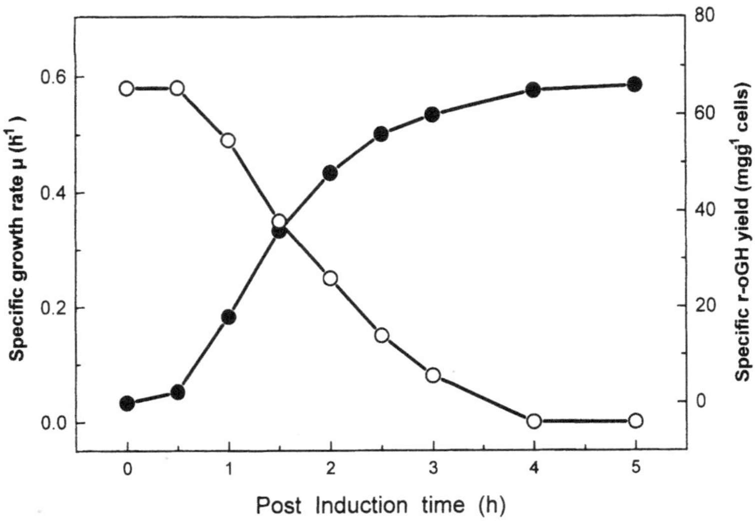

Fig. 2. Kinetics of inclusion body formation and its effect on specific growth rate and specific r-oGH yield. Reduction in specific growth rate of *E. coli* cells (*open circle*) and increase in specific r-oGH yield (*black circle*) as a function of post-induction time during the expression of ovine growth hormone

with a reduction in the specific growth rate of the culture (Fig. 2), indicating that the metabolic burden exhibited on growing cells was due to gene expression. Such an effect of increasing protein expression on the reduction of the specific growth of the growing culture has been quantitatively shown recently in *E. coli* [85]. This reduction of the specific growth rate during expression of the recombinant protein is of prime importance, particularly during high cell density fermentation with controlled feeding of glucose. During such a process, if glucose feeding is not lowered in accordance with the specific growth rate of the culture, there will be a build-up of excess glucose resulting in the secretion of the acetic acid, which will affect the metabolic activity of the growing cells [86].

3.3.4
Effect of Yeast Extract During High Cell Density Fermentation

High cell density growth of recombinant *E. coli* is always associated with secretion of acetic acid and lowering of specific cellular protein yield due to sub-optimal operational conditions. It has been observed that addition of yeast extract not only helps in reducing the secretion of acetic acid during growth of *E. coli* [65] but also helps in utilization of acetic acid during carbon limitation [67]. Apart from this, organic nitrogen sources like yeast extract and soybean hydrolysate have been reported to enhance the specific cellular yield of the expressed protein, particularly during high cell density fermentation where the demand for the nitrogenous source become very high following induction [56,68,73]. In order to find out the effect of yeast extract on the specific cellular

Table 5. Effect of yeast extract to glucose concentration ratio on cell growth and oGH expression during fed-batch fermentation

Yeast extract/ Glucose ratio in feed	Cell OD at 600 nm	Acetic acid conc (g L^{-1})	oGH conc (g L^{-1})	Specific r- oGH yield (mg g^{-1} cells)
0	40	6	0.68	42.5
0.25	56	5.4	1.12	50.
0.5	55	3.5	1.40	63.6
0.75	60	2.6	1.60	66.6
1.	60	2.2	1.60	66.6

yield of recombinant oGH, a series of fed-batch fermentations was carried out with different combinations of glucose and yeast extract and, in each case, cells were grown up to 35 OD and then induced with 1 mM IPTG. Cultures were allowed to grow at moderately high growth rates (0.5 $h^{-1)}$ for two hours and then the growth rate was decreased to 0.25 h^{-1} by controlling the feeding of the glucose. The feeding rates of nutrients (glucose and yeast extract) were reduced after induction with IPTG in accordance with the decrease in specific growth rate of the culture. The effects of yeast extract concentration on the specific cellular yield of recombinant oGH during fed-batch fermentation are presented in Table 5. It was observed that addition of yeast extract has profound effects on both growth and the specific cellular yield of expressed oGH. As the ratio of glucose to yeast extract (C/N) in the feeding medium was increased from 0 to 1, the specific cellular oGH yield increased from 42.5 mg g^{-1} dry cell weight to 66 mg g^{-1} dry cell weight. At a C/N ratio of 0.75, a maximum of 1.6 g L^{-1} of the recombinant oGH was produced at 24 g L^{-1} of dry cell weight concentration. The specific cellular yield was around 66 mg g^{-1} of dry cell weight, which was achieved in a simple optimized batch fermentation. Acetic acid secretion was low in the presence of medium containing higher amounts of yeast extract (data not shown). This result indicated that, for the maintenance of specific cellular protein yield during high cell density fermentation, it is important that the cultures should be provided with the optimum amount of nitrogenous source to supplement the required amino acids during high level synthesis of the expressed protein and to reduce the acetic acid secretion. As the volumetric recombinant oGH yield is affected by both cell concentration and specific cellular protein yield, it was expected that maximization of both during high cell density fed-batch fermentation would result in high volumetric yields of the protein.

3.3.5
High Productive Fermentation Process for Ovine Growth Hormone

High cell density fed-batch fermentation was carried out by continuous feeding of the nutrients after the consumption of initial glucose (yeast extract to glucose ratio=0.75) in order to grow the cells at a particular growth rate. A series of fed-batch fermentations was carried out with increasing cell concentra-

Table 6. Effect of yeast extract feeding on cell growth and oGH expression during high cell density fermentation

Cell conc. at IPTG induction (OD 600 nm)	Final cell conc (OD at 600 nm)	oGH conc ($g\ L^{-1}$)	Specific r-oGH yield ($mg\ g^{-1}$of cells)
4	15	0.4	65.6
15	32	0.8	64.5
30	60	1.6	66.6
50	78	2.0	65.4
86	124	3.2	65.

tions and then suitably induced for oGH expression [113]. The results are presented in Table 6. It was observed that as long as the yeast extract to glucose ratio was maintained around 0.75 during the nutrient feeding, oGH expression increased concomitantly with cell mass. The specific oGH yield was around 65 mg g^{-1} of cells. The highest yield of 3.2 g of r-oGH was obtained at a cell OD of 124 in 16 hours of batch time. The kinetics of oGH expression in optimized high cell density fermentation is presented in Fig. 3. A cell concentration of 86 OD (34 g L^{-1} dry cell weight) was achieved in 11 hours of fed-batch fermentation, after which the culture was induced with 2 mM IPTG (optimum IPTG for induction was 0.02 mM/L/OD culture). More than 95% of the culture sample has plasmids before the IPTG induction, indicating the high stability of the recombinant plasmid during extended fed-batch fermentation. The cultures were grown for another 6 hours and the batch was terminated at a cell OD of 124. The nutrient feeding rate during the expression period was decreased according to the fall of the specific growth rate of the culture due to gene expression. The specific growth rate of the culture declined during the induction period, as was experienced in batch process. The kinetics of inclusion body production also followed almost the same pattern as was observed during batch fermentation (data not shown). With the use of proper glucose feeding, particularly during the expression phase, along with the use of an optimal amount of yeast extract a maximum of 3.2 g L^{-1} of recombinant oGH was expressed in 16 hours time after which the expression plateaued. The cell OD was around 124 (49 g L^{-1}) and the specific cellular oGH yield was 65 mg g^{-1} dry cell weight. This specific cellular oGH yield was very close to that obtained in simple batch fermentation. A continuous supply of yeast extract and nutrient feeding according to the reduction in the specific growth rate following IPTG induction helped in maximizing the recombinant oGH production using fed-batch fermentation. The residual glucose concentration was maintained almost constant around 0.5 to 1 g L^{-1} throughout the fermentation period. The acetic acid concentration during the whole fermentation period was very low and apparently at such a low concentration that it did not have any detrimental effect on gene expression. In sixteen hours of fed-batch fermentation, a maximum of 3.2 g L^{-1} of oGH was expressed at a cell concentration of 124 OD, using the *E. coli*-based expression system. In conclusion, with proper analysis of bioprocessing parameters the volumetric productivity of a recombinant protein ex-

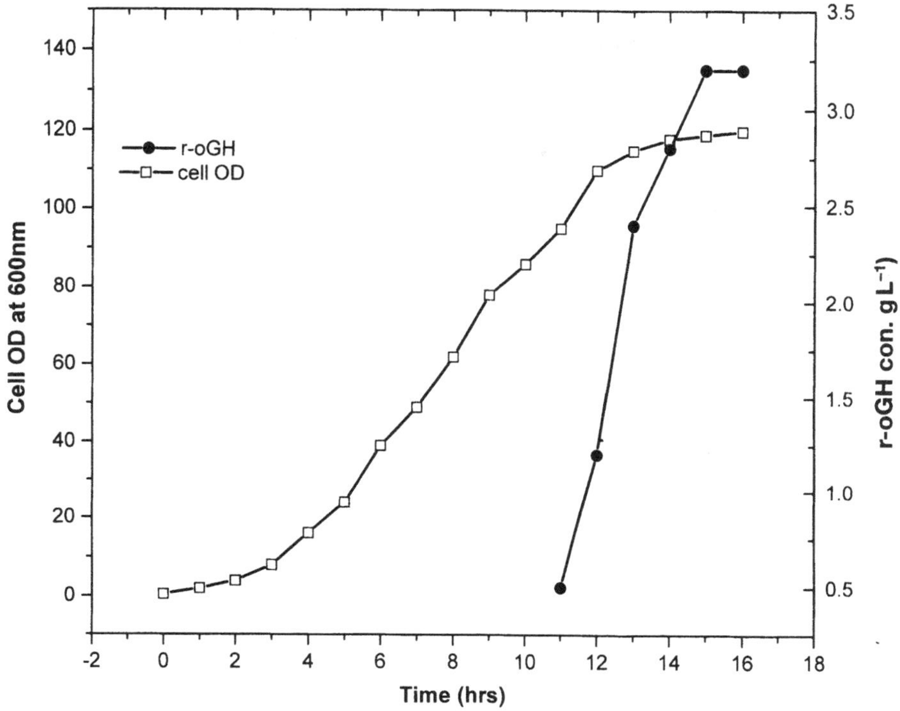

Fig. 3. High cell density fermentation of *E. coli* for the expression of r-oGH. On reaching an OD value of 86, the culture was induced with 2 mM IPTG. Following induction, glucose and nutrient feeding were reduced according to the fall in specific growth rate of the culture

pressed in *E. coli* was maximized by maintaining the specific cellular protein yield during high cell density fermentation.

The use of yeast extract along with glucose feeding helped in maintaining a higher growth rate during fermentation with very little acetic acid secretion. The presence of yeast extract in the medium provides nutrients and precursors for the synthesis of building blocks of cells and, thus, glucose is used mainly as an energy source. This helps in lowering the glucose uptake during the growth of cells and thereby leads to a lower secretion of acetic acid [32]. Lowering of the glucose uptake does not affect the cell growth rate in a complex medium as glucose is used only as an energy source. The presence of optimal amount of yeast extract during the entire period of fed-batch fermentation lowered the glucose uptake without compromising cell growth, produced less acetic acid and helped in maintaining the specific cellular protein yield. The amount of alkali used during the process to control the pH was also low due to the extensive utilization of the complex nitrogenous source. This is because utilization of the complex nitrogenous source for cellular metabolism releases ammonia, which is used to control the pH of the medium, thus avoiding excess addition of alkali whose ions has negative effects on gene expression [50]. The presence of yeast extract in the medium also helps in lowering the inhibitory effect of acetic acid and also works as a better physiological buffer in comparison to the minimal medium [113]. Therefore, its use in the feeding medium along with glucose helped in avoiding the need of complex genetic manipulation to lower the acetic acid secretion [68, 69]. Considering the fact that acetic acid dissipates proton motive force, and works as an uncoupler in *E. coli* cells resulting in growth inhibition, it is suggested that use of yeast extract in the medium will be a better and simpler approach than genetic manipulation to control the secretion of acetic acid during the aerobic growth in glucose. This indicates that, with the use of yeast extract in the feeding medium, not only the specific yield of protein can be maintained in high cell density fermentation but also the duration of the process can be reduced resulting in high volumetric productivity of the expressed protein.

It was observed that during gene expression, particularly from a strong promoter, the specific growth rate of the culture becomes an intrinsic property of the cells, which is reduced in a programmed manner during the post-induction period. Hence it is essential to design the feeding strategy of nutrients according to the reduction in specific growth rate of the culture during the post-induction period for maximizing expression of recombinant ovine growth hormone. A maximum of 3.2 g L^{-1} of recombinant ovine growth hormone was produced in fed-batch operation in sixteen hours time which is the highest value ever reported for this hormone using an *E. coli*-based expression system.

Using a similar nutrient feeding strategy, high volumetric yields of recombinant bonnet monkey zona pellucida glycoprotein bMZPC [114] and human growth hormone [115] were achieved. In the case of the zona pellucida glycoprotein, even though the level of expression was low, high cell density fed-batch fermentation resulted in improving the volumetric yield of the protein considerably. A recombinant bmzPC yield of 3–5 mg L^{-1} in shaker flask culture was enhanced up to 160 mg L^{-1} using high cell density fed-batch fermentation taking care of the most of the bioprocess engineering considerations described in the present text. Similarly, a high volumetric yield of recombinant human growth hormone was also achieved. What is more important is that the development of such bioprocessing resulted in high cell growth while maintaining the specific protein yield. In a simple fed-batch fermentation, 1.6 g L^{-1} of hGH was expressed as inclusion bodies in 10 hours of batch time at a cell concentration of 25 g L^{-1} dry cell weight [115]. Increasing the cell mass while maintaining the specific recombinant protein yield in both cases resulted in a high volum etric productivity of the recombinant protein. Such a nutrient feeding strategy in combination with other bioprocessing criteria like acetate inhibition, metabolic burden due to gene expression, plasmid stability and supply of nitrogenous nutrients to meet the demand of amino acid supply during high-level expression of gene can be used for enhancing the volumetric productivity of recombinant proteins expressed in *E. coli*.

4
High Throughput Purification

Refolding and purification of bioactive protein is the major bioprocessing parameter for the efficient production of therapeutic proteins from *E. coli* [41, 116].

The levels of expression for soluble proteins are in general low in *E. coli* and most of the time soluble protein accumulates in the periplasmic space [117]. *E. coli* rarely expresses soluble protein in a large amount, the exceptions being IGF and leptin [21, 27]. For recovering soluble protein, standard chromatography procedures are employed after the cell lysis. As the yields of soluble protein are very low, such process are rarely employed for large-scale production of therapeutic proteins. High-level expressions of protein in the form of inclusion bodies are generally used for large-scale production of therapeutic protein [4, 30]. The formation of inclusion bodies in *E. coli* facilitates the isolation of the protein of interest from the cytoplasm at the cost of its native structure. In general, inclusion bodies are solubilized by the use of high concentrations of denaturants such as urea or guanidine hydrochloride, along with a reducing agent such as DTT or β-mercaptoethanol [41, 118]. The solubilized proteins are then refolded by slow removal of the denaturant in the presence of an oxidizing agent [30, 119, 120]. The solubilization of protein from the inclusion body by high concentrations of chaotropic reagents results in the loss of its secondary structure, leading to the random coil formation of the protein structure and exposure of hydrophobic surfaces [40]. Loss of the secondary structure during solubilization and the interactions among the denatured protein molecules leading to their aggregation are considered to be the main reasons for the poor recovery of bioactive proteins from the inclusion bodies. In most of the therapeutic recombinant protein cases, the yield of bioactive protein from the inclusion bodies is around 15–25% of the total expressed protein. The overall process yield and economic viability of the recombinant *E. coil* fermentation process mostly depend on the efficient recovery of bioactive protein from the inclusion bodies [8]. It is thus necessary to have information about the formation, characteristics, solubilization and refolding of inclusion body protein from *E. coli.*

4.1
Recombinant Protein as Inclusion Bodies

The formation of inclusion bodies is mainly attributed to the overexpression of recombinant proteins in the cell lacking the required accessories for its folding to the native form [28]. Endogenous proteins, when overexpressed in *E. coli*, also accumulate as inclusion bodies [29]. There is no direct correlation between the propensity of the inclusion body formation of a certain protein and its intrinsic properties, such as molecular weight, hydrophobicity, folding pathways and so on [120]. In case of proteins having disulfide bonds, the formation of protein aggregation as inclusion bodies is anticipated since the reducing environment of bacterial cytosol inhibits the formation of disulfide bonds. The consequences are improper folding of the protein in aggregation to inclusion bodies. It is generally assumed that high-level expressions of non-native protein (higher than 2% of cellular protein) and highly hydrophobic protein are more prone to accumulate as inclusion bodies in *E. coli.* High-level expression of recombinant protein results in aggregates not only in *E. coli* as host, similar deposits of protein aggregates have been reported in several other host systems, for example, bacillus, yeast, insect cells and in higher eukaryotic cells [121]. In the human system the aggregation of protein to insoluble aggregates results in many diseases, some important ones being cystic fibrosis, Alzheimer's disease [122]. Understanding the protein aggregation thus may provide valuable tools for developing protein aggregation antagonist agents to control such diseases [123]. A great variety of experimental approaches indicates that the formation of inclusion bodies results from partially folded intermediates in the intracellular folding pathways of the protein and not from the totally unfolded or native protein. This indicates that the protein aggregates in the form of inclusion bodies and provides a model to study the nature of protein interactions involved in aggregation. Investigations of the basis of protein folding by studying the solubilization of P22-like protein have given valuable information about the intermediates of protein folding pathways [124, 125]. Although protein expression in the form of inclusion bodies is often considered undesirable, their formation can be advantageous, as their isolation from cell homogenates is a convenient and effective way of purifying the protein of interest [30, 120]. Converting this inactive misfolded inclusion body protein into the soluble active form can result in a high recovery of the therapeutic recombinant protein.

4.1.1
Inclusion Body Formation, Isolation and Characterization

Inclusion bodies are dense particles of aggregated protein found in both cytoplasmic and periplasmic spaces of *E. coli* during the high-level expression of foreign protein. These protein aggregates form electron-refracting particles in the cell that can be distinguished from other cell components and thus are called refractile bodies. Their size varies from 0.17–1.3 μm and the protein aggregation in inclusion body may have either an amorphous or paracrystalline nature depending on the localization [120, 126]. In many cases the inclusion body constitutes 20 to 50% of the total cellular protein of the cell. Growth conditions and expression systems have profound effects on the composition of the inclusion body protein [127]. In addition to the heterologous protein, inclusion bodies contain very low amounts of host protein, ribosomal components and DNA/RNA fragments [128]. It has also been reported that the presence of contaminants in inclusion bodies is mainly due to incomplete purification of the inclusion bodies following cell lysis. The presence of the above minor components could have been due to be their accidental trapping during the aggregation of polypeptides into inclusion bodies. These inclusion bodies often contain almost exclusively the overexpressed protein and aggregation in inclusion bodies has been reported to be reversible [129, 130].

The exact reason for protein aggregation into inclusion body formation is not clear yet [29]. There may be several possible reasons for the intracellular aggregation of the recombinant protein and the predominant ones could be due to (1) the high local concentration of protein synthesis in the cytoplasm, (2) the lack of cellular compartmentation and the oxidizing environment in the *E. coli* cell thus preventing formation of the S-S bond necessary for proper folding, (3) the lack of mammalian post-translational modifying enzymes and foldases (PDI, PPI and chaperone system) during high-level expression of protein, and

(4) the aggregation behavior of the intermediates of the protein folding pathways. The propensity to form insoluble aggregates does not correlate with other factors such as size of the expressed protein and use of fusion construct. Recombinant protein deposition in inclusion bodies is commonly observed with hydrophobic proteins as hydrophobic interactions among the partially folded protein molecules have been found to be responsible for aggregation into inclusion bodies. In conclusion, inclusion body formation during the high-level expression of a foreign gene is determined by a combination of effects such as rate of folding, aggregation and protein synthesis, the solubility and thermodynamic stability of the folding intermediate, and the native state, interaction with chaperone, and disulfide bond content of the protein [131].

The aggregation leading to inclusion body formation has also been reported to be due to specific intermolecular interactions among a single type of protein molecule [132]. Significant features of protein aggregates in inclusion bodies are the existence of the native-like secondary structure of the expressed protein and their resistance to proteolytic degradation [133, 134]. Analysis of the secondary structure of β-lactamase inclusion bodies from *E. coli* by Raman spectroscopy indicated the presence of an amide bond, thus indicating the existence of a native-like protein structure in the inclusion body protein. Structural characterization studies using ATR-FTIR have shown that the insoluble nature of the inclusion bodies may be due to their increased level of non-native β-sheet content compared with native and salt-precipitated protein. The formation of inclusion bodies thus facilitates the easy isolation and recovery of the expressed proteins in denatured form. As the inclusion bodies have a high density ($\sim$1.3 mg mL^{-1}), they are easily separated by high-speed centrifugation after cell disruption. The most efficient process for complete cell lysis is high-pressure disruption following a lysozyme treatment. Further purification of inclusion bodies can be achieved by washing with detergents, low concentration of salt and or urea [30, 39]. With proper isolation and washing processes, more than 95% pure inclusion body can be prepared from *E. coli* [135]. Sucrose gradient centrifugation can be used to realize very pure inclusion body preparation [127]. Recently ultrafiltration using membranes of different pore sizes has been used for the isolation of inclusion bodies from *E. coli* cells [136]; however, centrifugal isolation has been found to be the best method for separating the inclusion bodies from the membrane or cellular components.

4.1.2
Solubilization of Inclusion Body Proteins

The strategy used to recover bioactive protein from inclusion bodies involves four steps: isolation and purification of inclusion bodies from *E. coli* cells, solubilization of the protein aggregates, purification, and refolding of the solubilized protein [30, 137, 138]. Among these steps, refolding of the solubilized protein is the most crucial step and needs careful attention for a high recovery of protein. Most of the protein aggregation leading to low recovery of the recombinant protein occurs due to the use of a suboptimal refolding procedure. Considering that protein aggregation leading to inclusion body formation is highly specific, it can be expected that, by using optimal washing procedures, more 90% inclusion body purity can be achieved with the washing step [135]. This may help in reducing the number of steps for purification. In general, proteins expressed as inclusion bodies are solubilized by the use of high concentration (6–8 M) of chaotropic solvents. Chaotropic agents such as urea, guanidine hydrochloride, and thiocyanate salts [41, 139] detergents such as SDS [140], *N*-cetyltrimethylammonium chloride [141] and sarkosyl (sodium *N*-lauroylsarcosine) [142] along with reducing agents like β-mercaptoethanol, dithiothreitol or cysteine have been extensively used for solubilizing the inclusion body proteins. In many cases of inclusion body solubilization, use of a reducing agent like DTT, 2-mercaptoethonal improves the solubilization yield in the presence of chaotropic agents. This helps in maintaining the cysteine residue in a reduced state and thus prevents non-native intra- or interdisulfide bond formation in highly concentrated protein solutions at alkaline pH. Chelating agents like EDTA are frequently used in the solubilization buffer to prevent metal-catalyzed air oxidation of cysteines.

Use of extreme pH values in combination with a low concentration of denaturing agent or temperature has also been used for the solubilization of inclusion body protein [12, 135]. Very recently, solubilization of inclusion bodies while applying high hydrostatic pressure (1–2 kbar) along with a reducing agent has been reported [143]. The use of a different solubilization buffer can be examined to evaluate of the importance of the different protein interactions leading to the accumulation of inclusion bodies [115]. Such information can be then used judiciously to develop an ideal inclusion body solubilization method for the given protein. A sparse matrix-based solubilization approach has been described to solubilize inclusion body protein based on the understanding of the interactions involved in protein aggregation obtained by using different buffer compositions [144]. The idea is to understand the nature of protein aggregation leading to inclusion body formation. With this information in hand a suitable buffer can be employed to solubilize the inclusion body protein. In the case of a very high level of protein expression, in situ solubilization of inclusion bodies by directly adding denaturant to the fermentation broth has also been reported [145]. The main advantages of this method is the elimination of a mechanical disruption method with a centrifugation step for the recovery of inclusion bodies and thus may help in increasing the overall yield of the recombinant protein. The choice of the solubilizing agent greatly affects the refolding yield and cost of the overall process. Developing an efficient and economic denaturant-based solubilization step is thus a key step in achieving high throughput and cost effective purification.

4.1.3 Renaturation of Solubilized Recombinant Proteins

The soluble proteins in general are refolded into their native state after removing the chaotropic agents or other salts by dialyzing the proteins in buffers containing reducing and oxidizing agents [30, 41]. Purification of the recombinant protein either can be carried out before renaturation under denaturing conditions or after refolding of the solubilized protein. Refolding followed by purification is generally preferable as some of the high molecular aggregates along with the contaminants can be copurified in single step. In spite of several protocols available for protein solubilization and refolding, the overall recovery of bioactive proteins from inclusion bodies is often very low. Dilution of the solubilized protein directly into the renaturation buffer is the most commonly used method for small-scale refolding of recombinant proteins. This helps in reducing protein aggregation. In general, the protein concentration is kept around 20–50 μg mL^{-1} to achieve the best refolding yield. At high initial protein concentrations, aggregation of the folding intermediates (high-order reaction) predominates over the first-order refolding, leading to lower yields of the refolded protein. However, the major limitation of the dilution method is the problems of scale-up. Refolding large amounts of recombinant protein using the dilution method needs a large refolding vessel, a huge amount of buffer and additional concentration steps after protein renaturation and thus adds to the high cost of protein production [8, 13]. Pulse renaturation involving the addition of a small amount of solubilized protein to the renaturation buffer at successive time intervals helps in reducing the volume of buffer and thus improves the overall performance of the refolding process [137, 146] The success of this process is based on the fact that once a small amount of denatured protein is refolded into the native from it does not form aggregates, thus by choosing the protein concentration and time of successive addition of solubilized protein, large quantities of the solubilized protein can be refolded in the same buffer tank. This method can be operated either in continuous mode or fed-batch mode to achieve high yields of the recombinant protein. Pulse renaturation processes have been successfully tried for the recovery of gamma-interferon and lysozyme [147, 148].

Renaturation of disulfide-bonded protein needs an oxidizing agent for the formation of the disulfide bond. Air oxidation in the presence of a metal catalyst is the simplest way of oxidizing protein but is highly empirical. Oxidation can also be achieved by adding a mixture of oxidized and reduced thiol reagents such as glutathione, cysteine and cystamine. The most widely used thiol reagents are reduced/oxidized glutathione (GSH/GSSH), DTT/GSSH, cysteine/cystine, cysteamine/cystamine at a total concentration of 5–15 mM with a molar ratio of reduced to oxidized species of 1:1 to 5:1, respectively [41, 131, 139]. Because thiol disulfide exchange proceeds via a nucleophilic attack of thiolate ion, mild alkaline conditions are generally used for thio-disulfide exchange. Renaturation with the help of mixed disulfide bond formation using oxidized glutathione also helps in the high recovery of native protein. This involves the formation of a disulfide bond between the glutathione and the denatured protein followed by renaturation in the presence of a catalytic amount of reduced glutathione [146]. The advantages of this process are that a mixed disulfide increases the solubility of the protein during refolding and reduces the incorrect disulfide bond formation. In vitro folding of human tissue plasminogen activator from inclusion bodies of *E. coli*, which contains 17 disulfide bonds in its native form has been successfully achieved by formation of a mixed disulfide using oxidized glutathione [131, 149].

Apart from the above two methods, sulfonation is also used to improve the refolding yield of disulfide bond-containing proteins [120, 131]. Essentially it involves the treatment of denatured protein with sodium sulfite and sodium tetrathionate resulting in sulfonated protein. During refolding, the sulfonate group is displaced by sulfhydryl to from a mixed disulfide, which is further converted into disulfide by intramolecular displacement [120]. Improved refolding of protein using sulfonation has been successfully reported for the recombinant β-subunit of human chorionic gonadotropin hormone [150].

Renaturation of bioactive protein with little aggregation has also been achieved by addition of low molecular weight additives [137]. Often, additives such as acetone, acetoamide, urea, detergents, sugar, short-chain alcohols, DMSO and PEG are used to enhance the yield of folded bioactive protein [151]. The most commonly used low molecular weight additives have been L-arginine, low concentrations (1–2 M) of urea or guanidine hydrochloride, and detergents. A detailed list of additives used for refolding has been published recently [146]. Among the additives, the positive effects of L-arginine/HCl in reducing aggregation have been demonstrated on various proteins like tissue plasminogen activator, single chain immunotoxin, fab fragments [146], human gamma-interferon [147] and for fish growth hormone [152]. Despite the beneficial effects of arginine, its exact function in reducing protein aggregation is still not clear. Probably arginine, because of its guanidino structure, helps in a better solubilization of folding intermediates without destabilizing them. The exact mechanisms of action of the low molecular weight additives are also not known but they have been found suitable for many refolding procedures. These additives may influence both the solubility and stability of the unfolded protein, folding intermediates as well as the final intermediate state. These are easy to remove from solubilization buffer, with the exception of detergents, which need special treatment after protein refolding. In most of the refolding methods, addition of renaturing agent or optimal buffer or conditions (usually low temperature) helps in reducing the aggregation of the protein intermediate. As the propensity of aggregation decreases with lower concentrations of the protein, most of the time refolding is carried out in dilute conditions, which result in high operational process costs.

4.1.4
Improved Methods of Protein Refolding from Inclusion Bodies

One of the major problems associated with the low recovery of the refolded protein from the solubilization mixture is its aggregation [30, 138]. Aggregation is a higher order reaction whereas refolding is a first order reaction. Thus, the rate of aggregation is more than the rate of folding at high initial protein concentrations [28, 153]. Because of this kinetic competition, the yield of correctly folded protein decreases at increasing protein concentrations. Protein concentrations in the range of 10–100 $\mu g\ mL^{-1}$ are typically used during protein refolding [139, 146]. A lower extent of protein aggregation and thus improvements in the yield of the refolded protein have been achieved by use of monoclonal antibody [116] as well as by use of chaperone and foldase during refolding [154–156]. In recent years many novel high throughput protein-refolding methods have been developed. These methods, in general, are carried out at high initial protein concentrations thus resulting in high recovery of the refolded protein. These process essentially involve physical separation of the partially folded protein molecules during buffer exchange thus avoiding protein-protein interactions. Reduction in protein-protein interactions, particularly during the refolding stage, improves the renaturation yield due to reduced aggregation of the folding intermediates. High throughput refolding of inclusion body solubilized protein at high protein concentrations has been reported by using diafiltration and immobilized minichaperones [156], reversed micelles [157] and immobilized liposome chromatography [158].

Simultaneous buffer exchange, refolding and purification of inclusion body solubilized protein can be achieved by use of ion exchange chromatography where the denatured proteins of interest bind to the matrix [159]. Intermolecular interaction leading to aggregation is minimized when the folding molecules are isolated through binding to the support matrix. Simultaneous use of denaturant free buffer and optimization of elution condition leads to the purification of protein in bioactive form. Such chromatographic matrixes have been successfully used for the refolding of interleukin-2, retroviral v-myb oncoprotein [159] and also for single chain cellulose binding domain [160] expressed as inclusion bodies in *E. coli*. Such simultaneous renaturation and purification of inclusion body protein using nickel-chelating chromatography has also been reported for *E. coli* membrane proteins [161]. Using similar methodology, reduced lysozymes at a very high concentration (9 mg mL^{-1}) have been successfully refolded into the bioactive form with almost 100% recovery using immobilized liposome chromatography [158]. This methodology is highly efficient in reducing protein aggregation and thus provides maximum recovery of the refolded protein.

Protein refolding can also be achieved by the use of diafiltration and dialysis using ultrafiltration membranes [162, 163]. A high recovery of the renatured protein can be achieved by use of proper operating conditions at which the binding of the protein to the membranes is less. Huge volumes of solubilized protein material can be processed using such refolding procedures. Rapid buffer exchange and refolding of solubilized inclusion body protein can also be achieved by size exclusion chromatography [164]. Using size exclusion chromatography, reduced lysozyme and carbonic anhydrase at a very high concentration (80 mg/mL) were successfully refolded on Sephacryl S-100 columns. Removal of denaturant and renaturation of the reduced denatured protein were simultaneously achieved using gel filtration chromatography. Use of a proper size of gel filtration matrix helps in trapping the different form of folding intermediates depending upon their hydrodynamic radius, thus physically separating the individual protein molecules from interactions with each other. This results in reduced aggregation of the folding intermediates and thus improves the renaturation yield of the denatured protein. Purification of inclusion bodies through microfiltration followed by renaturation using gel filtration has improved the overall recovery of the recombinant hen egg white lysozyme. Reduced aggregation of proteins during refolding has also been achieved by direct loading of denatured protein and carrying out renaturation in a gel filtration column. Using such chromatographic methods platelet-derived growth factor has been successfully renatured to the monomeric form [165]. Using size exclusion chromatography, urokinase plasminogen activator has been successfully separated [166]. It was found that the use of chromatography results in a higher yield of the refolded protein in comparison to the dilution method. As size exclusion chromatography offers multiple advantages of buffer exchange, protein refolding and separation of monomer from aggregates, it provides an ideal method for large-scale refolding of inclusion body protein. Such chromatography has also been used for the refolding and purification of recombinant ovine growth hormone from the inclusion bodies [135].

5 Novel Method of Protein Solubilization from Inclusion Body of *E. coli*

There exists a growing body of evidence that proteins expressed as inclusion bodies in *E. coli* have extensive native-like secondary structures [133, 134] and that the formation of inclusion bodies is a result of specific aggregation between folding intermediates of the protein molecules [132]. The structure estimates of the protein expressed as inclusion bodies localized at different compartments of *E. coli* have been found to be similar, suggesting that inclusion body formation takes place at a late stage of the protein folding pathway, and thus proteins retain most of their secondary structure during aggregation into inclusion bodies [127, 133]. This observation is also indirectly supported by the fact that the formation of a secondary structure from the primary amino acid sequences takes place very rapidly, mostly within 10–20 milliseconds and this secondary structure stabilizes the protein structure, which renatures into its native conformation [167]. All the foregoing information suggests that the protein in the inclusion body already exists at an intermediate stage of the folding pathway and thereby has a considerable amount of secondary structure. We assumed that, if protein from inclusion bodies could be solubilized without disturbing its existing native-like secondary structure, the overall yield of the bioactive protein from the inclusion body would be much higher [135]. This can be achieved by manipulating the experimental conditions, such as pH and the use of different solubilizing agents in the presence of low concentrations of denaturants.

5.1 Solubilization and Refolding of Ovine Growth Hormone (oGH)

Despite an extensive literature on the expression, structure and function of different growth hormones, very little has been published on recombinant oGH expressed in *E. coli* [168]. Solubilization of inclusion body protein with restoration of its secondary structure by the use of a surfactant has been reported. However, the overall yield of the protein using such a process was low [169]. We have recently reported the expression and single-step purification of recombinant ovine growth hormone (r-oGH) containing a histidine tag at the N-terminus with the help of a Ni^{2+}-NTA affinity column [112]. Solubilization of the recombinant protein from inclusion bodies without disturbing its secondary structure, its purification and refolding conditions were optimized to achieve a higher yield of recombinant oGH using the above novel procedures.

5.1.1
Isolation and Purification of r-oGH Inclusion Body from E. coli

Induced *E. coli* cells from the fermenter were centrifuged at 6,000 rpm for 1 h and the cell pellet (60 g wet weight) was suspended in 200 mL of 50 mM Tris-HCl buffer (pH 8.0) containing 5 mM EDTA and 1 mM PMSF [90, 135]. The cell suspension was subjected to French press (18000 psi, 3 cycle) for disruption and further centrifuged at 12,000 rpm for 30 minutes to isolate the inclusion bodies from the cell debris. The inclusion bodies thus obtained were washed with 50 mM Tris-HCl buffer (pH 8.0) containing 5 mM EDTA and 2% deoxycholate solution. After extensive washing of the inclusion bodies in deoxycholate, they were further washed with 50 mM Tris-HCl buffer (pH 8). The inclusion bodies were again washed with distilled water to remove contaminating salt and detergent, centrifuged at 12,000 rpm for 30 minutes and the pellet containing proteins in the form of purified inclusion bodies was stored for further studies. At this stage the purity of the inclusion body preparation was around 90% and thus was directly used for solubilization and refolding without further purification [135].

5.1.2
Solubilization of r-oGH from Inclusion Body

The existence of native like structural features of proteins in inclusion bodies of IL-1β [134] and β-lactamase [133] indicates that the inclusion bodies form from a folding intermediate having a considerable amount of native-like secondary structure. The secondary structure content of aggregated IL-1β, formed either during in vitro refolding or by thermal denaturation, has also been reported to be identical to that of the inclusion bodies, indicating that the main part of the folding process of the protein occurs before the onset of aggregation [133]. Protein aggregation leading to inclusion body formation has been reported to be taking place through specific interactions of certain conformations of the folding intermediates [132]. These observations provide a rationale that, if the inclusion bodies first be purified to homogeneity and then solubilized with retention of the existing structure of the protein molecule, the overall yield of the bioactive proteins will be high [135]. To preserve the secondary structure of the protein while solubilizing the inclusion bodies, strong denaturing agents such as 8 M urea or 6 M guanidine hydrochloride were avoided. Solubilization was carried out in 2 M Tris buffer at different pH values from 4–12. Sodium acetate buffer (2 M) was used for the pH ranges from 4-6, whereas Tris-HCl buffer (2 M) was used for the pH range 7–12. Supernatant and the pellet were analyzed individually by SDS-PAGE and RIA to monitor the concentration and to calculate the percent solubilization. The secondary structure of the solubilized r-oGH was determined from the CD spectra at each pH. Maximum solubilization of r-oGH inclusion bodies was achieved at pH 12 with addition of 2 M urea. More than 90% of the r-oGH from the inclusion body was solubilized at pH 12 in the presence of a two molar concentration of urea with a high initial protein concentration (2.6 mg mL^{-1}). The solubilized r-oGH retained its secondary structure and was also found to be immunoreactive. The fluorescence and UV spectra of the protein solubilized at pH 12 were close to those of the native, pituitary-derived growth hormone. The spectral data indicated the existence of intermediate protein structures in the solubilized inclusion body protein. The solubilization of the inclusion body proteins with retention of their native-like secondary structure is perhaps the most important criterion for its successful refolding into the native conformation. Solubilization of the inclusion body protein with retention of the secondary structure has been reported by the use of a cationic surfactant [169]. However, the use of surfactants for solubilization of inclusion body protein necessitated complex washing procedures to remove the surfactant, resulting in a considerable loss of protein. The use of pH-induced solubilization retains the existing secondary structure, solubilizes the maximum amount of protein from the inclusion body, and also helps in dissociating the dimer into monomers as reported for the native oGH [170]. Solubilization of oGH inclusion bodies at high pH indicated the importance of ionic interactions in the aggregation of oGH to inclusion bodies.

5.1.3
Refolding and Characterization of Recombinant oGH

The solubilized r-oGH was refolded using a conventional dilution method. The concentrated protein after refolding was analyzed by CD spectroscopy to determine the content of secondary structure and the yield of the refolded protein was estimated by RIA. The overall yield of oGH was higher when the proteins were solubilized at pH 12 and refolded at pH 8. A maximum of 40% of the solubilized protein could be refolded using the above refolding procedure. The loss during refolding was mainly due to the formation of insoluble aggregates, which appeared during the buffer exchange and concentration [135].

To improve the yield of the refolded protein, refolding was also carried out in a gel filtration column (Sephacryl-100, Pharmacia, Sweden). The main purpose was to allow the solubilized protein to enter the gel filtration matrix and then carry out the buffer exchange, so that intermolecular aggregation could be avoided [135]. Refolding experiments were carried out within two hours of the solubilization. Elution from S-100 matrix not only separated the monomeric protein from aggregates, but also helped in buffer exchange. During refolding in the gel filtration column, around 35% of the protein aggregated and the aggregates were separated during elution. The yield of the refolded protein using gel filtration was around 60–70%. The protein eluted from the gel filtration column migrated as a single band in SDS-PAGE, native PAGE and Western blot analysis. The fluorescence and CD spectra of the r-oGH were quite similar to those of the native protein suggesting the formation of the correctly folded form of the protein by this method.

Refolding of the denatured proteins is often hindered due to protein aggregation during the removal of salt and denaturants [153]. In general, aggregation of the protein during refolding is minimized by carrying out the process at very high dilution. To optimize the yield of refolded protein from the solubilizing buffer, refolding of the r-oGH was carried out in a gel filtration column. The solubilized protein molecules, while passing through the gel filtration matrix, are physically separated from each other (condition equivalent to an Anfinsen cage), which reduces the chance of protein-protein interactions during the buffer exchange process. Aggregation of the r-oGH was much less in comparison to that seen during refolding by simple dilution followed by concentration. The yield of purified refolded protein during the gel filtration process was 60–70%, which is higher than the 40% achieved by the dilution method. In addition to achieving a higher yield of purified refolded protein, separation from aggregates and buffer exchange was accomplished in a single-step procedure by refolding within a gel filtration matrix. This novel procedure resulted in recovery of a high yield of bioactive ovine growth hormone with minimum use of urea and water and thus substantially reduced the overall cost of protein recovery from the inclusion bodies of *E. coli*.

5.2 Solubilization and Refolding of Human Growth Hormone (hGH)

This novel inclusion body solubilization and refolding method was applied for the high throughput recovery of recombinant human growth hormone from the inclusion bodies of *E. coli*. Human growth hormone (hGH), a single chain polypeptide containing 191 amino acid residues, apart from stimulating cell growth, plays an important role in a variety of metabolic, physiologic and anatomic processes [171]. The protein folds into a four-helix bundle structure with two disulfide bridges, one connecting distant parts of the molecule involving the 53rd and 165th amino acid residues (large loop) and another between residues 182 and 189 (small loop). The large-scale requirement for r-hGH necessitates its high-level expression in *E. coli* as inclusion bodies [172]. However, expression of the protein along with a fusion tag and subsequent use of high concentrations of chaotropic reagents for solubilization and purification make the overall process more complex and expensive as these steps lower the yield of bioactive r-hGH [108]. Solubilization of r-hGH from the purified inclusion bodies was carried out without disturbing the existing native-like secondary structure as discussed above for ovine growth hormone.

5.2.1 *Solubilization of r-hGH from Inclusion Body*

Induced *E. coli* cells expressing human growth hormone were used for the solubilization and refolding of human growth hormone [115]. The r-hGH in the form of inclusion bodies was separated and purified as carried out for ovine growth hormone [135]. The notable difference in purification of inclusion body of hGH was that it was always associated with high molecular aggregates. It has

Table 7. Effect of pH on the solubility of r-hGH inclusion body. Inclusion body were solubilized in the buffer and the protein concentration was determined by taking absorbance at 280 nm

pH	Percent solubilization of inclusion body	
	100 mM Tris buffer (No urea)	100 mM Tris buffer with 2 M urea
5	–	5
6	–	5
7	–	5
8	5	10
9	5	10
10	12	15
11	20	25
12	40	85
12.5	50	95

been widely reported that growth hormone inclusion bodies of different species expressed in *E. coli* can be solubilized at alkaline pH [12, 108, 135]. Thus, the purified r-hGH inclusion bodies were solubilized at different pH values in 100 mM Tris buffer (pH 3–13) and the percent solubilization of r-hGH was monitored as shown in Table 7. Solubilization of r-hGH from inclusion bodies was observed while increasing the pH from 8 to 12.5. High alkaline pH (>12.5), even though it helped in solubilizing r-hGH from inclusion bodies, resulted in extensive degradation of r-hGH (SDS-PAGE data not shown). A maximum of 2 mg mL^{-1} of r-hGH was solubilized in 100 mM Tris buffer at pH 12.5 without the addition of urea or guanidine hydrochloride. Higher solubilization of r-hGH from inclusion bodies was observed by incorporating 2 M urea in 100 mM Tris buffer at pH 12.5 (Table 7). Furthermore, the addition of urea in 100 mM Tris buffer at pH 12.5 did not help in solubilizing a higher amount of r-hGH from the inclusion bodies. In 100 mM Tris buffer at pH 12.5 containing 2 M urea, a maximum of 6 mg mL^{-1} of r-hGH were solubilized from the inclusion bodies. The solubility of r-hGH was comparable to that of 8 M urea in Tris buffer at pH 8 [115].

5.2.2
Effect of β-Mercaptoethanol

The isolation of r-hGH and its subsequent solubilization at alkaline pH was always associated with the presence of dimer (44 kDa, in SDS-PAGE), which constituted around 5–8% of the total inclusion body protein. These r-hGH dimers were also observed in the SDS-PAGE in spite of a high reducing and denaturing environment. Addition of increasing amounts of β-mercaptoethanol (2–20 mM) in 100 mM Tris buffer at pH 12.5 containing 2 M urea had very little effect on dissociating the oligomers into monomeric r-hGH. Dissociation of oligomers and dimers to monomeric r-hGH was also not observed in 8 M urea solution containing 100 to 200 mM of β-mercaptoethanol (SDS-PAGE data not shown).

Expression of r-hGH in *E. coli* as inclusion bodies was always associated with the formation of dimers and oligomers, which were also observed at the early stage of protein synthesis. Bacterial cytosol being very reducing in nature does not allow the formation of disulfide bonds, so the aggregates could not be due to intermolecular disulfide bonds. Possibilities for mixed disulfide bond formation leading to high molecular weight aggregates may occur during cell lysis. However, as the r-hGH oligomers could not be dissociated in to monomers in the presence of a strong reducing agent such as β-mercaptoethanol or reduced:oxidized glutathione, it can be concluded that non-disulfide covalent bonding is responsible for oligomer formation in r-hGH inclusion bodies. The existence of such non-disulfide covalent bonds in r-bGH oligomers has been reported through Raman spectroscopic studies by Thamann [173].

Thus in conclusion, r-hGH was solubilized from the inclusion bodies without its existing native-like secondary structure being disturbed. The extent of solubilization in 100 mM Tris buffer pH 12.5 containing 2 M urea was found to be comparable with that of 8 M urea and a maximum amount of 6.5 mg mL^{-1} of r-hGH could be solubilized from the inclusion bodies. The increased solubility of r-hGH in the above buffer could be an effect of both urea and pH, indicating the existence of both ionic and hydrophobic interactions in the inclusion bodies.

5.2.3
Purification and Refolding of Recombinant hGH

For the large-scale purification of r-hGH, pure inclusion bodies were isolated (~104 mg protein isolated from 65 mL of high cell density fermentation broth) and solubilized in 16 mL of 100 mM Tris buffer pH 12.5, containing 2 M urea [115]. The solubilized r-hGH was diluted 5 times and pH of the buffer was adjusted to 8.5. No aggregation of the solubilized r-hGH was observed during dilution and buffer exchange. Solubilized r-hGH was passed through a DEAE-Sepharose column for purification. The recombinant human growth hormone which eluted between the conductivity range of 14 to 16 mS/cm was found to be homogeneous and represented 40% of the total protein. However, some amount of r-hGH was co-eluted along with r-hGH dimer between conductivities of 22 and 25 mS/cm, which constituted about 25–30% of the total protein. The overall recovery of r-hGH from the ion-exchange matrix was around 65%. Pure r-hGH containing dimers/oligomers was passed through the size exclusion chromatography column for further purification. The dimeric or higher forms of the proteins were removed through gel filtration. The overall yield of the purified refolded r-hGH from the inclusion bodies of *E. coli* was thus ~50% (Table 8).

The authenticity of the purified r-hGH was further confirmed from the N-terminal analysis of r-hGH and from spectroscopic analysis. The UV spectrum of the purified r-hGH showed an absorbance maxima at 276.8 nm, and a shoulder at 283 nm, which was comparable to that of native human growth hormone. The fluorescence spectrum of refolded r-hGH was found to be identical to that of the native hGH, which gave a peak at 340 nm. Growth kinetics of the prolactin-dependent Nb2 lymphoma cell line were monitored to evaluate the

Table 8. Purification of r-hGH from inclusion bodies of *E. coli*. 1 g of *E. coli* cells was processed for recovery of pure r-hGH. Purity is defined as percentage of monomeric r-hGH in the final protein

Steps	Total protein (mg)	Overall yield (%)	Purity (%)
Cell lysate	500	–	13
Pure inclusion body	65	100	90
Solubilization	61	94	92
Chromatography (ion exchange and gel filtration)	33	50	98

bioactivity of purified r-hGH. Addition of prolactin, commercial hGH and r-hGH promoted growth of Nb2 cells arrested at the Go/G1 phase by serum deprivation (Fig. 4). Growth of Nb2 cells in the presence of different concentrations of r-hGH was found to be comparable to that observed for the commercial hGH, indicating the bioactivity of the preparation [115]. The overall yield of r-hGH from the inclusion bodies was ~50% in comparison to 20 to 25% achieved when solubilizing the inclusion bodies in high concentrations of chaotropic reagents. Solubilization of the r-hGH from inclusion bodies while retaining the native-like secondary structures helped in lowering the extent of protein aggregation during buffer exchange and dilution. In spite of the presence of two disulfide bonds, extensive protein aggregation during refolding due to incorrect disulfide bond formation was not observed for r-hGH. Recombinant human growth hormone showed efficient growth promoting activity in the Nb2 cell proliferation assay, indicating that the purified and refolded r-hGH has a biologically active conformation. The high recovery of bioactive protein from the inclusion bodies of *E. coli* further substantiated the usefulness of using the novel solubilization procedure followed by a suitable refolding method for high throughput recovery of bioactive protein.

5.3 Solubilization and Refolding of Bonnet Monkey Zona Pellucida Glycoprotein C (bmZPC)

The high pH inclusion body protein solubilization was applied for a complex mammalian protein (bonnet monkey zona pellucida glycoprotein C) expressed as inclusion bodies in *E. coli* [114]. The mammalian oocyte is surrounded by a non-cellular envelope termed as the zona pellucida (ZP). ZP is composed of three distinct glycoproteins which, on the basis of the size of mRNA transcripts, have been classified as ZPA, ZPB and ZPC [174]. Functionally, in mouse, hamster and human, ZPC serves as the putative sperm receptor. Recombinant bonnet monkey (*Macaca radiata*) zona pellucida glycoprotein C (r-bmZPC) was expressed as inclusion bodies in *E. coli* under the *lac* operator control in the pQE-30 vector (QIA express; Qiagen GmbH, Hilden, Germany) as described previously [175]. The r-bmZPC was expressed as a polyhistidine (His_6) fusion protein in the BL-21(pLysS) strain of *E. coli* deficient in *ompT* and *lon* proteases.

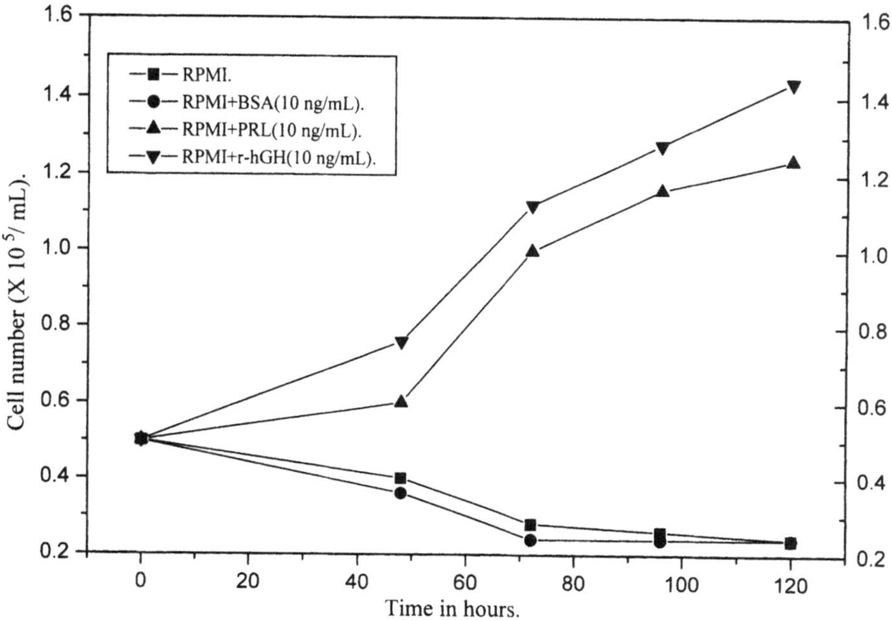

Fig. 4. Receptor-mediated growth promoting activity of purified r-hGH on Nb2 cell lines. Growth arrested cells were supplemented with BSA (*black circle*) prolactin (*black upright triangle*) and r-hGH (*black inverted triangle*) for cell proliferation. Purified r-hGH stimulated cell growth comparable to that of prolactin

For large-scale production of r-bmZPC, transformed cells were grown in a 3.5 L fermenter (2 L working volume) and induced with 1 mM isopropyl β-D-thio-galactopyranoside (IPTG) and used for purification studies [114].

5.3.1
Purification and Solubilization of Inclusion Body

Induced *E. coli* cells were lysed by a French press at 18000 psi and inclusion bodies were recovered by centrifugation at 8000 g for 30 min at 4°C. The inclusion body pellet thus obtained was washed with 50 mM Tris-HCl buffer (pH 8.5) containing 5 mM EDTA and 2% deoxycholate and was purified as described previously [135]. The purified inclusion bodies were solubilized in 100 mM Tris-HCl buffer at different pH values (4–12). Solubilization of r-bmZPC from inclusion bodies was further optimized by adding different concentrations of urea (2–8 M).

Recombinant bmZPC expressed as inclusion bodies in *E. coli* was purified to near homogeneity by extensive washing with detergent [176]. In order to protect the native-like secondary structure, r-bmZPC inclusion bodies were solubilized at alkaline pH, containing a moderate concentration of a chaotropic agent (2 M urea). A six-fold increase (45%) in the solubility at pH 12 as compared to the neutral pH suggested that the charge distribution provided by the highly alkaline pH along the protein chain could be responsible for solubilization of r-bmZPC from the inclusion bodies. Further, the presence of 2 M urea at alkaline pH of 12 improved r-bmZPC solubilization from inclusion bodies that may be attributed to the effect of the diminishing of hydrophobic interactions by urea [177]. The amount of solubilized r-bmZPC in Tris buffer at pH 12 containing 2 M urea was comparable to that obtained in 8 M urea solution. Since the protein could not be denatured at a 2 M urea concentration, the above buffer was used for solubilization of r-bmZPC inclusion bodies.

5.3.2
Refolding, Purification and Characterization of r-bmZPC

Purified r-bmZPC inclusion bodies were solubilized in optimized conditions comprised of 100 mM Tris buffer (pH 12) containing 2 M urea for 30 min at room temperature. The pH of the solubilized protein was brought down to 8.5 by adding 1 N HCl followed by extensive dialysis at 4 °C against renaturation buffer (20 mM Tris buffer pH 8.5 containing 1 mM EDTA, 1 mM reduced glutathione, 0.1 mM oxidized glutathione and 10% sucrose). Refolded r-bmZPC was further purified using ion-exchange chromatography [176]. Very little aggregation was observed during refolding. Recombinant bmZPC was eluted from a Q-Sepharose column at a conductivity of 40–44 mS/cm and was more than 90% homogenous with minor high molecular weight aggregates that were separated by size exclusion chromatography using Sephacryl S-200HR. The purified r-bmZPC showed a dominant band of 43 kDa in SDS-PAGE (data not shown). The overall yield of pure bmZPC was around 28% (Table 9).

Recombinant bmZPC when purified by Ni-NTA affinity chromatography followed by removal of urea resulted in loss of protein due to aggregation [114, 175]. This problem was overcome by dialysis of the solubilized proteins in buffer containing reduced and oxidized forms of glutathione, which have a profound effect on the refolding of r-bmZPC, and helped in reducing the protein aggregation. The r-bmZPC contains 14 cysteine residues that might be getting stabilized due to enhanced thiol-disulfide exchange by the additives and restricting the conformational flexibility of the unfolded state. The efficiency of forming correct disulfide bonds was increased not only due to the presence of oxidized and reduced forms of glutathione but also due to the preservation of the native-like protein conformations of r-bmZPC during solubilization. Refolded r-bmZPC did not bind to Ni-NTA resin under non-denaturing condi-

Table 9. Purification efficiency of r-bmZPC from the inclusion bodies of *E. coli.* Purified r-bmZPC inclusion bodies after solubilization were refolded in refolding buffer and then purified using ion exchange chromatography

Purification steps	BmZPC (mg)	% Recovery
Inclusion body pellet	80	100
Solubilization	70	87
Refolding	37	46
Chromatography	23	28

tions, indicating the absence of a His tag on the surface of the protein molecule (data not shown). Thus, in spite of the presence of a His tag at the N-terminus, r-bmZPC was purified using ion-exchange chromatography followed by gel filtration. The purified refolded r-bmZPC also gave characteristic spectra in CD, UV and fluorescence indicating the presence of the native-like protein structure. The CD spectrum, in particular, indicated the presence of both an α-helical structure and a β-sheet conformation in the r-bmZPC [176]. The purified refolded r-bmZPC reacted with to bonnet monkey spermatozoa, suggesting the presence of a native-like conformation in the protein molecule. In a mouse model, it has been demonstrated that the presence of O-linked oligosaccharides on ZPC is critical for its binding to spermatozoa. However, in the present studies, r-bmZPC expressed in *E. coli* and hence presumably deficient in carbohydrate moieties binds to spermatozoa, suggesting that presence of oligosaccharides is not an absolute requirement for binding of ZPC to spermatozoa in non-human primates. This is in agreement with the observed binding of human ZPC expressed in *E. coli* to human spermatozoa [178].

In conclusion, an improved method for purification of r-bmZPC in folded form from the inclusion bodies of *E. coli* is described. In spite of the absence of carbohydrate moieties, the r-bmZPC can be expressed and refolded from the inclusion bodies of *E. coli* to a native-like structure. The efficiency of recovery of this complex protein was also high, indicating the suitability of the novel solubilization procedures for high recovery of denatured protein from the inclusion bodies of *E. coli*.

5.4
Ideal Method for Solublization and Refolding of Inclusion Body Protein

The use of novel solubilization and refolding processes for inclusion body protein of oGH, hGH and bMZPC resulted in high recovery of bioactive proteins. Use of a low amount of urea at high pH not only helped in reducing the urea consumption during inclusion body solubilization but also protected the native-like secondary structure of the expressed protein. This helped in a better recovery of the solubilized protein in the bioactive form. Protein aggregation is a major stumbling block during the refolding of inclusion body protein. However using these novel procedures, the extent of aggregation was also reduced which resulted in improved yields of the recombinant proteins. Proteins are more prone to aggregation in the non-native state than in the partially folded conformation. The solubilization of inclusion body protein without the use of high concentrations of chaotropic substances helped in retention of the native-like secondary structure, and thus reduced the extent of protein aggregation during refolding. This method also resulted in the use of lower amount of buffer for refolding. This helps in lowering the overall cost of refolding. In conclusion, the novel method of solubilization increases the recovery of protein with less use of urea and buffer. This will be a significant development towards bioprocessing of recombinant proteins, as urea and buffer constitute a major cost during the protein refolding process. With the use of different solvents for solubilization of inclusion body protein, it is possible to determine the major interactions leading to protein aggregation during high level expression of recombinant protein in *E. coli.* Once the major interactive forces which lead to aggregation of protein into inclusion bodies are known such information can be exploited without using a high molar concentration chaotropic solvent for solubilization of protein with retention of the native-like secondary structure. As the existence of the native-like secondary structure helped in better refolding of inclusion body proteins, novel solubilization procedures hold the key for successful high throughput purification and refolding of inclusion body protein. Once the solubilization of inclusion body protein with low molar urea is achieved, subsequent refolding procedures can be developed to enable high throughput recovery of bioactive therapeutic protein from the inclusion bodies of *E. coli.*

6
Conclusion

E. coli continues to be the ideal host for the large-scale expression and purification of recombinant protein. The successful production of heterologous bioactive protein from *E. coli* needs a multidisciplinary approach considering the impact of cell and molecular biology, fermentation process development, high throughput purification and stable formulation for better application of the final product. However, the success of commercialization of the product mostly depends on the volumetric productivity of the purified protein. In this context, a high productive fermentation process and a high throughput purification become the most important bioprocess engineering parameters for the successful production of therapeutic protein from *E. coli.* With the cell biology information available for *E. coli* and the development of high density fed-batch culture, it has been possible to achieve very high cell densities, which was otherwise impossible to achieve. This has helped in reducing the size of the reactors and thus has had a very big impact on the overall cost of production. Cell densities up to 200 g L^{-1} have been achieved. If protein expression can be achieved at such high cell densities then the volumetric yield will be very high. This, coupled with understanding of protein structure, its aggregation behavior and, particularly, the nature of protein in the form of inclusion body has tremendously helped in improving the volumetric yields of recombinant proteins expressed in *E. coli.* Developments of new and novel protein refolding procedures have not only improved the recovery of bioactive proteins but have also provided new information about the behavior of proteins, particularly at high concentrations. In this review the uses of high cell density fermentation and high throughput refolding of inclusion body proteins have been discussed. Taking the examples of three different proteins, it has been demonstrated that by proper bioprocess analysis of the expression system, the volumetric yields of the recombinant proteins and their recovery can be improved substantially. Particularly with ovine growth hormone, the volumetric yield was improved almost 100 times using a fed-batch fermentation process. Using the novel solubilization and refolding procedures, the recovery of bioactive ovine growth hormone was improved considerably. Similar yield improvements were observed for human growth hormone and for a complex protein like zona pellucida glycoprotein C. It is expected that such a bioprocessing approach can be applied to many other proteins which are expressed as inclusion bodies in *E. coli*. By analyzing the impact of high cell density fermentation and inclusion body solubilization in an interactive way, the overall yield of therapeutic proteins expressed in *E. coli* can be improved. Such methodologies can be applied for bioprocess development of therapeutic proteins expressed as inclusion bodies in *E. coli*.

Acknowledgement. This work is supported by the financial grant from the Department of Biotechnology, Government of India and core funds of National Institute of Immunology (NII) New Delhi. The author would like to acknowledge the help rendered by Dr. S. M. Totey, Dr. L. C. Garg, and Dr. S. K. Gupta for their collaboration and help in various recombinant gene product projects. Thanks to Dr. S. K. Basu, Director NII for his help and encouragement.

7
References

1. Hodgson J (1993) Bio/Technology 11:887
2. Baneyx F (1999) Curr Opin Biotechnol 10:411
3. Bucknel P (1996) Trends Pharma Sci 17:450
4. Swartz JR (2001) Curr Opin Biotechnol 12:195
5. Kane JF, Hartley DL (1988) Trends Biotechnol 6:95
6. Yee L, Blanch HW (1992) Bio/Technology 10:1550
7. Kleman GL, Strohl, W R (1994) Curr Opin Biotechnol 15:180
8. Petrides D, Sapidou E, Calandranis J (1995) Biotechnol Bioeng 48:529
9. Makrides SC (1996) Microbiological Reviews 60:512
10. Friehs K, Reardon KF (1993) Adv Biochem Eng Biotechnol 48:53
11. Kroeff EP, Owens RA, Campbell EL, Johnson RD, Marks HI (1989) J Chromatogr 461:45
12. Storrs SB, Przybycien TM (1991) In: Georgiou G, De Bernardez-Clark E (eds), Protein Refolding. ACS Symposium Series No. 470, American Chemical Society, Washington, DC, p 197
13. Datar RV, Cartwright T, Rosen C-G (1993) Bio/Technology 11:349
14. Hockney RC (1994) Trends Biotechnol 12:456
15. Hannig G, Makrides SC (1998) Trends Biotechnol 16:54
16. Georgiou G (1988) AIChE Journal 34:1233
17. Goldstein MA, Doi RH (1995) Biotechnol Ann Rev 1:105
18. Chao YP, Chiang CJ, Hung WB (2002) Biotechnol Prog 18:394
19. Lim HK, Jung KH, Chung SL (2000) Appl Microbiol Biotechnol 53:201
20. Wilms B, Hauck A, Reuss M, Syldakt C, Mattes R, Siemann M, Altenbuchner J (2001) Biotechnol Bioeng 73:95
21. Joly JC, Leung WS, Swartz JR (1998) Proc Natl Acad Sci USA 95:2773
22. Wood TK. Peretti SW (1991) Biotechnol Bioeng 38:397
23. Mc Carthy JEG (1991) Adv Gene Technology 2:145
24. Coburn GA, Mackie GA (1999) Prog Nucleic Acid Res Mol Biol 62:55
25. Carrier TA, Keasling JD (1999) Biotechnol Prog 15:58
26. Wang G, Liu N, Yang K (1995) Protein Expr Purif 6:284
27. Jeong KJ, Lee SY (2000) Biotechnol Bioeng 67:398
28. Mitraki A, King J (1989) Bio/Technology 7:690
29. Schein CH (1989) Bio/Technology 7:1141
30. De Bernardez Clark E (2001) Curr Opin Biotechnol 12:202
31. LaVallie ER, McCoy JM (1995) Curr Opin Biotechnol 6:501
32. Chou C-H, Bennett GN, San K-Y (1994) Biotechnol Bioeng 44:952

33. Farmer WR, Liao J (1997) Appl Environ Microbiol 63:3205
34. Nilsson M, Kallio PT, Bailey JE, Bulow L, Wahlund KG (1999) Biotechnol Prog 15:158
35. Riesenberg, D (1991) Curr Opin Biotechnol 2:380
36. Lee SY (1996) Trends in Biotechnol 14:98
37. Chio L-I, Lee SY (1999) Appl Environ Microbiol 65:4363
38. Van Wegen RJ, Ling Y, Middleberg APJ (1998) Trans IChemE 76:417
39. Rudolph R, Bohm G, Lilie H, Jaenicke R (1997) In: Creighton TE (ed), Protein Function, A practical approach. IRL press at Oxford University Press, New York, p 57
40. Dill KA, Shortle D (1991) Ann Rev Biochem 60:795
41. Fischer B, Sumner I, Goodenough P (1993) Biotechnol Bioeng 41:3
42. Datar R, Rosen CG (1990) In: Asenjo JA (ed), Separation Process in Biotechnology. Marcel Dekker, New York, p 741
43. Strandberg L, Enfors, S-O (1991) Biotechnol Lett 13:609
44. Yee L, Blanch HW (1993) Biotechnol Bioeng 41:781
45. Ingrahm JL, Maaloe O, Neidhardt FC (1983) Growth of bacterial cells. Sinauer Associates Inc, Sunderland, MA
46. Bermer H, Dennis PP (1996) In: Neidhartd FC, Curtiss R, Ingraham JL, Lin ECC, Low KBB, Magasanik B, Reznikogg WS, Riley M, Schaechter M, Umbarger HE (eds), *Escherichia coli* and *Salmonella*. Cellular and Molecular Biology, ASM press, Washington, DC, p 1553
47. Holms WH (1986) Curr Top Cell Reg 28:69
48. Vila P, Corchero JL, Benito A, Villaverde A (1994) Biotechnol Prog 10:648
49. Mori H, Yano T, Kobayashi T, Shimizu S (1979) J Chem Eng Jpn 12:313
50. Bech Jensen E, Carlsen S (1990) Biotechnol Bioeng 36:1
51. Lee YL, Chang HM (1990) Biotechnol Bioeng 36:330
52. Riesenberg D, Schulz V, Knorre WA, Phol HD, Korz D, Sanders EA, Ross A, Deckwer D (1991) J Biotechnol 20:17
53. Yoon SK, Kang WK, Parl TH (1994) Biotechnol Bioeng 43:995
54. Korz DJ, Rinas U, Hellmuth K, Sanders EA, Decker WD (1995) J Biotechnol 39:59
55. Nakano K, Rischke S, Sato S, Markl H (1997) Appl Microbiol Biotechnol 48:597
56. Shin CS, Hong MS, Bae CS, Lee J (1997) Biotechnol Prog 13:49
57. Lim H-K, Jung K-H (1998) Biotechnol Prog 14:548
58. Ahn WS, Park SJ, Lee SY (2000) Appl Environ Microbiol 66:3624
59. Fuchs C, Koster D, Weibusch S, Mahr K, Eisbrenner G, Markl H (2002) J Biotechnol 93:243
60. Guest JR (1992) J Gen Microbiol 138:2253
61. Konz JO, King J, Cooney CL (1998) Biotechnol Prog 14:393
62. El-Mansi EMT, Holms WH (1989) J Gen Microbiol 135:2875
63. Luli GW, Strohl WR (1994) Appl Environ Microbiol 60:3952
64. Axe DD, Bailey JE (1995) Biotechnol Bioeng 47:8
65. Han K, Lim HC, Hong J (1992) Biotechnol Bioeng 39:663
66. Koh BT, Nakashimada U, Pfeiffer M, Yap MGS (1992) Biotech Lett 14:1115
67. Nancib N, Mosrati R, Boudrant J (1993) Chem Eng J 52:B35
68. Chou C-H, Bennett GN, San K-Y (1994) Biotechnol Prog 10:644
69. Aristidou AA, San K-Y, Bennett GN (1994) Biotechnol Bioeng 44:944
70. Aristidou AA, San K-Y, Bennett GN (1999) Biotechnol Prog 15:140
71. Shiloach J, Kaufman J, Guillard AS, Fass R (1996) Biotechnol Bioeng 49:421
72. van de Walle M, Shiloach J (1998) Biotechnol Bioeng 57:71
73. Jung G, Denefle P, Becquart J, Mayaux JF (1988) Ann Inst Pasteur/Microbiol 139:129
74. Konstantinov KB, Nishio N, Seki T, Yoshida T (1991) J Ferm Bioeng 5:350
75. Pan JG, Rhee JS, Lebeult JM (1987) Biotech Lett 9:89
76. Seeger A., Schneppe B, McCarthy JEG, Deckwer W-D, Rinas U (1995) Enz Microb Technol 17:947
77. Shimizu N, Fukuzone S, Harada Y, Fujimori K, Gotoh K, Yamzaki Y (1991) Biotechnol Bioeng 38:37

78. Chae, HJ, DeLisa MP, Cha HJ, Weigand WA, Rao G, Bentley WE (2000) Biotechnol Bioeng 69:275
79. Markl H, Zenneck C, Dubach A, Ogbana JC (1993) Appl Microbiol Biotechnol 39:48
80. Summers D (1998) Mol Microbiol 29:1137
81. Kumar PK, Maschke HE, Fries K, Schugerl K (1991) Trends Biotechnol 8:279
82. Chen HC, Hwang CF, Mou DG (1992) Enz Microb Technol 14:321
83. Shimiju M, Ijima S, Kobayashi T(1992) J Ferm Bioeng 74:163
84. Ryan W, Collier P, Loredo L, Pope J, Sachdev R (1996) Biotechnol Prog 12:596
85. Dong H, Nilsson L, Kurland CG (1995) J Bacteriol 177:1497
86. Neubauer P, Haggstrom L, Enfors SO (1995) Biotechnol Bioeng 47:139
87. Ramirez DM, Bentley WE (1993) Biotechnol Bioeng 41:557
88. Anderson L, Yang S, Neubar P, Enfors SO (1996) J Biotechnology 46:255
89. Kurland CG, Dong H (1996) Mol Microbiol 21:1
90. Panda AK, Khan RH, Appa Rao KBC, Totey SM (1999) J Biotechnol 75:161
91. Wong HH, Kim YC, Lee SY, Chang HN (1998) Biotechnol Bioeng 60:271
92. Namdev PK, Irwin N, Thompson BG, Gray MR (1993) Biotechnol Bioeng 41:666
93. Hopkins DJ, Betenbaugh MJ, Dhujati P (1987) Biotechnol Bioeng 29:85
94. Farr SB, Kogoma T (1991) Microbiol Rev 55:561
95. Davies KJA, Lin SW (1998) Free Radical Biol Med 5:215
96. Bylund F, Castan A, Mikkola R, Veide A, Larsson G (2000) Biotechnol Bioeng 69:119
97. Olins PO, Lee SC (1993) Curr Opin Biotechnol 4:520
98. Kurland C, Gallant J (1996) Curr Opin Biotechnol 7:489
99. Rosenberg AH, Goldman E, Dunn JJ, Studier FW, Zubay G (1993) J Bacteriol 175:716
100. Kane JF, Violand BN, Curran DF, Staten NR, Diffn KL, Bogosain G (1992) Nucleic Acid Res 20:6707
101. Budisa N, Steipe B, Dermange P, Eckerskorn C, Kellermann J, Huber R (1995) Eur J Biochem 230:788
102. Lu HS, Rausset Pr, Sotos LS, Clogston CL, Rhode MF, Stoney Ks, Herman AC (1993) Protein Expr Purif 4:465
103. Calderone TL, Stevans RD, Oas TG (1996) J Mol Biol 262:407
104. Scorer CA, Carrier MJ, Rosenberger RF (1991) Nucleic Acid Res 19:3511
105. Barak Z, Lindsey D, Gallant J (1996) J Mol Biol 256:676
106. Bogosian G, Violand BN, Dorward-King EJ, Workman WE (1989) J Biol Chem 264:531
107. Del Tito BJ, Ward JM, Hodgson J, Gershater CJ, Edwards W, Wysocki LA, Watson FA, Sathe G, Kane JF (1995) J Bacteriol 177:7086
108. Shin NK, Kim DY, Shin CS, Hong MS, Lee J, Shin HC (1998) J Biotechnol 62:143
109. Schmidt M, Babu KR, Khanna N, Marten S, Rinas U (1999) J Biotechnol 68:71
110. Jeong KJ, Lee SY (1999) App Environ Microbiol 65:3027
111. Li CH, Godon D, Knorr J (1973) Arch Biochem Biophy 156:493
112. Appa Rao KBC, Garg LC, Panda AK, Totey SM (1997) Protein Expr Puri 11:201
113. Panda AK, Khan RH, Mishra S, Appa Rao KBC, Totey SM (2000) Bioproces Engg 22: 379
114. Srivastava M, Nayak J, Mehrotra V, Kaul R, Sheela P, Gupta SK, Panda AK (200) Process Biochem 35:451
115. Patra AK, Mukhopadhya R, Makhija R, Garg LC, Panda AK (2000) Protein Expr Puri 18:182
116. Sadana A (1995) Biotechnol Bioeng 48:481
117. Wulfing C, Pluckthun A (1994) Mol Microbiol 12:685
118. Creighton TE, Darby NJ, Kemmink J (1996) FASEB Journal 10:110
119. Fischer B, Perry B, Sumner I, Goodenough P (1992) Protein Eng 5:593
120. Mukhopadhyay A (1997) Adv Biochem Eng Biotechnol 56:61
121. Yon JM (1996) In: Meyers RA (ed), Encyclopedia of Molecular Biology and Molecular Medicine. Volume 5, VCH, Weinheim, p 73
122. Thomas PJ, Qu BH, Pederson PL (1995) Trends Biochem Sci 20:456
123. Wetzel R (1994) Trends Biotechnol 12:193

124. Sather SK, King J (1994) J Biol Chem 269:25268
125. Speed MA, King J, Wang DIC (1997) Biotechnol Bioeng 54:333
126. Taylor G, Hoare M, Gray DR, Marston FAO (1986) Bio/Technology 4:553
127. Bowden GA, Paredes AM, Georgiou G (1991) Bio/Technology 9:725
128. Valax P, Georgiou G(1993) Biotechnol Prog 9:539
129. Carrio MM, Cubarsi R, Villaverde A (2001) FEBS Lett 471:7
130. Mar Carrio M, Villaverde A (1998) FEBS Lett 489:29
131. Rudolph R (1996) In: Cleland JE, Craik CS (eds), Protein Engineering: Principles and Practice. Wiley-Liss, New York, p283
132. Speed MA, Wang DIC, King J (1996) Nature Biotechnology 14:1283
133. Przybycien, TM, Dunn JP, Valax P, Georgiou G (1994) Protein Eng 7:131
134. Oberg K, Chrunyk BA, Wetzel R, Fink AL(1994) Biochemistry 33:2628
135. Khan RH, Appa Rao KBC, Eshwari ANS, Totey SM, Panda AK (1998) Biotechnol Prog 14:722
136. Batas B, Schiraldi C, Chaudhuri JB (1999) J Biotechnol 68:149
137. Lilie H, Schwarz E, Rudolph R (1998) Curr Opin Biotechnol 9:479
138. De Bernardez Clark E (1998) Curr Opin Biotechnol 9:157
139. Rudolph R, Lilie H (1996) FASEB J 10:49
140. Stockel J, Doring K, Malotka J, Jahnig F, Dornmair K (1997) Eur J Biochem 248:684
141. Cardmone M, Puri NK, Brandon MR (1995) Biochemistry 34:5773
142. Burgess RR (1996) Methods Enzymol 273:145
143. St John RJ, Carpenter JF, Balny C, Randolph TW (2001) J Biol Chem 276:46856
144. Lindwall G, Chau MF, Gardner SR, Kohlstaedt LA (2000) Protein Engg 13:67
145. Falconer RJ, O'Neill BK, Middelberg AP (1999) Biotechnol Bioeng 62:455
146. De Bernardez Clark E Schwarz E, Rudolph R (1999) Methods Enzymol 309:217
147. Arora D, Khanna N (1996) J Biotechnol 52:127
148. Katoh S, Katoh Y (2000) Process Biochem 35:1119
149. Rudolph R, Buckner J, Lenz H (1991) US Patent 5,077,392
150. Mukhopadhyay A (2000) Vaccine 18:1802
151. Yasuda M, Murakami Y, Sowa A, Ogino H, Ishikawa H (1998) Biotechnol Prog 14:601
152. Hsih MH, Kuo JC, Tsai HJ (1997) Appl Microbiol Biotechnol 48:66
153. Goldberg ME, Rudolph R, Jaenicke RA (1991) Biochemistry 30:2790
154. Altamirano MM, Garcia C, Possani LD, Fersht AR (1999) Nature Biotechnology 17:187
155. Kohler RJ, Preuss M, Miller AD (2000) Biotechnol Prog 16:671
156. Altamirano MM, Golbik R, Zahn R, Buckle AM, Fersht AR (1997) Pro Natl Acad Sci USA 94:3576
157. Cleland JL, Wang DIC (1990) Biotechnology 8:1274
158. Yashimoto M, Kubio R (1999) Biotechnol Prog 15:480
159. Hoess A, Arthur AK, Wanner G, Fanning E (1988) Bio/Technology 6:1214
160. Berdichevsky Y, Lamed R, Frenkel D, Gophna U, Bayer EA, Yaron S, Shoham Y, Benhar I (1999) Protein Expr Puri 17:249
161. Rogol H, Kosemund K, Kuhlbrandt W, Collinson I (1998) FEBS Lett 432:21
162. West SM, Chaudhuri JB, Howell JA (1998) Biotechnol Bioeng 57:509
163. Hamaker KH, Liu J, Seely RJ, Ladisch CM, Ladisch MR (1996) Biotechnol Prog 12:184
164. Batas B, Chaudhuri JB (1996) Biotechnol Bioeng 50:16
165. Muller C, Rinas U. (1999) J Chromatography A 855:203
166. Fahey EM, Chaudhuri JB, Binding P (2000) Sep Sci Technol 35:1734
167. Ptitsyn O B (1994) Protein Eng 7:593
168. Adams TE, MacIntosh B, Brandon MR, Wordsworth P, Puri NK (1992) Gene 122:371
169. Puri NK, Crivelli E, Cardamone M, Fiddes R, Bertolin J, Ninham B, Brandon MR (1992) Biochem J 285:871
170. Bewley TA, Li CH (1972) Biochemistry 11:927
171. Isaksson OG, Eden S, Jansson JO (1985) Ann Rev Physiol 47:483
172. Mukhija R, Rupa P, Pillai D, Garg LC (1995) Gene 165:303
173. Thamann TJ (1998) Anal Biochem 265:202

174. Harris JD, Hibler DW, Fontenot GK, Hsu KT, Yurewicz EC, Sacco AG (1994) DNA Sequence 4:361
175. Kaul R, Afzalpurkar A, Gupta SK (1997) Mol Reprod Dev 47:140
176. Patra AK, Gahlay GK, Reddy BVV, Gupta SK, Panda AK (2000) Eur J Biochemistry 267:7075
177. Vanzi F, Madan B Sharp K (1998) J Am Chem Soc 120:10748
178. Chapman N, Kessopoulou E, Andrews P, Hornby D, Barratt CR (1998) Biochem J 330:839

Received: May 2002

Adv Biochem Engin/Biotechnol (2003) 85: 95–124
DOI 10.1007/b11046

Research and Application of Microbial Enzymes – India's Contribution

Subhash Chand[1] · Prashant Mishra[2]

[1] Department of Biochemical Engineering and Biotechnology, Indian Institute of Technology Delhi, Hauz Khas, New Delhi-10016, India. *E-mail: subhashc@dbeb.iitd.ernet.in*
[2] Department of Biochemical Engineering and Biotechnology, Indian Institute of Technology Delhi, Hauz Khas, New Delhi-10016, India. *E-mail: pmishra@dbeb.iitd.ernet.in*

Enzymes have attracted the attention of scientists world over due to their wide range of physiological, analytical and industrial applications. Although enzymes have been isolated, purified and studied from microbial, animal and plant sources, microorganisms represent the most common source of enzymes due to their broad biochemical diversity, feasibility of mass culture and ease of genetic manipulation. With the advent of molecular biology techniques, a number of genes of industrially important enzymes has been cloned and expressed in order to improve the production of enzymes, substrate utilization and other commercially useful properties. Special attention has been focused on enzymes isolated from thermophiles due to their inherent stability and industrial applications. In addition, a variety of methods have been employed to modify enzymes for their industrial usage including strain improvement, chemical modifications, modification of reaction environment, immobilization and protein engineering. A wide range of applications of enzymes in different bioprocess industries is discussed.

Keywords. Microbial enzymes, Gene cloning, Purification, Modification, Applications

Abbreviations

AIDS	Acquired immune deficiency syndrome
6-APA	6-Aminopenicillanic acid
ATCC	American Type Culture Collection
ATPS	Aqueous two-phase systems
BHU	Banaras Hindu University
CCMB	Centre for Cellular and Molecular Biology
CDRI	Central Drug Research Institute
CFTRI	Central Food Technological Research Institute
CGAs	Colloidal gas aphrons
CNBr	Cyanogen bromide
CM-Cellulase	Carboxymethyl cellulase
CTAB	*N*-Cetyl-*N,N,N*-trimethylammonium bromide
DEAE	Diethylaminoethyl
DNA	Deoxyribnucleic acid
DMF	Dimethylformamide
DMSO	Dimethyl sulfoxide
GRAS	Generally regarded as safe
HAL	Hindustan Antibiotics Limited
IIT	Indian Institute of Technology
IU	International units
IUPAC	International Union of Pure and Applied Chemistry
IME	Immobilized enzymes
MKU	Madurai Kamraj University

NADH	Nicotinamide adenine dinucleotide (reduced)
NCIM	National Collection of Industrial Microorganisms
NCL	National Chemical Laboratory
NTG	*N*-Methyl-*N'*-nitro-*N*-nitrosoguanidine
OT	Sodium bis(2-ethylhexyl)sulfosuccinate
PCR	Polymerase chain reaction
PEG	Polyethylene glycol
RNA	Ribonucleic acid
RNAse	Ribonuclease
RRL	Regional Research Laboratory
SSF	Solid state fermentation
TERI	Tata Energy Research Institute
UV	Ultraviolet

1 Introduction

The current demand for better utilization of renewable resources and pressure on industry to operate within environmentally compatible limits has been a stimulus to the development of new enzyme-catalyzed industrial processes, leading to a steady growth of enzyme market in India during the last 30 years. The consumption pattern of industrial enzymes in India during the last decade is depicted in Fig. 1 [1]. Although a large number of academic institutions are involved in basic research in enzymology, applied research in this sector has been limited. In India, the potential of immobilized enzymes (IME) was realized at an early stage and research and development activities in this area started at Biochemical Engineering Research Centre, Indian Institute of Technology, Delhi. Some of the enzyme systems studied during this early phase include glucose isomerase [2, 3], and urease [4]. The use of non-living whole cells as a source of enzyme for producing stable immobilized enzyme preparations was a novel approach for the application of enzymes. Subsequently, similar studies were also initiated at NCL, Pune; CFTRI, Mysore and CDRI, Lucknow.

Numerous applications of enzymes have emerged in the food, medical (diagnostics), biochemical processing and pharmaceutical industries. By far the greatest commercial user of enzymes has been the food and beverage industry largely due to historical and safety reasons. In industry where chemical and enzymatic catalysis can be directly compared (such as starch hydrolysis, inversion of sugar, cellulose hydrolysis etc.), biocatalysis has maintained an overwhelming advantage. Diagnostics has been largely a biocatalyst-driven market and enzymes have a proven record of accomplishment in this area. This industry requires highly selective and stable biocatalysts in order to develop the accurate and sensitive sensors that are necessary for diagnostic purposes. The commercial use of enzymes in the chemical and pharmaceutical industries has so far been limited, with a possible exception of penicillin acylase for the synthesis of penicillin derivatives. While a number of processes can use enzymes, their large-scale availability at reasonable cost has often limited their scope. However,

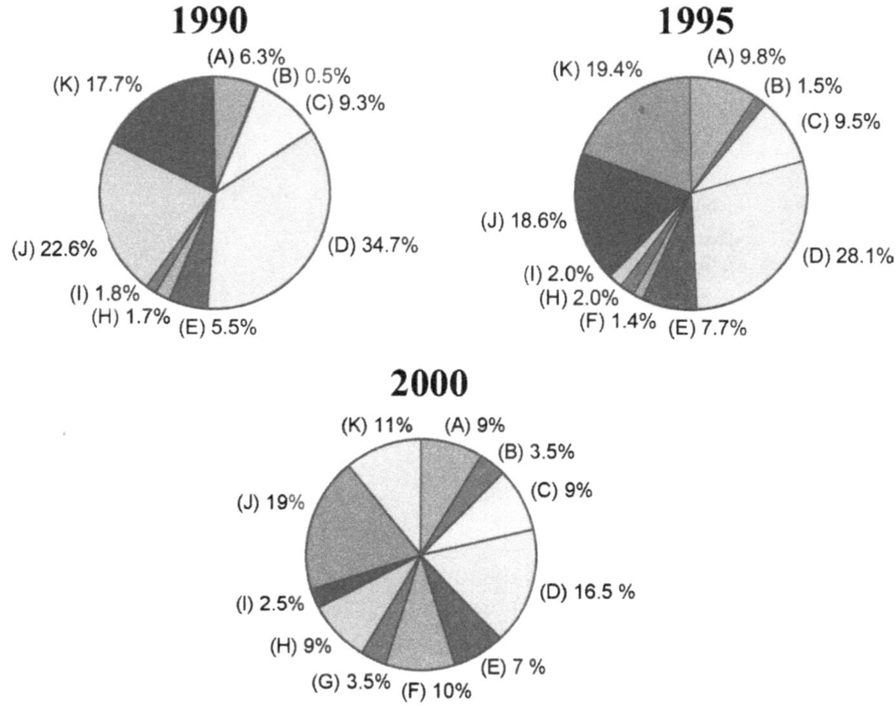

Fig. 1. Consumption of industrial enzymes in India during the past ten years (1990–2000) in various industrial and other sectors [1]. Starch processing (A), Sugars and alcohol (B), Leather processing (C), Pharmaceutical and diagnostics (D), Textile and paper (E), Detergents (F), Amino acids (G), Food processing (H), Research and Development (I), Export (J), Misc. (K)

a significant potential for novel uses of enzymes in textile processing, organic synthesis, drugs and pharmaceuticals, and polymer synthesis will emerge in the near future, which will necessitate large-scale enzyme production facilities.

The growing knowledge of the genetic make-up of various microorganisms and the ease with which they can be tailored for specific applications have made microbial enzymes more attractive for industrial purposes. Microorganisms can be tailored using the tools of genetic engineering to enable higher yields of enzymes, for their ability to utilize cheaper carbon sources as substrates, for their stability at higher temperatures and their resistance to altered pH. Knowledge of metabolic regulation has also helped a lot in adopting strategies for better enzyme production. For industrial applications of enzymes, it is necessary to have less expensive methods for their purification on a large scale. In view of this, methods have been developed to isolate and purify enzymes depending on their applications. The usefulness of enzymes specifically for industry has been improved by the design of enzymes for their robustness. Enzymes have been made reusable by immobilization, and thus more stable and less costly. On the other hand, the tools of protein engineering have also helped in developing stable enzymes. In addition, new bioreactor designs with various controls have helped to meet the challenges for scaling-up the processes for enzyme production. The present review will summarize various aspects of enzyme production, microbial sources employed, genetic manipulations, strategies for enzyme isolation and purification, various aspects of modifications and applications of microbial enzymes in the context to the contributions made by Indian scientists mainly during the last decade.

2 Sources and Production of Microbial Enzymes

Enzymes can be obtained from all living organisms but due to the fact that microorganisms can be easily and quickly grown on a large scale and, in many cases, produce extracellular enzymes; they are placed in a favorable position for enzyme production. Microorganisms that are non-pathogenic, produce no toxins and have a well established record of safety are categorized as "generally regarded as safe" (GRAS) and are preferred for the production of enzymes for application in food processing and health care products [1]. Due to overwhelming diversity of the microorganisms present in nature, the process of selection and isolation of new microorganisms as a source of enzyme production is still an interesting area of research. Thus at present, most of the work related to enzymes is devoted to finding a new microbial source of enzymes, to improving the yield of these enzymes and to finding out new industrial applications for them.

The production of microbial enzymes is generally achieved through either aerobic submerged culture or solid-state fermentation. Aerobic submerged culture has been most commonly employed for growing microbial cultures for enzyme production. The major enzyme systems studied for their production by various microbial sources using submerged aerobic culture include amylases, cellulases, chitinase, β-galactosidase, glucose isomerase, lipase, alkaline protease, penicillin acylase, tannase and xylanases. A number of strategies to improve the production of enzymes, like the effect of medium supplementation or carbon and nitrogen sources, has been generally studied. In many studies a cheaper source of carbon and nitrogen (mainly agro wastes) has been used to lower the production cost of enzymes. Different operating conditions, like temperature, pH, aeration rate, agitation, and fermentation mode, have been studied for microbial enzyme production. Response surface methodology has also been employed to optimize the microbial enzyme production (discussed in Sect. 8).

Bajpai et al. [5–7] compared various carbon and nitrogen sources for the production of α-amylase using *Bacillus* sp. and optimized a process using cheese whey as substrate. Numerous other microbial sources including *Cellulomonas* sp. [8], *Humicola paecilomyces* [9], *Mucor* sp. [10] and *Streptomyces megasporus* [11] have also been used as a source of amylolytic enzymes. Ghose and coworkers [12–15] at IIT, Delhi carried out extensive studies on the production of cellulase from *Trichoderma reesei* during the 1970s and 1980s. They used various cell culture strategies for optimizing the enzyme productivity. *Humicola fuswatra* [16] and coculture of *Trichoderma reesei* and *Aspergillus niger* [17] provided other important sources of cellulase. Panda et al. [18–20] from IIT, Madras have reported chitinase production by *Trichoderma harzianum*. These studies involved optimization of nutrients and environmental parameters as well as characterization of the enzyme chitinase. Saini et al. [21] studied the productivity of a stable β-galactosidase from *Bacillus* sp. The microbial technology developed for the production of β-galactosidase from *Klyuveromyces* sp. at IIT, Delhi has been transferred to a pharmaceutical company under a consultancy assignment.

The intracellular glucose isomerase activity demonstrated from *Streptomyces* sp. [3] and *Actinoplanes* [22] is, perhaps, still the highest reported activity in India so far. Similar studies on glucose isomerase production using *Streptomyces kanamyceticus* [23] and *Arthobacter* sp. [24] have been reported. However, none of these systems have yet found industrial application in India due to availability of sucrose as the bulk-sweetening agent.

Lipases have been most extensively studied for their production and application. A number of microbial cultures has been isolated for production of lipases at IIT, Delhi [25]. Lipase production from *Aspergillus* sp. and *Penicillium* sp. [26] showed some interesting applications in bio-organic syntheses. Several groups are involved in the production of alkaline proteases for their bulk process application, particularly in the detergent and leather industries. A high activity alkali-tolerant protease from *Conidiobolus coronatus* with interesting applications has been reported from NCL, Pune [27–30]. Penicillin acylase used for the hydrolysis of penicillin to 6-APA in the synthesis of penicillin derivatives is another industrially important enzyme. A process for the production of penicillin V acylase from *Fusarium* sp. has been developed at HAL, Pimpri [31]. The purified enzyme was used after immobilization. *Aspergillus japonicus* [32], *Aspergillus awamori* [33] and *Rhizopus oryzae* [34] have been used for optimizing the submerged production of tannase which has potential applications in the hydrolysis of tannins and the synthesis of alkyl gallates. Tannase from *Aspergillus awamori*, being intracellular, offers an advantage of being used as an immobilized non-living, whole cell system. Three major groups including NCL, Pune [35]; IIT, Delhi [36] and TERI, New Delhi [37–38] have been working on the production of xylanases, mainly for prebleaching of kraft pulp to reduce the conventional use of bleaching agent and thereby reduce the pollution load in the effluent.

Solid-state fermentation (SSF) has been employed mainly to grow fungi to produce extracellular enzymes like amylases [39–42], alkaline proteases [43–46], xylanase [47] and cellulase [48]. In the SSF process microbes are grown on a porous solid substrate generally in the absence of free water. The water and nutrient absorbed on the substrate supports the growth of cells. The growth and secretion of product occurs both on the surface of the solid support and within the support matrix. Major parameters studied using SSF are moisture content, pH, dissolved oxygen and substrate loading. SSF has exhibited advantages of higher enzyme concentration, lower capital and operational cost; lesser wastewater output and is convenient at small-scale operations. But due to the lack of process control options, lower productivity (based on per unit substrate) and the difficulty in using recombinant strains, widespread commercial use of SSF has been limited. CFTRI, Mysore; IIT, Kharagpur and RRL, Trivandrum have developed good research and development facilities for SSF.

Current industrial production of enzymes in India is restricted to hydrolytic enzymes mainly by SSF, with the exception of a few high-value restriction endonucleases. Major facilities for enzyme production are mainly located in Bangalore, Thane, Ahemdabad and Himanchal Pradesh.

3 Extremophiles as a Source of Microbial Enzymes

Extremophiles ranging from psychrophiles to hyperthermophiles constitute a valuable source of industrial enzymes. Major research has been focused on the screening and isolation of enzymes from thermophiles. Enzymes from thermophilic sources offer significant advantages in terms of thermal stability – a problem area for the application of some of the enzymes. α-Amylases represent the most thermotolerant enzymes used commercially for the solubilization of starch gel. Most of the thermotolerant amylolytic enzymes have been obtained from *Bacillus* sp. [49–53]. However, their thermal stabilites are far below those of some of the commercially available enzymes. Many microbial cultures have been reported to produce alkalophilic and thermotolerant xylanases. Raj and Chandra from IIT, Madras [54] have reported xylanase from alkalophilic *Aspergillus fischeri* that has an optimum pH of 6.0, an optimum temperature of 60°C and exhibited stability over the pH range 5.0–9.5. A thermostable extracellular xylanase from *Bacillus* sp. has shown an optimum temperature of 70°C and dual pH optima at 7.0 and 8.4 [55]. Thermotolerant xylanases have also been reported from *Cephalosporium* sp. [56] and *Thermonospora* sp. [57]. The enzyme from *Thermonospora* sp. was active over a pH range from 5–9 (with pH optimum at 7) and temperature range of 40–90°C with an optimum activity at 70°C. Sarkar and Upadhyay [58] have described a cellulase obtained from a thermophilic as well as an alkalophilic *Bacillus thermoalkalophilus* isolated from termite mound. The enzyme showed optimum activity at 70°C and at pH 8.5, which has potential applications for detergents. Other enzymes from extremophilic sources include lipases [59–61], thermolysin [62], chitinase [63] etc.

In addition to thermophilic, alkalophilic, and acidophilic microbes, there is a need to exploit barophilic (resistant to high pressure) and hyperthermophilic (which are resistant to extremes of temperatures) microorganisms for the production of enzymes.

4 Molecular Cloning and Expression of Genes

Due to rapid progress in techniques of molecular biology, our understanding of genetic systems of various industrially important microorganisms has improved. At present more than 50% of the industrially important enzymes are produced world-wide through recombinant technology [64]. Keeping pace with this many genes coding for industrially useful enzymes have been cloned in

India. Expression of recombinant DNA has been aimed at: (i) overproducing the enzymes and (ii) improving the characteristics of the enzymes with respect to their heat stability, pH optima, and susceptibility to protease. Efforts have also been made to improve the stability of plasmids during fermentation.

Most of the genes coding for enzymes have been isolated from bacterial sources and expressed in *E. coli*. Table 1 provides a summary of genes cloned, their sources and vectors employed to construct recombinant DNA. Srivastava et al. [66, 67], at CDRI, Lucknow cloned the cellulase gene from *Bacillus subtilis* into *E. coli* using PUC8/PUC18 as vectors. The recombinant (pBcelR) containing an insert of 1.2 Kb expressed both β-glucosidase and cellulase activity. In this study the enzyme activities were localized in the periplasmic space. Extensive studies on cloning and expression of the β-glucosidase gene (*βglu1*) encoding β glucosidase I or BglI [69–70] and the β-glucosidase gene (*βglu2*) encoding β gluclucosidase II or BglII [71] from the thermotolerant yeast *Pichia etchellsii* into *E. coli* have been carried out at IIT, Delhi. While Bgl1 has been reported as an intracellular enzyme, BglII was localized to the periplasmic space. The pH and temperature optima of the two enzymes have also been reported to be different. The BglII has been shown to have relatively high activity for hydrolyzing sophorose and has potential application for the synthesis of unique oligosaccharides [71].

Cloning and expression of the xylanase gene from *Bacillus circulans* into *Bacillus subtilis* have been carried out at TERI, New Delhi. The results demonstrated a 14-fold increase in extracellular xylanase activity over the corresponding expression in *E. coli* [72]. The recombinant enzyme also exhibited better thermal stability. Shendye et al. [73] at, NCL Pune also reported expression of xylanase activity by a recombinant clone of *Bacillus subtilis* to the extent of 6-fold compared to the original *Bacillus* sp. The xylanase gene has also been cloned and expressed from *Chainia* sp. [74] at NCL, Pune. The gene for amylase has been cloned from *Bacillus laterosporous* [75] and *Cellulomonas* sp. [76]. In *Cellulomonas* sp. multiple amylases are encoded by independent genes [76]. Various species of *Streptomyces* have been employed for the isolation of genes coding for glucose isomerase [77, 78], nitroaryl reductase [79], actinokinase [80] and chitinase [81]. The penicillin amidase gene has been isolated from *E. coli* [82], whereas the lipase gene and β-lactamase gene have been isolated from *Xanthomonas campestris* [83] and *Zymomonas mobilis* [84], respectively. Cloning and expression of a mercuric ion-resistant gene from *Zymomonas mobilis* has also been reported [85].

In order to enhance the β-galactosidase production in *E. coli*, phage T7 RNA polymerase gene under the control of heat stable inducible lambda PL promoter has been employed [86]. Studies have shown the effect of specific amino acids on the synthesis of recombinant glucose isomerase of *Streptomyces* expressed in *E. coli* [77]. The stability of recombinant shuttle plasmid pCPPS-31 encoding CM-cellulase in *E. coli* and *Bacillus subtilis* has been studied in batch, fed-batch and continuous culture [87] and efforts have been made to improve its stability and expression during fed-batch cultivation [88].

Table 1. Genes coding for industrial enzymes

Source of gene	Enzyme coded	Vector used	Recombinant plasmid	Expressed in host	Ref.
Acinetobacter calcoaceticus S19	Esterase B	pRS5	–	*E. coli*	[65]
Bacillus subtilis	Cellulase	pUC8 pUC18	pBcelR	*E. coli*	[66, 67]
Rumino-coccus sp.	Cellulase	pBR322	pSK1 pSK211	*E. coli*	[68]
Pichia etchellsii	*β*-glucosidase	Yep13 pUC19	–	*E. coli*	[69–71]
Bacillus cir-culans teri-42	Xylanase	pUC19 pUB110	pAQA pBA7	*E. coli* *Bacillus subtilis*	[72]
Bacillus sp. NCIM 59	Xylanase	–	pLPX6.5	*Bacillus subtilis*	[73]
Chainia sp.	Xylanase	pUC8	–	*E. coli*	[74]
Bacillus latero-sporous P3	*α*-Amylase	pKT240	–	*E. coli*	[75]
Cellulomonas sp. NCIM 2353	Amylase	pACs2 pACs10 pACs16 pACs20	–	*E. coli*	[76]
Streptomyces sp. NCIM 2730	Glucose/ Xylose isomerase	pUC8	pMSG27	*E. coli*	[77, 78]
Streptomyces aminophillus MCMB411	Nitroaryl reductase	pUC18 pIJ 702	pSD103 pSD105	*E. coli* *Streptomyces lividans*	[79]
Streptomyces megasporus SD5	Actinokinase	pBR322	pSR500 pT7–7acks	*E. coli*	[80]
Streptomyces peucetius SPV1	Chitinase	pUC18	pEMU723	*E. coli*	[81]
E. coli ATCC 1105	Penicillin amidase	–	pUSAD2	*E. coli*	[82]
Xanthomonas campestris sesami	Lipase	pUC18	pLMS1	*E. coli*	[83]
Zymomonas mobilis B-806	*β*-lactamase	pACYC184	pGMV2	*E. coli*	[84]

5
Isolation and Large Scale Purification of Enzymes

A number of industrial enzymes such as amylases, tannase, β-glucosidase and lipase has been purified using series of purification steps from a number of microbial sources. Most of these purification protocols involve ammonium sulfate precipitation as the first step of purification followed by ion exchange chromatography and or gel filtration. For many industrial applications, partially purified enzyme preparations are good enough whereas for analytical and medical applications enzymes must be highly purified and free from undesirable contaminants. Cyclomaltodextrin glucanotransferase has been purified from *Bacillus firmus* by Gawande et al. at NCL, Pune using ultrafiltration, affinity chromatography and gel filtration and resulted in a 400-fold purification with 64% yield [89]. β-Glucosidase from *Clostridium papyrosolvens* has been purified using alcohol precipitation and DEAE ion-exchange chromatography [90]. Saxena et al. [91] have reviewed the isolation and purification of lipases from various microbial sources. Recently at IIT, Delhi, tannase has been isolated and purified from *Aspergillus awamori* using ultrafiltration, precipitation and sorption of contaminant proteins [92]. The enzyme was purified 134-fold with 68% yield. Dextransucrase has been purified from *Leuconostoc mesentroides* in three successive steps of fractionation using PEG 400 [93]. Studies on the interactions of protein with polyelectrolyte resulting in protein precipitation offers an interesting approach [94–95]. Some of the methods that hold potential for purifying industrial enzymes, including partitioning in aqueous two-phase system and affinity-based methods and their application in isolation/purification of enzymes are described here.

5.1
Aqueous Two-Phase Systems (ATPS)

Aqueous two-phase systems have been used by many workers to purify lipases [96, 97], xylanases [98,99], proteases [100] and glucoamylase [101] from microbial sources. It involves partitioning of the desired enzyme in one of the phases of ATPS as a result of combination of electrostatic interactions, hydrophobicity, molecular size and other physico-chemical properties [102, 103]. Advantages of aqueous two-phase separation for protein purification include negligible mass transfer resistance, amenability to continuous processing, a high capacity for handling of solids and lower investment cost. But aqueous two-phase systems demix slowly due to their similar properties, thus posing a problem for large-scale enzyme purification. A method to accelerate the demixing rate of these systems using a traveling acoustic wave field has been developed. The acoustically assisted method reduced the demixing time by two-fold compared to gravity alone in the PEG/ maltodextrin system [104]. Another approach employed to intensify the conventional aqueous two-phase system is by conversion of the dispersed phase into colloidal gas aphrons (CGAs). In this method, the PEG phase was converted to CGAs by adding surfactants and stirring at 6000 rpm for 15 seconds. The effects of the surfactant and its concentration, dispersed phase velocity, phase composition and type of sparger on the dispersed phase mass transfer coefficient, $K_d a$, were investigated. A multi-orifice sparger and 0.33 g/L cetyltrimethylammonium chloride gave optimal results [105]. Studies suggest that highly hydrophobic surfactants such as cetyltrimethylammonium bromide (CTAB) and tetradecyldimethylamine oxide partition preferably towards the top PEG phase in the PEG/salt system, whereas tetraalkylammonium halide organic salts (that carried the shielded, charged head group) preferred the salt phase [106]. In another study, the effects of surface active additives like sodium butylbenzenesulfonate, sodium butylmonoglycol sulfate, tetrabutylammonium bromide and Tween 20 on the partitioning of albumin, lysozyme, glucose oxidase and *β*-lactaglobulin were investigated in the PEG/dextran system. The partitioning of enzymes or proteins was dependent upon their structure and charge. Due to electrostatic and hydrophobic interactions the amphiphiles partition unevenly between the two phases. The hydrophobic effect contributed to partitioning of the proteins if the proteins had a significant number of surface hydrophobic amino acids [107]. Recently, polyelectrolyte-based ATPS has been described where the formation of two phases occurs at a low polymer concentration (as low as 0.5% as compared to the traditional 10–12% concentration). The partitioning of proteins in polyelectrolyte-based ATPS is predominantly controlled by electrostatic interactions [102, 103].

5.2 Affinity Based Methods

In general, many purification protocols include affinity chromatography as a final step to obtain purified enzyme preparations [108, 109]. Affinity based purifications have also been used as a highly specific method for single-step enzyme purification. Affinity precipitation using alginate as a macroaffinity ligand to purify lipase from *Chromobacterium viscosum* has been reported. This method yielded 87% enzyme [110]. Affinity precipitation using the anionic polymer eudragit S100 yielded 85% enzyme with 10-fold purification of *endo* 1,4-*β*-D-xylanase from a crude culture of *Aspergillus* sp. [111]. Using yeast enzyme alcohol dehydrogenase as a ligand, affinity ultrafiltration was used to purify coenzyme NADH from permeabilized cells of *Saccharomyces cerevisiae* [112].

Milk clotting aspartyl protease was purified from the culture filtrate of *Rhizomucor miehei* using two-step purification involving ion-exchange chromatography followed by affinity separation, where benzamidamine was used as an affinity ligand. This two-step purification protocol has resulted in 6.7-fold purification and 22% recovery [113]. 2,4-Dichlorophenol hydroxylase has been purified from *Pseudomonas cepacia* using single-step affinity chromatography employing 2,4 dichlorophenol (DCP)-sepharose CL-4B. This method has yielded 60.5% enzyme with 34-fold purification [114]. Immunoaffinity purification of glucose/xylose isomerase from *Streptomyces* sp. NCIM 2730 was carried out. The polyclonal antibodies against xylose isomerase were raised in rabbits. Antibodies coupled with divinylsulfone active sepharose CL-4B were used for purifying the enzyme, which yielded 75% purification [115]. Fluidized-bed affinity chromatography using alginate and chitosan is yet another approach employed for the separation of cellulase from *Aspergillus niger* and has resulted in 30-fold purification with 80% recovery [116].

In addition to partitioning in ATPS and affinity-based methods, another novel method used for enzyme purification includes a reversed micellar system in organic solvent for extraction of acid phosphatase from *Aspergillus niger* [117]. Protease has been purified using adsorption on activated charcoal followed by treatment with hydrogen peroxide and acetone precipitation. This has resulted in 95% recovery with 1.6-fold purification [118]. Extracellular tannase has been purified from the culture filtrate of *Aspergillus japonicus* using tannic acid and PEG 600 to precipitate the enzyme at low pH values [119].

6 Modification and Improvement of Enzymes

A number of approaches has been opted for designing enzymes for their optimal stability with respect to their reaction environment such as temperature, pH, solvents etc. The approach used for obtaining such an enzyme depends on the particular application and cost effectiveness. A combination of various approaches for designing an enzyme can also be used. In this section, the efforts of Indian scientists to modify enzymes for their stability and improved activity in aqueous and non-aqueous environments by selection of microorganisms after mutagenesis, chemical modification, medium engineering and the novel approach of protein engineering are discussed.

6.1 Selection of Microbial Strains and Genetic Improvement

Microorganisms can survive in a wide range of temperatures, pH values and other extremes of environment. The microbes selected from these environments have intrinsic ability to produce enzymes that are resistant to such extremes of environment and have potential industrial applications. Microbial strains obtained from extremes of environment such as temperature and pH have been discussed in Sect. 3.

Microorganisms usually synthesize enzymes that are suitable for aqueous environment and physiological pH and temperature conditions. This is based on organism's genome. Thus, one of the approaches to improve functioning of enzymes is to modify the genome of the organism. In addition to gene cloning and expression of enzymatic activity in host (as discussed in Sect. 4), mutation and protoplast fusion have been widely employed for obtaining stable enzyme.

Normally, mutation (inheritable changes in gene) occurs during the process of evolution. Exposing the microbial culture to UV radiation and or various chemicals can also cause the mutation. Such an approach has been opted for improving strains for better enzyme production, improved thermostability, required pH optima, better substrate utilization or resistance to catabolite repression.

Aspergillus niger has been mutagenized using UV for amylase overproduction at CFTRI, Mysore [120]. This enzyme has potential applications for the starch processing industries. Using UV or NTG mutagenesis of *Thermomyces lanuginosus*, α-amylase hyperproducing mutants [121] have also been reported. Extracellular β-amylase was produced from *Bacillus megaterium* B 6 mutants, isolated by UV radiation and NTG mutagenesis at Bose Institute, Kolkata [122]. The mutant strain was capable of growing on starch wastes like those from arrowroot, maize, potato etc. obtained from market dumps. Thus the enzyme can be produced at low cost with effective waste utilization.

A program at IIT, Delhi on mutagenesis of *Trichoderma reesei* using UV/NTG resulted in several hypercellulase producing strains that were used for further studies [13]. A cellulase hyperproducing strain from *Aspergillus nidulans* ATCC using UV mutagenesis has been reported [123]. Using UV or NTG mutagenesis of *Thermomyces lanuginosus,* xylanase hyperproducing mutant [124] has been reported. At IIT, Delhi *Fusarium oxysporum* was mutagenized using UV and NTG to select hyperxylanolytic mutants that secreted high levels of xylanolytic enzymes on commercial xylan and several agricultural residues of which wheat bran supported maximum enzyme yield [125]. Another study employed *Melanocarpous albomyces* to select xylanolytic mutants [126]. Bakshi et al. [127] mutagenized *Bacillus cereus* using UV radiation to increase pullulanase production that showed a potential application for starch saccharification.

Corynebacterium and *Rhizopus arrhizus* were mutagenized to obtain respective mutants with high lipase activity at Jadavpur University and University of Kolkata [128, 129]. *Streptomyces peucetius* ATCC 29050 was mutagenized and mutants defective in daunorubicin biosynthesis were selected and were found to overproduce chitinase at MKU, Madurai [130]. Another group at MKU, Madurai has mutagenized *Zymomonas mobilis* to obtain better ethanol-producing strains [131] and poor levan-forming activity [132]. Using transposon mutagenesis, *Anabena* sp. was mutagenized to enhance nitrogenase activity and excrete ammonia in the medium at BHU, Varanasi [133, 134].

The protoplast fusion technique has increased the prospects of combining characteristics found in two different strains. Both intrageneric and intergeneric protoplast fusion has been employed. The fusants are grown to regenerate their cell wall.

The intergeneric protoplast fusion between xylanase producing *Bacillus subtilis* LYT (rifampcin-resistant mutant) capable of degrading xylan and pullulan and *Corynebacterium acetoacidophilum* (a lysine producer) was carried out at IIT, Delhi. Some of the selected fusants exhibited both xylanase- and lysine-producing abilities [135]. Protoplast fusion of *Trichoderma reesei* QM 9414 mutants has been employed for improving the cellulase activity [136]. Mutants showed enhanced yields of cellobiohydrolase whereas cellulase and β-glucosidase yields were intermediate as compared to their respective parents. The production of CM-cellulase from the two intergeneric fusants M14 and M62 of *Trichoderma reesei* QM 9414 has been demonstrated at IIT, Madras [137]. Increased β-glucosidase activity was also noted in fusants obtained from *Sporotrichum thermophile* at Pantnagar Agricultural and Technological University, Pantnagar [138]. NCL, Pune has been working on intergeneric protoplast fusion between

Cellulomonas sp. and *Bacillus subtilis* that has resulted in fusants. These fusants produced extracellular aryl β-glucosidase upon growing them on insoluble substrates. The enzyme exhibited a broad pH range (pH 5.5–7.0) and showed better thermotolerance compared to the parent strains [139]. The same group at NCL obtained amylase hyperproducing stable intergeneric hybrids of *Bacillus subtilis* and *Zymomonas mobilis* [140]. Amylase hyperproducing and catabolite-resistant fusants of *Thermomyces lanuginosus* have also been obtained by intraspecific protoplast fusion [141]. CFTRI, Mysore has been working on protoplast fusion between *Aspergillus carbonarius* and *Aspergillus niger* that has resulted in fusants with higher α-glucosidase activity and increased polygalacturonase production during solid-state fermentation [142].

6.2 Chemical Modification of Enzymes

Various groups have carried out chemical modifications of the surface of enzyme molecules to achieve stabilization of enzymes. As such, any modification in the primary structure of the enzyme is likely to alter the conformation of the enzyme and that may alter the stability. Due to the lack of our detailed understanding of the physical relationship between structure and stability, these approaches have mostly been empirical. Nevertheless, chemical modification has resulted in improved stability of some of the enzymes.

IIT, Delhi has been working on cross-linking of enzymes using glutaraldehyde to chemically modify enzymes. Using this approach cellulase obtained from *Aspergillus niger* has been modified [143]. A number of other enzymes such as catechol oxidase, acid phosphatase, β-galactosidase and trypsin have been chemically modified using cattle serum albumin and cross-linking with glutaraldehyde to form aggregates of the enzymes. The amorphous aggregates of these enzymes exhibit stability towards heat and organic cosolvents [144]. Studies at Centre for Protein Engineering and Biomedical Research, Madaras have been carried out to stabilize enzymes using glutaladehyde. β-Galactosidase from *Aspergillus oryzae* [145], cholesterol esterase from *Pseudomonas* sp. and trypsin [146] have been stabilized using this approach. The degree of cross-linking required for stabilization of an enzyme was found to be dependent on the type of enzyme. Trypsin cross-linked with glutaraldehyde at a ratio of 1:250 showed modifications of 80% of the NH_2 groups and the resultant enzyme exhibited stability against temperature, denaturants and organic solvents [146]. On the other hand, cross-linking of β-galactosidase from *Aspergillus oryzae* at an enzyme:glutaraldehyde ratio of 1:10 produced a stable enzyme [145]. Polymeric sucrose has also been used to stabilize cholesterol oxidase [147] and papain [148]. At University Department of Chemical Technology, Bombay papain was chemically modified by succinylation, which showed a shift in the pH optima of enzyme from pH 6 to 8. This modification has resulted in an enzyme suitable for detergent formulations [149].

Chemical modification of carboxylic groups of the xylanase enzyme from *Bacillus* sp. has been carried out at NCL, Pune using Woodward's reagent K and it was observed that one carboxylic group was essential for the activity of enzyme. Kinetic analysis of the modified enzyme showed similar K_m but lower K_{cat} values than the native enzyme, indicating that the chemical modification of carboxylic groups affected catalytic hydrolysis but not the substrate binding [150]. At MKU, Madurai the β-D-fructo-furanosidase enzyme obtained from *Thermomyces lanuginosus* was incubated with exogenous protein [151] or phospholipid [152] and showed enhancement of its enzymatic activity. This activation was specific to the thermophilic enzyme.

6.3 Modification of Reaction Environment

Enzymatic reactions are susceptible to factors of the reaction environment such as pH, temperature, ionic concentrations, solvents and soluble additives. A number of studies has been reported on the factors affecting enzyme catalysis due to changes in these environmental conditions.

6.3.1 *Effect of Organic Solvents*

Until recently enzymes have been employed predominantly in an aqueous environment. However, due to the poor solubility of many organic compounds (like fats, oils, aromatic compounds, steroids etc. in water), undesirable side reactions and unfavorable thermodynamic equilibrium in water, water proved to be a poor reaction medium for many biotransformations. To overcome such problems, enzymes have been employed in organic solvents.

Sundaram and his group at Centre for Protein Engineering and Biomedical Research, Chennai have shown that the stability of β-glucosidase obtained from *Aspergillus oryzae* is maximum in water miscible organic solvents, e.g., acetone, ethanol, dimethylformamide (DMF), dimethyl sulfoxide (DMSO), acetonitrile and dioxane at pH 4.6. Addition of polyhydroxy compounds such as ethylene glycol, glycerol, mannitol, sorbitol stabilized the enzyme in borate buffer (at pH 8.0) against the solvents. It appears that the binding of borate to the carbohydrate moiety of the glycoprotein alters the conformation of the enzyme which then facilitates the interaction of organic solvents leading to inactivation of the enzyme. However, a small amount of carbohydrates such as mannitol prevented this inactivation by competing with the enzyme binding the borate ions and thus preventing the conformational change [153].

Work carried out at NCL, Pune; CFTRI, Mysore and University of Bombay has elucidated the role of solvent hydrophobicity on the activity of different enzymes. Properties of various glycosidases like β-glucosidase, β-xylosidase and α-L-arabino-furanosidase from *Sclerotium rolfsii* were studied in organic solvents. A correlation between hydrophobicity (log P), molecular weight of solvents and enzymatic activity was established [154]. Solvents with log P values of 0.49–2.20 and molecular weight 80–155 supported the activity and thermostability of enzymes [154]. Similarly, a higher activity of lipozyme from *Rhizomucor miehei* was noted in the presence of highly non-polar solvents (solvents with log P values greater than 2) [155]. *Mucor miehei* lipase showed temperature-induced thermodeactivation of enzyme, which was aggravated on exposure to butanol and reduced in the presence of lauryl alcohol. Non-polar solvents such as isooctane conferred stability to the thermodenaturation of this enzyme [156].

6.3.2
Role of Soluble Additives

The use of external additives to modify the microenvironment of enzyme reactions is a promising approach for stabilizing enzymes. Studies at NCL, Pune have demonstrated that, in the presence of xylitol and sorbitol, the hydrolytic activity of pullulanase was stabilized against temperature, while ethylene glycol showed no protective effect. The stabilizing effect of polyhydric alcohol was related to their concentration and the number of -OH groups present per molecule [157]. Additions of sorbitol, mannitol and glycerol were found to increase the thermostability of *endo*-1,4 β-xylanase, whereas protection against thermoinactivation was produced by glycine [158]. The melting temperatures of lysozyme, RNAse and other enzymes have been increased using sugars and polyols. Thermolabile enzyme preparations have been stabilized by high ionic strength solutions. Based on these observations it was inferred that possibly electrostriction leads to a hydrophobic environment in aqueous solution and enhances the shelf-life of an enzyme [159]. A significant increase in the activity of β-xylanase from *Bacillus* sp. was observed on addition of sugars like glucose, xylose, mannose, arabinose, rhamnose, sucrose, non-ionic detergents and bile salts [160]. Enzymes like restriction endonucleases and modifying enzymes T_4 DNA ligase and T_7 DNA polymerase showed extraordinary stability when dried at 37°C or ambient temperature with trehalose. The enzyme was stable up to 70°C and remained intact up to 9 months at 37°C as reported from Gauhati University [161].

A report published from Pantnagar Agricultural and Technological University, Pantnagar suggested that cations like Ca^{2+}, Mg^{2+} and Mn^{2+} apparently protect the enzyme alkaline protease obtained from *Bacillus* sp. against thermal denaturation and this effect was pronounced in presence of a Ca^{2+} concentration above 5 mM [162].

6.3.3
Role of Reverse Micelles

Various studies have been performed to modify the bulk phase of enzymatic reactions by entrapping enzymes in reverse micelles. Katiyar and Dey [163] have shown that glucose 6-phosphate dehydrogenase from yeast can be solubilized in a reverse-micellar solution of mixed surfactant of sodium bis(2-ethylhexyl)sulfosuccinate and polyoxyethylene (5) octylphenol in *n*-heptane. Interestingly, using this approach large and complex enzymes like glucose 6-phosphate dehydrogenase can be solubilized in apolar solvents where the enzyme retained its conformational integrity as well as optimal interactions between subunits and showed high activity [163]. Another study has demonstrated that the enzyme glucose oxidase obtained from *Aspergillus niger* entrapped in reverse micelles showed a more than two-fold increase in activity compared to aqueous solution [164]. When entrapped in reverse micelles using various surfactants, the increase in enzymatic activity has been also shown by the enzymes cellulase, α-amylase, β-D-fructofuranosidase, in addition to glucose oxidase obtained from *Aspergillus niger* [165]. Lower K_m values of the substrate in reverse micelles than in aqueous systems have been observed. In addition to increased activity of enzyme, an improved stability of enzymes in the presence of polyols in reverse micelles has been observed [166]. Work at University of Delhi and Hamdard University demonstrated 100% increases in the enzymatic activity of penicillinase enzyme from *Bacillus cereus* entrapped in aerosol OT reverse micelles as compared to aqueous solution [167]. Also, it has been shown that the activity and stability of *S. cerevisiae* alcohol dehydrogenase entrapped in aerosol OT reverse micelles were higher than those in aqueous buffer [168].

6.4 Immobilization

Immobilization of enzymes and non-living cells to insoluble matrices gave a serious impetus to the use of enzymes as catalysts in a number of process industries. The use of immobilized enzymes/cells essentially involved the use of heterogeneous catalysis, which has been well established in chemical process industries. Although the use of immobilized enzyme has potential difficulties such as loss of enzyme activity and operational stability, the success of immobilization is dependent on minimizing these problems. During the 1970s many institutions in India including IIT, Delhi; HAL, Pimpri; CFTRI, Mysore; RRL, Trivandrum and NCL, Pune started activities on enzyme immobilization. Entrapment of glucose isomerase containing heat-treated *Streptomyces/Actinoplanes* cells in polyester sacs (pore size ≤25 μ) was developed by Ghose and Chand at IIT, Delhi [3, 22]. This interesting system was used in the development of enzyme technology for the production of glucose/fructose syrup from cellulose hydrolyzate. Another notable development from the same Institute (Ghose and Kannan) resulted in the immobilization of urease containing bacterial cells by entrapment in cellulose acetate fibers [4]. A suspension of cellulose acetate containing heat-treated bacterial cells were forced through a nozzle and the mixture was drawn in the form of fiber through a trough containing a non-polar solvent so as to obtain insoluble fibers containing urease activity. These fibers were then used for hydrolysis of urea in a packed-bed reactor.

During the last decade several studies have been published using physical and chemical methods of enzyme/cell immobilization but they lack detailed studies on design and analysis of enzyme reactors for their industrial applications. These studies include immobilization of aldolase from *Haloferax mediterranei* by cross-linking in a proteinic matrix [169], entrapment of amylase containing yeast cells in reversed micelles [170], cyclomaltodextrin glucanotransferase on sepharose 6B [171] and endonuclease *EcoR1* by an immunoaffinity-based method on sepharose 4B [172]. A number of support matrices like agar, alginate, polyacrylamide have been used to immobilize whole cell [173–176]. For bioconversion of gallotannin to gallic acid, whole cells and spores of various fungi were entrapped in calcium alginate [177]. Similarly, bioconversion of progesterone to11α-hydroxyprogesterone was carried out using immobilized conidia of *Aspergillus ochraceus* [178]. *Saccharomyces cerevisiae* immobilized on jute fabric through adhesion using polyethyleneimine has been used for sucrose syrup inversion [179].

A number of enzymes has been immobilized using chemical methods, for example, covalent binding of fructosyl transferase from *Aureobasidium pullulans* has been carried out on CNBr-activated agarose for the preparation of a non-digestible sweetener fructo-oligosaccharide [180]. Similarly, aminated silica gel has been used as a support for covalent immobilization of polygalacturonase from *Aspergillus ustus*. Alkaline protease has been immobilized by cross-linking with activated nylon support using trichlorotriazine [181]. The *Fusarium oxysporum* enzyme innulinase was immobilized by cross-linking with a water-insoluble protein support prepared from soybean flour, mung bean or egg white [182].

Entrapment of enzymes or whole cells using polyacrylamide, alginate or carragenan has been the most common approach for immobilization. This method has been used for immobilization of penicillin amidase from *E. coli* [183], penicillin acylase from *Chainia* sp. [184] and α-amylase from *Halobacterium halobium* [185]. Physical adsorption, being the simplest and cheapest method, has been method of choice for the immobilization of industrial enzymes at large-scale operations. Using this method, immobilization of protease from *Bacillus* sp. on a vermiculite support [186], lipase from *Candida lipolytica* on activated alumina beads [187] and penicillin acylase on kieselguhr [184] has been carried out.

6.5 Protein Engineering

Various strategies involving protein-engineering tools for stabilization of enzymes have been used. These include disulfide bond formation, substitution to decrease the configurational entropy of chain unfolding or to remove aspartic acid residues, improvement of helix formation, creation of appropriate charges at the end of helices, addition of new interactions in rigid parts of the structure, elimination of unsatisfied hydrogen bonding groups and interior cavities, increase in internal hydrophobicity and reduction of the solvent-accessible hydrophobic surfaces [188].

Thermal inactivation of oligomeric enzymes is often irreversible and leads to precipitation of the enzyme. Balaram and his group at Indian Institute of Science, Bangalore is working on engineering of *Lactobacillus casei* thymidylate synthetase, a dimeric enzyme. This enzyme forms aggregates at higher temperatures and in the presence of urea. This enzyme was engineered across the subunit interface having symmetrically related disulfide bridges. For site-directed mutagenesis, sites were chosen on the basis of an algorithm for introduction of stereochemically unstrained bridges into proteins. A marked enhancement in the thermal stability of the covalently cross-linked double disulfide-containing dimeric enzyme was demonstrated. The mutant enzyme remained soluble and retained its secondary structure even at 90°C in contrast to the wild-type enzyme, which precipitated at 52°C [189]. The enzyme was also engineered for its stability in the presence of urea. Engineered mutants of TS were obtained in which 2 subunits were cross-linked between residues 155–188′ and 188–155′ and this has resulted in resistance to urea [190]. The crystal structure of the covalently cross-linked enzyme was determined at 2.8 Å resolution. The remarkable increase in the thermostabilty of the mutant enzyme has been attributed to a covalent reinforcement of the relatively fragile protein-protein interface [191]. Another group at Indian Institute of Science, Bangalore has reported on model-based design and engineering of biocatalysts [192]. In this work, the Asp-His-Ser triad of serine protease has been incorporated into a non-protease scaffold by rational design to obtain a new biocatalyst. The work was based on knowledge-based computer modeling, which identifies sites in non-protease that are suitable for modeling the protease triad. A hydrogen-bonded triad that mimics protease catalytic triad geometry was modeled at these sites. Immunoglobulins containing this triad have shown catalytic activity [192]. Reddy and his group at CCMB, Hyderabad have used a novel approach for predicting the in vivo stability of a protein from its primary sequence. They have established a correlation between the stability of a protein and its dipeptide composition [193]. They have also studied the effect of substitution mutations on protein stability [194].

At Institute of Microbial Technology, Chandigarh work is being carried out in order to understand the basic structure-function relationship of proteins and to improve their functions. Sahni and coworkers, using tools of protein engineering, have designed and developed a new, second-generation version of streptokinase with improved clot specificity and fibrin affinity [195]. Another group has worked to understand the molecular mechanism of staphylokinase by isolating site-specific, modified and deletion mutants of staphylokinase. The studies have suggested that the Lys 11 residue of staphylokinase participates in modulating the interaction of the substrate during the staphylokinase-mediated plasminogen activation process [196].

At Centre for Protein Engineering and Biomedical Research, Madras, the stability of proteases like trypsin, chymotrypsin and papain was studied [197]. The stability of the enzyme was improved using various cross-linkers without loss in activity using distance mapping as a tool in protein engineering. Theoretical energy calculations were utilized to study the effect of protein modification on protein structure and activity. Based on the model calculations using the program AMBER 3.0 revision A on papain, differences in energy between native and chemically modified forms were calculated. Distance mapping of proteins was used to select cross-linkers to produce stable proteins [197].

7
Application of Enzymes

Although the enzyme industry has grown progressively in India, the pace of growth has been rather slow. Most of the enzymes used for various applications belong to the hydrolases class of enzymes. Amongst the hydrolases, extensive studies have been carried out to explore the possible applications of cellulases, proteases, amylases and lipases. Very few enzymes belonging to the transferase, oxido-reductase and lyase classes have been studied for their potential applications.

Some of the conventional and established applications of amylases in process industry include starch processing, maltose production and desizing in the textiles and paper industry. A major effort was undertaken during the 1970s and 1980s at IIT, Delhi to establish lignocellulosics derivative from agricultural residues (e.g., rice/wheat straw, bagasse etc.) as feedstock for organic chemicals, liquid fuel and food, under the leadership of Prof. T. K. Ghose, that involved the use of cellulases for hydrolysis of cellulose to produce glucose. Glucose syrup was then used for various bioconversion processes including fermentation to ethanol, acetone-butanol and fructose-containing syrup [198]. In view of the prevailing difficulties to compare the data on cellulase assays reported by many researchers world-wide, the Commission of Biotechnology, IUPAC, appointed an international committee chaired by Prof. T. K. Ghose to produce a document of standardized methods of cellulase assay [199].

In this program, extensive studies were conducted for the bulk production of cellulase [12–15] and saccharification of cellulosics [200, 201]. Based on these studies, a demonstration facility was created for techno-economic validation of laboratory/bench scale results. The technology package was found to be technically feasible but had economic constraints. The economic scenario is likely to change with the availability of cellulases/β-glucosidase in bulk quantities and alternative applications of lignin as by-product.

Some of the major applications of enzymes that have emerged from R and D activities in Indian laboratories over the last decade are described below.

7.1 Paper and Pulp

Applications of enzymes, especially hemicellulases [202] and xylanases [203] in the paper and pulp industry has been reviewed earlier. Hemicellulases and lignolytic enzymes have been used in bleaching, deinking, pulping and debarking [202]. White rot fungi, ligninase and hemicellulase enzymes have been used for biobleaching of kraft pulp and it was observed that enzymatic treatment reduces the need for active chlorine and thus provides an environment-friendly process. Xylanases have been extensively studied for biobleaching of kraft pulp [204–206] and bagasse pulp [207]. Cellulase has also been used for newspaper pulp deinking and decolorization [208].

7.2 Textile

Alkaline and themostable poygalacturonase (pectinase) from *Bacillus* sp. has been used for degumming of ramie (*Boehmeria nivea*) and sunn hemp (*Crotalaria juncea*) blast fibers for use in the textile and packaging industries [209]. The role of pectinases in the textile industry has been reviewed recently [210]. Extensive studies on the application of cellulase for biopolishing of cotton fiber

[211] and alkaline protease for degumming of silk [212] have been reported from IIT, Delhi.

7.3
Leather

To avoid tannery waste, enzymes are viable alternates in pre-tanning operations such as soaking, dehairing, bating and degreasing. The role of enzymes in the leather industry has been reviewed [213]. The roles of lipase, protease and tannase [213, 214] from various microbial sources have been studied in leather processing.

7.4
Laundry

Lipases [215] and alkaline serine protease from various *Bacillus* species [216] have been used as an additive in detergents.

7.5
Food, Beverage and Processing of Fruits

Both α- and β-amylases have been used in the food and beverage industry [217–219]. α-Amylases have been used for saccharification of dextrin, starch processing [217], starch liquefaction [220] and maltose production [221]. Fructosyl transferases have been used in preparation of fructo-oligosaccharide, a non-digestible sweetener [222], lipases have been used for cheese processing [223] and the synthesis of modified lipids [224–226]. Pectinases, naringinase and tannases have been employed for the processing of fruit juice and wine. Naringinase has been used for debittering of Kinnow juice [227]. Pectinases find their application in fruit juice extraction, debittering and removal of cloudiness [210, 228] and clarification of grape juice [209]. Tannase has been used in beer chill proofing and wine making [229].

7.6
Pharmaceutical

A number of enzymes find their application in the synthesis of pharmaceuticals. However, the major pharmaceutical products produced using enzymes are semi-synthetic penicillins. Penicillin V acylase has been used for the production of 6-aminopenicillanic acid (6-APA) [230]. D-Amino acid oxidase has been used for bioconversion of cephalosporin [231]. Lipases have shown potential for catalyzing reactions useful in the synthesis of pharmaceuticals or their resolution. Lipase from *Candida cylinderacea* has been used for deacylation of different classes of polyphenols that can be used as building blocks in the synthesis of anti-AIDS and anticancer agents [232]. Lipase from *Pseudomonas cepacia* has been used for preparation of chiral *R*- and *S*-atenolal (an antihypertensive drug) from phenylacetic acid [233]. The resolution of *R*- and *S*-dropropizine, an antitussive and sedative agent, has been achieved using *Pseudomonas cepacia* lipase [234]. *Candida rugosa* lipase has been used for the stereospecific esterification of ibuprofen [235].

7.7
Bio-Organic Synthesis

In non-aqueous environments lipases have been used to synthesize a number of stereospecific compounds useful as fine chemical, esters having specific properties such as those used in perfumery [236, 237] and chiral compounds [233, 235, 238, 239] used in the pharmaceutical industry.

7.8
Analytical

The use of enzymes in the development of biosensors has opened up a number of analytical applications of enzymes. Alcohol dehydrogenase has been used to develop an enzyme electrode for monitoring ethanol concentrations [240]. Glucose oxidase from *Aspergillus niger* was used to develop an interference-free, flow injection-based biosensor for monitoring micromolar concentrations of glucose [241]. Polyaniline microtubular biosensors were constructed using immobilized glucose oxidase, lipase and urease on three closely spaced devices, allowing the simultaneous analysis of a mixture of glucose, urea and triglycerides [242]. Recently, an ascorbate oxidase-based biosensor has been developed for monitoring organophosphorus pesticides [243].

8
Reactor Design

The design of bioreactors for microbial enzyme production involves a detailed analysis of the parameters associated with cell growth and enzyme biosynthesis. Once a suitable organism has been found and genetically improved, the operational and physiological parameters like substrate concentration, pH, temperature etc. need to be optimized for maximum enzyme productivity. In kinetic terms, optimization of the synthesis rate means finding the highest specific cell growth rate that favors the desired enzyme synthesis. The relationship between the specific enzyme synthesis rate and the specific growth rate can vary for different enzyme systems. However, once such a relationship is established, a limiting factor can be used to control the growth rate and thereby the synthesis rate of the enzyme. Such a strategy is often absent in the current literature. Another approach – statistical optimization by response surface methodology – has been commonly used for process optimization of various industrial enzymes, e.g., tannase [33], pullulanase [244] etc.

Of the three modes of microbial cell cultivation (batch, fed-batch and continuous flow), most enzyme production systems reported (Sect. 2) have followed batch cultures both for submerged and solid-state fermentation. A departure from this can be cited for the design of a packed-bed column bioreactor for the production of glucoamylase by SSF [245]. Another significant development relates to the optimization of a fed-batch bioreactor using a neural network model for the production of β-D-fructo-furanosidase from recombinant yeast [246]. A fed-batch strategy has also been reported for the overexpression of recombinant streptokinase [247].

The design of industrial bioreactors for enzyme production also requires an analysis of their transient behavior. Process noise during reactor operation associated with the environment needs to be minimized. Recombinants that have been studied for transient behavior via mathematical models of the cell culture system include β-galactosidase [248] and streptokinase [249] producers. The fermentation performance has been significantly improved by coupling of the neural filter with a neural controller for the production of β-galactosidase by recombinant *E. coli* fed-batch culture [250]. Similarly, streptokinase production was improved by controlled filtration of process noise (variation in substrate inflow rate) in a fed-batch culture [251].

The bioreactor design for enzyme applications requires detailed information on the mode of reactor operation, enzyme loading in the reactor, mass transfer limitations, if any, microenvironment effects and an operational strategy for constant conversion and productivity. Such studies, although reported in early 1980s for enzymes like glucose isomerase [3, 22] and urease [4], are largely lacking in the current literature. Studies on immobilized glucose isomerase-containing cells at IIT, Delhi resulted in an immobilized enzyme technology to produce glucose-fructose syrup. *Actinoplanes* cells were grown on the hydrolyzed hemicellulose fraction of lignocellulosic residue, heat-treated (70°C, 1 hour) and entrapped in polyester sacs. The sacs containing 484 IU/g dry cell weight of glucose isomerase activity were used in a packed-bed reactor for the continuous isomerization of concentrated cellulose hydrolyzate (2.5 M glucose). The immobilized cell system was shown to be reaction kinetics-controlled and possessed a half-life of 21 days at an operating temperature of 65°C. A reactor operational strategy for constant conversion and productivity was also proposed. Similarly, detailed bioreactor design and analysis was carried out for an immobilized-cell, packed-bed reactor for the hydrolysis of urea.

The possibility of a rapid rate of saccharification of cellulose in fairly high concentrations was demonstrated in a novel membrane bioreactor that removes glucose syrup in high concentration ~ 14% with simultaneous rejection of the cellulase enzyme into the reactor continuously [252]. This was the first report on the use of a membrane reactor for a soluble enzyme-catalyzed reaction and provided a novel approach for the simultaneous removal of the product and recycling of the enzyme for reuse.

Applications of lipases for bio-organic synthesis have been extensively studied [235–239]. A major additional requirement for these reactions is the control of water activity during the reaction. A new design of bioreactor with a water activity control system has been proposed by Kaur et al. [253, 254] that involves recirculation of saturated salt solution through silicon tubing immersed in the reaction mixture and is perhaps one of the important approaches reported. The system has been used for studying the kinetics of water transfer for biocatalysis in organic media. Such an approach also provides a reactor scale-up strategy for reverse hydrolytic reactions. Chand et al. [255] also reported a packed-bed reactor design for the lipase-catalyzed esterification of diols, where the two cosubstrates are immiscible.

9 Conclusion

Advances in enzyme technology have been possible due to the multitude of new knowledge in various areas of biochemical engineering and biotechnology. Tools of genetic engineering like gene cloning and PCR have made it possible to improve the yield of an enzyme from a specific microorganism. The development of powerful computers has helped in establishing the structure and function relationship of proteins and thus the design of novel enzymes with the help of site-directed mutagenesis (protein engineering). Thermophilic microorganisms have extended the use of enzymes in the harsher environments (like high temperature) required in many industrial processes. In addition, the development of newer approaches for enzyme isolation using ATPS and affinity-based methods has helped in reducing the cost of enzyme production. Immobilization techniques have helped in the reuse of enzymes in continuous processes. Similarly, process engineering parameters using a neural filter and a neural controller as well as a number of mathematical models for optimization of enzyme production have made it possible to produce enzymes in a large scale. However, when natural enzymes are recruited for industrial applications they are often not well suited due to poor substrate solubility, breakdown of unstable products and the conditions for the enzyme reaction may not be suitable for large-scale applications. Thus, the main challenge in enzyme technology is the design of enzymes for such applications.

India, being a rich source of biodiversity, has great potential for exploring new microorganisms with unique biocatalytic properties. Although a large number of thermophilic microorganisms has been isolated for obtaining thermotolerant enzymes, there is a need for the screening of barophilic and hyperthermophilic microorganisms for enzyme production.

In recent years, possibilities are emerging for new approaches that allow us to explore enzyme functions lacking in the natural environment and for which the molecular basis is poorly understood. Recent advances in the ability to create genetic diversity and screening for selected functions of enzymes have attracted the attention of scientists world-wide and has resulted in the development of tailor-made enzymes for specific functions. Such an in vitro evolution of enzymes has not been explored in India and requires the special attention of enzyme technologists. In addition, there are few data on reactor design for enzyme-catalyzed reactions and the development of novel bioreactor designs and more strategies for their scale-up and operation are needed.

Although the research and development base in enzyme science and technology has now expanded to a large number of government-funded institutions, its impact on industrial growth in this sector of biotechnology has been very limited. Most of the work related to enzymes in these institutions is devoted to finding new microbial sources of enzymes, improving the yield and studying potential applications in laboratories. There is a need for engineering of the processes developed for their commercialization. Some of the interesting and novel developments made at IIT, Delhi offer commercial potential in near future.

Acknowledgement. The authors wish to thank Prof. T. K. Ghose for his critical comments and suggestions. Technical help provided by Ms. Santosh Yadav and Ms. Mili Prabhakar during the preparation of the manuscript is also acknowledged.

10 References

1. Technomarket survey report on enzymes (1992) Department of Science and Technology, New Delhi
2. Ghose TK, Chand S (1977) In: Overmira TG (ed), Proc Reg Sem on Microbial Conversion Systems for Food and Fodder Production and Waste Management. KSIR, Kuwait
3. Ghose TK, Chand S (1978) J Ferm Technol 56:315
4. Ghose TK, Kannan V (1979) Enz Microb Technol 1:47
5. Bajpai P, Bajpai PK (1991) Appl Biochem Biotechnol 31:159
6. Bajpai P, Gera RK, Bajpai PK (1992) Enzyme Microb Technol 14:679
7. Bajpai P, Verma N, Neer, Bajpai PK (1991) J Biotechnol 18:265
8. Kumar NN, Bhide A, Gokhale DV, Deobagkar DN (1995) Biotechnol Appl Biochem 22:345
9. Gaur R, Garg SK, Singh SP, Verma J (1993) Bioresource Technol 462:13
10. Mohapatra BR, Banerjee UC, Bapuji M (1998) J Biotechnol 60:113
11. Dey S, Agarwal SO (1999) Indian J Biochem Biophys 36:150
12. Ghose TK, Sahai V (1979) Biotechnol Bioeng 21:283
13. Ghosh VK, Ghose TK, Gopalkrishanan KS (1982) Biotechnol Bioeng 24:141
14. Mishra S, Gopalkrishanan KS, Ghose TK (1982) Biotechnol Bioeng 24:251
15. Ghose TK, Panda T, Bisaria VS (1985) Biotechnol Bioeng 27:1353
16. Rajendran A, Gunasekaran P, Lakshmanan M (1994) Indian J Microbiol 34:289
17. Madamwar D, Patel S (1992) World J Microbiol Biotechnol 8:183
18. Felse PA, Panda T(1999) Process Biochem 34:563
19. Kapat A, Panda T (1997) Bioprocess Eng 16:269
20. Kapat A, Rakshit SK, Panda T (1996) Bioprocess Eng 15:13
21. Saini RK, Chakraborti S, Sobti RC, Patnaik PR, Banerjee UC (1999) Folia Microbiol 44:367
22. Venkataramani ES, Chand S, Ghose TK (1980) In: Ghose TK (ed), Proceedings IInd International Symposium on Bioconversion and Biochemical Engineering. Delhi, India, p 407
23. Debnath M, Majumdar SK (1992) J Microb Biotechnol 6:12
24. Prabhakar G, Raju DC (1993) Bioprocess Eng 8:283
25. Tayal S, Chand S (1994) In: 34th Annual Conference of AMI. Ludhiana, India
26. Yadav RP, Saxena RK, Gupta R, Davidson S (1998) Folia Microbiol 43:373
27. Phadatare SU, Deshpande VV, Srinivasan MC (1993) Enzyme Microb Technol 15:72
28. Sutar II, Srinivasan MC (1991) Biotechnol Lett 13:119
29. Sutar II, Srinivasan MC, Vartak HG (1992) World J Microbiol Biotechnol 8:254
30. Bhosale SH Rao MB, Deshpande VV, Srinivasan MC (1995) Enzyme Microb Technol 17:136
31. Sudhakaran VK, Shewale JG (1993) World J Microbiol Biotechnol 9:233
32. Bradoo S, Gupta R, Saxena RK (1997) Process Biochem 32:135
33. Seth M, Chand S (2000) Process Biochem 36:39
34. Hadi TA, Banerjee R, Bhattacharyya BC (1994) Bioprocess Eng 11:239

35. Balakrishnan H, Srinivasan MC, Rele MV, Chaundhari K, Chandwadkar AJ (2000) Curr Sci 79:95
36. Saraswat V, Bisaria VS (1997) J Ferment Bioeng 83:352
37. Dhillon A, Khanna S (2000) World J Microbiol Biotechnol 16:325
38. Dhillon A, Gupta JK, Jauhari BM, Khanna S (2000) Bioresource Technol 73:273
39. Ramesh MV, Lonsane BK (1991) Appl Microbiol Biotechnol 35:591
40. Ramesh MV, Lonsane BK (1991) Biotechnol Lett 13:355
41. Selvakumar P, Ashakumary L, Pandey A (1998) Bioresource Technol (1998) 65:83
42. Pandey A, Selvakumar P, Ashakumary L (1994) World J Microbiol Biotechnol 10:348
43. Aikat K, Bhattacharya BC (2001) Process Biochem 36:1059
44. Tunga R, Banerjee R, Bhattacharyya BC (1998) Bioprocess Eng 19:187
45. Tunga R, Banerjee R, Bhattacharyya BC (1999) Bioprocess Eng 21:107
46. Chakraborty R, Srinivasan M (1993) J Microb Biotechnol 8:7
47. Narang S, Sahai V, Bisaria VS (2001) J Ferment Bioeng 91:425
48. Kuhad RC, Singh A (1993) World J Microbiol Biotechnol 9:100
49. Goyal N, Sidhu GS, Chakrabarti T, Gupta JK (1995) World J Microbiol Biotechnol 11: 593
50. Narang S, Satayanarayana T (2001) Lett Appl Microbiol 32:31
51. Malhotra R, Noorwez SM, Satayanarayana T (2000) Lett Appl Microbiol 31:378
52. Jana M, Chattopadhyay DJ, Pati BR (1997) Acta Microbiol Immunol Hung 44:281
53. Chakraborty K, Bhattacharyya BK, Sen SK (2000) Folia Microbiol 45:207
54. Raj KC, Chandra TS (1996) FEMS Microbiol Lett 145:457
55. Gupta N, Vohra RM, Hoondal GS (1992) Biotechnol Lett 14:1045
56. Bansod SM, Dutta Choudhary M, Srinivasan MC, Rele MV (1993) Biotechnol Lett 15:965
57. George SP, Ahmad A, Rao MB (2001) Bioresource Technol 78:221
58. Sarkar A, Upadhyay SN (1993) Folia Microbiol 38:29
59. Sidhu P, Sharma R, Soni SK, Gupta JK (1998) Folia Microbiol 43:51
60. Nayak DP Vyas VV (1999) World J Microbiol Biotechnol 15:73
61. Bradoo S, Saxena RK, Gupta R (1999) World J Microbiol Biotechnol 15: 97
62. Reddy AV (1991) Indian J Biochem Biophys 28:10
63. Bhushan B, Hoondal GS (1998) Biotechnol Lett 20:157
64. Hodgson J (1994) Bio/Technology 12:789
65. Shaif R, Ali A, Khanna S (1997) Indian J Microbiol 37:59
66. Srivastava KK, Verma PK, Srivastava R (1999) Biotechnol Lett 21:293
67. Srivastava R, Kumar GP, Srivastava KK (1995) Gene 164:185
68. Srivastava SK, Khanna S (1994) Indian J Microbiol 34:265
69. Pandey M, Mishra S (1995) J Ferment Bioeng 80:446
70. Pandey M, Mishra S (1997) Gene 190:45
71. Sethi B, Jain M, Chowdhary M, Soni Y, Bhatia Y, Sahai V, Mishra S (2002) Biotechnol Bioprocess Eng 7:43
72. Qureshy AF, Khan LA, Khanna S (2000) Enzyme Microb Technol 27:227
73. Shendye A, Gaikaiwari R, Rao M (1994)) World J Microb Biotechnol 10:414
74. Bandivadekar K, Deshpande V (1996) Enzyme Microbiol Technol 18:439
75. Manonmani HK, Kunhi AAM (1999) World J Microbiol Biotechnol 15:475
76. Kumar NN, Deobagkar DN (1995) Biotechnol Lett 17:797
77. Paul A, Bhosale SH, Maity TK, Deshpande VV (1998) Enzyme Microb Technol 23:506
78. Bhosale SH, Ghatge MS, Deshpande VV (1996) FEMS Microbiol Lett 145:95
79. Dey S (1999) Indian J Exp Biol 37:787
80. Chitte RR, Dey S (2001) Indian J Exp Biol 39:410
81. Vetrivel KS, Pandian SK, Chaudhary U, Daramlingam K (2001) Can J Microbiol 47:179
82. Bhattacharya S, Gupta VS, Prabhune AA, Sivaaman H, Debnath M, Ranjekar PK (1993) Enzyme Microb Technol 15:1070
83. Sheela P, Amuthan G, Mahadevan A (1996) Indian J Exp Biol 34:27
84. Mukundan AG, Kannan TR Gunasekaran P (1994) Indian J Microbiol 34:187
85. Karunakaran T, Gunasekaran P (1991) J Biotechnol 19:287

86. Gupta JC, Jaisani M, Pandey G, Mukherjee KJ (1999) J Biotechnol 68:125
87. Vyas VV, Gupta S, Sharma P (1994) Enzyme Microb Technol 16:240
88. Nayak DP, Vyas VV (1999) World J Microbiol Biotechnol 15:73
89. Gawande BN, Goel A, Patkar AY, Nene SN (1999) Appl Microbiol Biotechnol 51:504
90. Sharmila T, Sreemulu G, Nand K (1998) Biotechnol Appl Biochem 27:175
91. Saxena RK, Davidson WS, Gupta R (1999) In: Chand S, Jain S (eds), Fermentation Biotechnology Industrial Perspectives. All India Biotechnology Association, India
92. Seth M, Chand S (2002) Bioseparation (In Press)
93. Goyal A, Katiyar SS (1994) J Microbial Methods 20:225
94. Nath S (1995) J Chem Technol Biotechnol 62:295
95. Nath S, Patrickios CS, Hatton TA (1995) Biotechnol Prog 11:99
96. Boominathan R, Mishra P, Chand S (1995) Bioseparation 5:235
97. Gulati R, Saxena RK, Gupta R (2000) Process Biochem 36:149
98. Jain A, Johri B (1999) Bioresource Technol 67:205
99. Kulkarni N, Vaidya A, Rao M (1999) Biochem Biophys Res Commun 255:274
100. Sinha R, Singh RP, Ahmed S, Garg SK (1996) Bioresource Technol 55:163
101. Tamija S, Srinivas ND, Rao KSMSR, Gowthaman MK (1997) Process Biochem 32:635
102. Gupta V, Nath S, Chand S (2002) Indian J Biotechnol 1:87
103. Gupta V, Nath S, Chand S (2002) Polymers 43:3387
104. Srinivas ND, Barhate RS, Raghavarao KSMS, Todd P (2000) Appl Microbiol Biotechnol 53:650
105. Save SV, Pangarkar VG, Kumar SV (1993) Biotechnol Bioeng 41:72
106. Gaikar VG, Bodhankar SS, Latha V (1996) J Chem Technol Biotechnol 67:329
107. Bodhankar S, Gaikar VG (1998) J Chem Technol Biotechnol 73:251
108. Madyastha KM, Gururaja TL (1995) Biochem Biophys Res Commun 211:540
109. Phadtare SU, Rawat UB, Rao MB (1977) FEMS Microbiol Lett 146:79
110. Sharma S, Gupta MN (2001) Biotechnol Appl Biochem 33:161
111. Gawande PV, Kamat MY (1999) Process Biochem 34:577
112. Goyal A, Chand S (1996) Indian J Exp Biol 34:999
113. Preetha S, Boopathy R (1997) World J Microbiol Biotechnol 13:573
114. Rajendirane V, Bhat MA, Vadyanathan CS (1991) Arch Biochem Biophys 288:169
115. Ghatge M, Mawal Y, Gaikwad S, Deshpande V (1991) Appl Biochem Biotechnol 31:11
116. Roy I, Sardar M, Gupta MN (2000) Enzyme Microb Technol 27:53
117. Soni K, Madamwar D (2000) Process Biochem 36:311
118. Tunga R, Banerjee R, Bhattacharyya BC (1999) Bioprocess Eng 21:447
119. Gupta R, Bradoo S, Saxena RK (1997) Lett Appl Microbiol 24:253
120. Suresh C, Dubey AK, Srikanta S, Kumar SU, Karanth NG (1999) Appl Microbiol Biotechnol 51:673
121. Singh CB, Arvind SS, Singh SH (1996) Acta Microbiol Pol 45:31
122. Ray RR, Jana SC, Nanda G (1997) Indian J Exp Biol 35:285
123. Kaur A, Sandhu DK (1997) Acta Microbiol Immunol Hung 44:187
124. Chadha BS, Jaswinder K, Rubinder K, Saini HS, Singh S (1999) World J Microbiol Biotechnol 15:195
125. Singh A, Kuhad RC, Kumar M (1995) Enzyme Microb Technol 17:551
126. Jethro J, Ganesh R, Goel R, Johri BN (1993) J Microb Biotechnol 8:17
127. Bakshi A, Patnaik PR, Gupta JK (1992) Biotechnol Lett 14:689
128. Ray N, Ray L, Srimani BN, Chattopadhyay P (1999) Indian J Exp Biol 37:818
129. Chattopadhyay M, Banik AK, Raychaudhuri S (1999) Folia Microbiol 44:37
130. Vetrivel KS, Dharmlingam K (2000) Can J Microbiol 46:956
131. Kannan TR, Mukundan AG, Gunasekaran P (1993) J Ferment Bioeng 75:265
132. Sangiliyandi G, Gunasekaran P (2000) Process Biochem 36:543
133. Singh AP, Tiwari DN (1998) World J Microbiol Biotechnol 14:591
134. Tiwari DN, Singh AP (1997) J Plant Physiol 150:9
135. Deb JK, Malik S, Ghosh VK, Mathai S, Sethi R (1990) FEMS Microbiol Lett 71:287
136. Sandhu DK, Bawa S (1992) Appl Biochem Biotechnol 34:175

137. Srinivas R, Panda T (1998) J Ind Microbiol Biotechnol 21:178
138. Jaitly AK, Johri BN, Goel R (1993) Indian J Microbiol 33:175
139. Gokhale D, Deobagkar DN (1990) Appl Biochem Biotechnol 26:207
140. Gokhale D, Deobagkar DN (1994) Biotechnol Appl Biochem 20:109
141. Rubinder K, Chadha BS, Singh S, Saini HS (2000) Can J Microbiol 46:669
142. Kavitha R, Umesh Kumar S (2000) Biotechnol Bioeng 67:121
143. Sharma A, Khare SK, Gupta MN (2001) Bioresource Technol 78:281
144. Tyagi R, Batra R, Gupta MN (1999) Enzyme Microb Technol 24:348
145. Venkatesh R, Arun S, Rajalakshmi N, Subramanian L, Meenakshi V, Venugopalan K, Sundaram PV (1995) Ann NY Acad Sci 750:130
146. Venkatesh R, Sundaram PV (1993) Protein Eng 6:118
147. Venkatesh R, Sundaram PV (1996) Ann NY Acad Sci 799:569
148. Rajalakshmi N, Sundaram PV (1995) Protein Eng 8:1039
149. Kharparde SS, Singhal RS (2001) Bioresource Technol78:1
150. Chauthaiwae J, Rao M (1994) Biochim Biophys Acta 1204:164
151. Basha SY, Palanivelu P (1998) World J Microbiol Biotechnol 14:603
152. Basha SY, Palanivelu P (2000) Indian J Microbiol 39:217
153. Shubhada S, Sundaram PV, (1993) Enzyme Microb Technol 15:881
154. Chinnathambi S, Lachke A (1997) Biotech Tech 11:589
155. Hamsaveni DR, Praplla SG, Divakar S (2001) Process Biochem 36:1103
156. Gandhi NN, Sawant SB, Joshi JB, Mukesh D (1995) Enzyme Microb Technol 17:373
157. Kelkar HS, Deshpande MV (1991) Biotechnol Lett 13:901
158. George SP, Ahmad A, Rao MB (2001) Bioresource Technol 78:221
159. Kumar A (1997) Curr Sci 72:10
160. Capalash N, Gupta KG, Sharma P (1991) Lett Appl Microbiol 12:31
161. Handique AK (1994) Curr Sci 66:103
162. Paliwal N, Singh SP, Garg SK (1994) Bioresource Technol 50:209
163. Katiyar SS, De TK (1990) Biochem Int 20:1127
164. Madamwar DB Patel SJ, Jain N (1991) Indian J Microbiol 31:77
165. Madamwar DB Patel SJ, Jain N (1991) Fungal Biotechnol 75
166. Subramani S, Shah C, Madamwar D (1996) Appl Biochem Biotechnol 60:33
167. Chakravarty K, Varshney M, Maitra A (1995) Indian J Biochem Biophys 32:100
168. Sarkar S, Jain TK, Maitra A (1992) Biotechnol Bioeng 39:474
169. D'-Souza SE, Altekar W, D'Souza SF (1997) World J Microbiol Biotechnol 13:561
170. Gajjar L, Singh A, Dubey RS, Srivastava RC (1997) Appl Biochem Biotechnol 66:159
171. Abraham TE (1995) Biocatalysis Biotransform 12:137
172. Varshney H, Iqbal J, Saleedmuddin M (1999) Enzyme Microb Technol 25:172
173. Jamuna R, Ramakrishna SV (1992) Enzyme Microb Technol 14:36
174. Krishna KB, Varma A (1990) Physiol Immobilized Cells 449
175. Nandi R, Bhattacharyya PK, Bhaduri AN, Sengupta S (1992) Biotechnol Bioeng 39:775
176. Hemachander C, Bose N, Puvanakrishnan R (2001) Process Biochem 36:629
177. Bajpai B, Banerjee T, Patil S (1999) Indian J Exp Biol 37:94
178. Dutta TK, Samanta TB (1999) Curr Microbiol 39:309
179. D'Souza SF, Melo JS (2001) Process Biochem 36:677
180. Patil VB, Patil NB (1999) Indian J Exp Biol 37:830
181. Chellapandian M, Sastry CA (1996) Bioprocess Eng 15:95
182. Gupta AK, Kaur M, Kaur N, Singh R (1992) J Chem Technol Biotechnol 53:293
183. Prabhune A, SivaRaman H (1991) Appl Biochem Biotechnol 30:254
184. Chauhan S, Nichkawade A, Iyengar MRS, Chattoo BB (1998) Curr Microbiol 37:186
185. Patel S, Bagai R, Madamwar D (1996) Biocatalysis Biotransform 14:147
186. Chellapandian M, Sastry CA (1992) Bioprocess Eng 8:33
187. Padmini P, Rakshit SK, Baradarajan A (1993) Bioprocess Eng 9:43
188. Gupta MN (1991) Biotechnol Appl Biochem 14:1
189. Gokhale RS Agarwalla S, Francis VS, Santi V, Balaram P (1994) J Mol Biol 235:89
190. Agarwalla S, Gokhale RS, Santi V, Balaram P (1996) Protein Sci 5:270

191. Velanker SS, Gokhale RS, Ray RS, Ray SS, Gopal B, Parthasarthy S, Santi DV, Balaram P, Murthy MRV (1999) Protein Sci 8:930
192. Iengar P, Ramakrishnan C (1999) Protein Eng 12:649
193. Guruprasad K, Reddy BVB, Pandit MW (1990) Protein Eng 4:155
194. Reddy BVB, Datta S, Tiwari S (1998) Protein Eng 11:1137
195. Nihalani D, Raghava GPS, Sahni G (1997) Protein Sci 6:1284
196. Rajamohan G, Dikshit KL (2000) FEBS Lett 474:151
197. Nagaragan V, Pattabhi V, Sundaram PV (1996) Ann NY Acad Sci 799:413
198. Chand S, Ghosh P (1982) In: Jain SC (ed), Proc National Symposium on Biotechnology. Chandigarh
199. Ghose TK (1987) Pure Appl Chem 59:257
200. Ghose TK, Roychoudhury PK, Ghosh P (1984) Biotechnol Bioeng 26:377
201. Roychoudhury PK, Ghose TK, Ghosh P, Chotani GK (1986) Biotechnol Bioeng 28:972
202. Bajpai P (1999) Biotechnol Prog 15:147
203. Srinivasan MC, Rele MV (1999) Curr Sci 77:137
204. Dhillon A, Gupta JK, Jauhari BM, Khanna S (2000) Bioresource Technol 73:273
205. Subramaniyan S, Prema P (2000) FEMS Microbiol Lett 183:1
206. Bajpai P, Bajpai PK (1997) Adv Biochem Eng/Biotechnol 56:1
207. Kulkarni N, Rao M (1996) J Biotechnol 51:167
208. Bajpai P (1997) Adv Appl Microbiol 45:241
209. Sreenath HK, Santhanam K (1992) J Ferment Bioeng 73:241
210. Kashyap DR, Vohra PK, Chopra S, Tewari R (2001) Bioresource Technol 77:215
211. Gulrajani ML, Roy P, Agarwal R, Chand S (1999) Ind J Fibres Textile Res 24:88
212. Gulrajani ML, Agarwal R, Chand S (2000) Ind J Fibres Textile Res 25:138
213. Kamini NR, Hemachander C, Geraldine Sandana Mala J, Puvanakrishnan R (1999) Curr Sci 77:80
214. Mondal KC, Banerjee R, Pati BR (2000) Biotechnol Lett 22:767
215. Hemachander C, Puvanakrishnan R (2000) Process Biochem 35:809
216. Pundir CS, Malik V, Bhargava AK, Thakur M, Kalia V, Singh S, Kuchhal NK (1999) J Plant Biochem Biotechnol 8:123
217. Suresh C, Dubey AK, Srikanta S, Kumar SU, Karanth NG (1999) Appl Microbiol Biotechnol 51:673
218. Reddy RM, Reddy PG, Seenayya G(1999) Process Biochem 34:87
219. Ray RR, Nanda G (1996) Acta Microbiol Pol 45:241
220. Soni SK, Sandhu IK, Bath KS, Banerjee UC, Patnaik PR (1997) Folia Microbiol 41:243
221. Thacker SP, Ramamurthy V, Kothari RM (1992) Starch 44:339
222. Mujawar SK, Kotha A, Rajan CR, Ponrathnam S, Shewale JG (1999) J Biotechnol 75:11
223. Neelakantan S, Mohanty AK, Kaushik JK (1999) Curr Sci 77:143
224. Sridhar R, Lakshminarayana G, Kaimal TNB (1991) J Agric Food Chem 39:2069
225. Mukesh D, Banerji AA, Newadkar R, Bevinakatti HS (1993) Biotechnol Lett 15:77
226. Ray S, Bhattacharyya DK (1995) J Am Oil Chem Soc 72:327
227. Puri M, Marwaha SS, Kothari RM (1996) Enzyme Microb Technol 18:281
228. Kashyap DR, Chandra S, Kaul A, Tewari R (2000) World J Microbiol Biotechnol 16: 277
229. Lekha PK, Lonsane BK (1997) Adv Appl Microbiol 44:215
230. Shewale JG, Shudhakaran VK (1997) Enzyme Microb Technol 20:402
231. Mujawar SK, Kotha A, Rajan CR, Ponrathnam S, Shewale JG (1999) J Biotechnol 75:11
232. Parmar VS, Prasad AK, Sharma NK, Bisht KS, Sinha R, Taneja P (1992) Pure Appl Chem 64:1135
233. Bevinakatti HS, Banerji AA (1992) J Org Chem 57:6003
234. Salunkhe MM, Nair RV (2001) Enzyme Microb Technol 28:333
235. Lakshmi BS, Kangueane P, Guo Y, Chen YZ, Gautam P (1999) Biocatalysis Biotransform 17:475
236. Chatterjee T, Bhattacharyya DK (1998) Biotechnol Lett 2:865
237. Mestri S, Pai JS (1995) Biotechnol Lett 17:459
238. Roy A (1999) J Agric Food Chem 47:5209

239. Pannu J, Chand S (1999) In: Chand S, Jain SC (eds), Fermentation Biotechnology Industrial Perspective. All India Biotech Association, New Delhi
240. Pandey PC, Upadhyay S, Tiwari I, Tripathi VS (2001) Anal Biochem 288:39
241. Sapre AG, Bedekar A, Despande AV, Lali AM (2000) Biotech Lett 22:569
242. Sukeerthi S, Contractor AQ (1999) Anal Chem 71:2231
243. Rekha K, Gouda MD, Thakur MS, Karanth NG (2000) Biosensors Bioelectron 15:499
244. Reddy PRM, Reddy G, Seenayya G (1999) Bioprocess Eng 21:497
245. Pandey A, Radhakrishnan S (1993) Process Biochem 28:305
246. Chaudhari B, Modak JM (1998) Bioprocess Eng 19:71
247. Yazdani SS, Mukherjee KJ (1998) 20:923
248. Patnaik PR (1999) Bioprocess Eng 20:31
249. Patnaik PR (2000) Biotechnol Lett 22:393
250. Patnaik PR (1999) Biotechnol Tech 13:735
251. Patnaik PR (1999) Process Biochem 35:309
252. Ghose TK (1972) US Patent 3,642,580
253. Kaur J, Ernst W, Adlercreuetz P, Chand S, Mattiasson B (1997) Enzyme Microbiol Technol 21:496
254. Ernst W, Kaur J, Adlercreuetz P, Chand S, Mattiasson B (1997) Enzyme Microb Technol 21:502
255. Chand S, Adlercreuetz P, Mattiasson B (1997) Enzyme Microb Technol 20:102

Received: June 2002

Adv Biochem Engin/Biotechnol (2003) 85: 125–180
DOI 10.1007/b11047

Molecular Mechanisms of Energy Transduction in Cells: Engineering Applications and Biological Implications

Sunil Nath

Department of Biochemical Engineering and Biotechnology, Indian Institute of Technology, Hauz Khas, New Delhi 110 016, India. *E-mail: sunath@dbeb.iitd.ernet.in*

Dedicated to Prof. Tarun K. Ghose on the occasion of his 78th birthday

"Every novel idea in science passes through three stages. First people say it isn't true. Then they say it's true but not important. And finally they say it's true and important, but not new".
Anon

"All acquired knowledge, all learning, consists of the modification (possibly the rejection) of some sort of knowledge. All growth of knowledge consists in the improvement of existing knowledge which is changed in the hope of approaching nearer to the truth". K. R. Popper

The synthesis of ATP from ADP and inorganic phosphate by F_1F_0-ATP synthase, the universal enzyme in biological energy conversion, using the energy of a transmembrane gradient of ions, and the use of ATP by the myosin-actin system to cause muscular contraction are among the most fundamental processes in biology. Both the ATP synthase and the myosin-actin may be looked upon as molecular machines. A detailed analysis of the molecular mechanisms of energy transduction by these molecular machines has been carried out in order to understand the means by which living cells produce and consume energy. These mechanisms have been compared with each other and their biological implications have been discussed. The thermodynamics of energy coupling in the oxidative phosphorylation process has been developed and the consistency of the mechanisms with the thermodynamics has been explored. Novel engineering applications that can result have been discussed in detail and several directions for future work have been pointed out.

Keywords. ATP synthesis, Oxidative phosphorylation, Muscle contraction, Molecular mechanism, Energy transduction, Molecular machines, Molecular engineering, Nanotechnology

Symbols and Abbreviations

A_O	affinity of oxidation
A_P	affinity of phosphorylation
A_P'	phosphorylation affinity due to anions
a, b	constants in the adsorption isotherm [Eq. (4)]
a, b, c	subunits of the F_0 portion of ATP synthase enzyme
$\alpha, \beta, \gamma, \delta, \varepsilon$	subunits of the F_1 portion of ATP synthase enzyme
$\beta_E, \beta_C, \beta_{TP}, \beta_{DP}$	open, closed, loose, tight conformations, respectively, of the catalytic site, as per the torsional mechanism of ATP synthesis
C	closed; constant in Eq. (7)
C_H	proton leak through the inner mitochondrial membrane
$\boldsymbol{C}_H$	3-dimensional couple
d	distance
E	effective energy available; electromotive force
F	Faraday
F_0	hydrophobic, membrane-bound portion of ATP synthase
F_1	hydrophilic, extra-membrane portion of ATP synthase
F_1F_0	complete ATP synthase enzyme
$\boldsymbol{F}_R$	3-dimensional force
F_z	force in the z-direction
f	fraction; final
f_{anti}	fraction of ATP synthase molecules involved in antisequence-port
f_{total}	total fraction of ATP synthase molecules carrying out ATP synthesis
ΔG	difference in free energy
ΔH	difference in enthalpy
H^+/O	proton to oxygen ratio
I	ion I
i	initial; i^{th} component; inside
in	inside
J_{ATP}	rate of ATP synthesis
J_H	rate of proton translocation
J_O	rate of oxidation
J_{OX}	rate of oxidation
J_P	rate of phosphorylation
K_d	dissociation constant
K_V	equilibrium constant in Eq. (5)
k_{K^+}	constant of proportionality between rate of K^+ transport and concentration gradient [Eq. (8)]
K^+/ATP	K^+ to ATP ratio
L	loose
l	length
L_{HH}	phenomenological coefficient for proton translocation
L_{OO}	phenomenological coefficient for oxidation
L_{OH}	coupling coefficient between oxidation and proton translocation

L_{PH}	coupling coefficient between proton translocation and phosphorylation
L_{PO}	coupling coefficient between oxidation and phosphorylation
L_{PP}	phenomenological coefficient for phosphorylation
L_{00}	overall phenomenological coefficient on the phosphorylation side
L_{01}	overall coupling coefficient (Table 4)
L_{11}	overall phenomenological coefficient on the oxidation side
$\Delta\mu$	chemical potential difference
$\Delta\tilde{\mu}$	electrochemical potential difference
$\Delta\tilde{\mu}_H$	electrochemical potential difference of protons
n	operating number of ATP synthase enzyme complexes or molecules
n_O	redox pump stoichiometry
n_P	ATPase pump stoichiometry
n_{total}	total number of ATP synthase enzyme complexes or molecules
η	efficiency
O	open
o	outside
out	outside
Δp	"protonmotive" force
ΔpA	pAnion difference
ΔpH	pH difference
P/O	ATP to oxygen ratio
$\Delta\psi$	electrical potential difference
$\Delta(\Delta\psi)$	change in electrical potential
q	degree of coupling
R	universal gas constant
dS	differential change in entropy
d_eS	differential change in exchange entropy
d_iS	differential change in entropy internal to the system
T	tight; temperature
τ	twisting moment
VK_m	K^+-valinomycin concentration
V_m	membrane valinomycin concentration
V_{max}	maximum rate
V_t	total valinomycin concentration
v	rate of ATP hydrolysis
v_{K^+}	rate of K^+ transport
v_{an}	rate of anion transport
v_{syn}	rate of ATP synthesis [Eq. (1)]
$v_{syn,an}$	rate of ATP synthesis due to anion transport
v_{syn,K^+}	rate of ATP synthesis due to K^+-valinomycin transport
x	affinity or thermodynamic force ratio; distance
Z	mechanistic stoichiometry in oxidative phosphorylation
ADP	adenosine diphosphate
anti	antisequenceport

ATP	adenosine triphosphate
CD	catalytic domain of myosin
HMM	heavy meromyosin
LMM	light meromyosin
P_i	inorganic phosphate
RD	regulatory domain of myosin
RTT	rotation-twist-tilt mechanism
S-1	S-1 region of myosin molecule
S-2	S-2 region of myosin molecule
T	tail of myosin molecule

1 Introduction

Adenosine triphosphate (ATP), the general energy currency of the cell, is synthesized by the universal enzyme F_1F_0-ATP synthase, which is present in abundance in the mitochondria of animals, the chloroplasts of plants and in bacteria [1, 2]. Since the ocean area and the amount of biomass is very large, the synthesis and use of ATP is the most prevalent chemical reaction occurring on the surface of the earth. It is a very important reaction for life and it is of great fundamental interest to understand how it occurs. The enzyme consists of a hydrophobic membrane-bound base-piece (F_0) and a hydrophilic extramembrane head-piece (F_1, with stoichiometry $\alpha_3\beta_3\gamma\delta\varepsilon$ in *Escherichia coli*) [1–13]. The F_0 and F_1 domains are linked by two slender stalks. The central stalk is formed by the ε-subunit and part of the γ-subunit, while the peripheral stalk is constituted by the hydrophilic portions of the two b-subunits of F_0 and the δ-subunit of F_1. The proton channel is formed by the interacting regions of a- and c-subunits in F_0, while the catalytic binding sites are predominantly in the β-subunits at the α-β interface. Great interest has been generated in this field after the direct observation of rotation of the central stalk in the hydrolysis mode by innovative techniques, making ATP synthase the smallest-known molecular nanomachine [14–16].

Force generation in muscle involves the interactions between actin, a helical protein, and myosin, a highly asymmetric protein molecule [17–22]. It is fundamentally important to elucidate how the hydrolysis of ATP is coupled to motion, and how force is generated by the actomyosin system of muscle. A detailed analysis of the molecular mechanisms of energy transduction by these molecular machines should help us in understanding the means by which living cells produce and consume energy. Insights obtained from such an investigation would be expected to have several biological implications and to lead to novel engineering applications. These aspects will be critically reviewed in the subsequent sections.

2 Molecular Mechanisms of Energy Transduction in the F_1 Portion of ATP Synthase

Two major candidate molecular mechanisms of ATP synthesis are Boyer's binding change mechanism [23–26] and the torsional mechanism of ion translocation, energy transduction and storage, and ATP synthesis proposed by Nath and coworkers [1, 9, 27–40, 69, 70]. The binding change mechanism was postulated in 1973 (when very little was known about the ATP synthase) and represented a milestone for that era. However, it is a gross mechanism that deals only with the F_1 portion of ATP synthase and ignores mechanistic aspects within the F_0 portion as well as the coupling between F_0 and F_1. It was proposed chiefly based on enzymological studies without any structural evidence or use of computational aids, which were lacking at that time. Moreover, most of the biochemical experiments were conducted in the hydrolysis mode, with the enzyme acting as a hydrolase, not as a synthase. Nonetheless, a molecular mechanism of ATP synthesis was postulated from these hydrolysis studies. This is, in the opinion of this researcher, a difficult proposition because (as is now gradually but surely being realized by a minority of researchers in the field), the driving forces for the two processes are different, and ATP synthesis is not a simple reversal of ATP hydrolysis [1, 2, 41]. Thus one cannot, in our view, propose a mechanism for ATP hydrolysis based on the action of the enzyme as a hydrolase and simply reverse the arrows to obtain the mechanism of ATP synthesis. Note, however, that this does not imply that microscopic reversibility is violated. The binding change mechanism also fails to explain recent structural, spectroscopic, and biochemical observations. Finally, the details of the ATP synthesis mechanism and the mechanical, molecular machine-like nature of ATP synthase have not been proposed in the binding change mechanism from 1973 till 2002.

On the other hand, the torsional mechanism of ion transport, energy transduction, energy storage and ATP synthesis is a complete mechanism that has several novel features and addresses the details of the molecular mechanism within F_0 [1, 30, 33, 35, 37–39, 69, 70], the molecular mechanism in F_1 [1, 9, 32, 36, 38, 40], and the molecular mechanism of coupling between F_0 to F_1 [1, 9, 31, 32, 35–40, 69, 70] and provides a detailed sequence of events and their causes. In this section, the major differences between the torsional mechanism and the binding change mechanism are presented.

2.1 Principal Differences between the Torsional Mechanism and the Binding Change Mechanism

First, according to the torsional mechanism, every elementary step requires energy [9, 30–32, 38]; this differs from the fundamental tenet of Boyer's binding change mechanism that energy of the proton gradient is used not to make ATP but primarily to release tightly bound ATP from the enzyme-ATP complex [23–26]. Second, the torsional mechanism clearly reveals the absence of site-site cooperativity in ATP synthase in the steady state physiological mode of functioning [1, 9, 32, 38]. This is different from the second fundamental tenet of the binding change mechanism. Third, "binding changes" "drive rotation of the γ-subunit" in the binding change mechanism while, according to the torsional mechanism, conformational changes are caused by Mg-nucleotide binding as well as by fundamental γ-β and ε-β interactions which arise from torsion and intersubunit rotation in ATP synthase. Possibilities include: a) energy of bound MgADP·P_i equals the energy of bound MgATP at the site, i.e., an equilibrium at the enzyme catalytic site as postulated by the original binding change mechanism; b) energy of enzyme catalytic site-bound MgADP·P_i is far greater than energy of bound MgATP because of the much tighter binding of ATP (compared to ADP) to the enzyme catalytic site and this drives the reaction, i.e., the large negative free energy of ATP binding makes the reaction go, which is the view of Penefsky and Boyer; c) the energies of bound forms are different, but, as per the torsional mechanism of ATP synthesis, *this does not drive the change/reaction.* Thus, in our view, one needs to alter the catalytic site to make it prefer ATP and achieve ATP synthesis. Finally, according to the binding change mechanism, *the binding energy released during the ATP binding step* performs useful work in the "user" molecule (e.g., the actin-myosin system in muscle [22]). According to the torsional mechanism, *the enthalpy change upon ATP hydrolysis* is transduced to useful work [1, 9, 22]. Thus, the elementary step whose energy is employed for the performance of useful work differs radically between the two mechanisms. The torsional mechanism and the binding change mechanism are thus completely different from each other. They may be regarded as two *poles* of ATP synthesis mechanisms in the F_1 portion of ATP synthase. The chief differences between the two mechanisms are summarized in Table 1.

Which of these two poles appears more likely (Table 1)? Which one (if any) appeals or convinces the discerning scientist-engineer? This is for the scientific community to debate and to find out by theory and experimentation. But perhaps, for now, it seems sufficient (an achievement?) that a complete, more detailed alternative molecular mechanism exists and that the differences stand clearly and unambiguously accentuated.

2.2 Structural Studies to Validate the Postulates of the Torsional Mechanism

The catalytic site of a β-subunit of ATP synthase contains three major sub-domains of interest. In our interpretation, the adenine-binding sub-domain consists of the amino acid residues Tyr 345, Phe 418, Ala 421, Phe 424, Thr 425, Pro 346, Val 164, and Gly 161 (the residue numbers refer to mitochondria). The phosphate binding sub-domain is made up of the following residues of the β subunit: Lys 162, Thr 163, Val 164, Leu 165, Gly 161, Val 160, Gly 159, and Arg 189. The amino acid residues Lys 162, Thr 163, Glu 188, Arg 189, Glu 192, and Asp 256 of the β subunit contribute to coordination with the Mg^{2+} and form the third sub-domain [9].

One of the major postulates of the torsional mechanism of ATP synthesis is that the nucleotide cannot bind (and stay bound) in the open conformation. We studied the Walker crystal structure to provide a quantitative basis for this postulate. We first determined all the atoms within a distance of 5 Å from any atom

Table 1. The major differences between the torsional mechanism of ATP synthesis and the binding change mechanism

Binding change mechanism	Torsional mechanism
Site-site cooperativity exists among catalytic sites	No site-site cooperativity among catalytic sites in the steady state physiological mode of operation
Different affinities of catalytic sites for Mg nucleotides in ATP synthase are explained by a negative cooperativity of binding	Different affinities of catalytic sites for MgADP or MgATP are explained by intrinsic asymmetry of the catalytic sites due to their asymmetric interactions with the single copy subunits of F_1 governed by the position of the γ-subunit within the $\alpha_3\beta_3$ cavity and the ε-subunit
A~10^5-fold positive cooperativity of catalysis takes place in transition from "uni-site" to "bi-site" catalysis	The rate enhancement during ATP synthesis is explained to be due to an increase in the fraction of the F_1F_0 enzyme population containing bound nucleotide in all three catalytic sites with increase in substrate concentration
Reversible catalysis	Irreversible mode of catalysis under physiological conditions and for a single enzyme molecule
ATP synthesis occurs spontaneously on the enzyme	Energy is needed for the synthesis elementary step
P_i binding is conceived to be spontaneous in diagrams depicting the mechanism	P_i binding requires energy
Substrate binding precedes product release or is simultaneous with it during V_{max} ATP synthesis	Product release precedes substrate binding in V_{max} physiological mode of functioning
The energy of substrate binding at one catalytic site is transmitted to another catalytic site and used for product release from that site	Substrate binding energy is used in situ to cause conformational changes at that catalytic site. The energy for product release comes from an interaction of a β with a subunit/agent outside, and not part of, the $\alpha_3\beta_3$ ring
Two catalytic sites only need to be filled by bound nucleotides for physiological rates of ATP synthesis	Three catalytic sites need to be filled by bound nucleotides to achieve physiological rates of ATP synthesis. Catalysis takes place in the three-nucleotide state
Free rotation of γ	Torsion of γ
Continuous	Discrete, quantized
No energy storage	Energy storage is crucial
No closed catalytic site in catalytic cycle. Substrate can bind to the catalytic site with the open, distorted conformation and remain bound.	Closed catalytic site, where the substrate can stay bound, is an intermediate in the catalytic cycle
Driving force is nucleotide binding	Driving force is ΔpH+ΔpAnion
Entropic	Enthalpic

Table 1 (continued)

Binding change mechanism	Torsional mechanism
One point of ^{18}O water entry; one pathway of oxygen exchange	Three points of ^{18}O water entry; two pathways of oxygen exchange
Binding changes are fundamental	Conformational; both conformational changes caused by nucleotide binding and by fundamental γ-β and ε-β interactions which arise from torsion and intersubunit rotation in ATP synthase are essential and help each other
In the hydrolysis mode, binding of substrate MgATP to a catalytic site provides the driving force for rotation of γ	In the hydrolysis mode, the 120° rotation of the γ-ε is driven by the energy of ATP hydrolysis occurring in the β_{TP} site (i.e., site 2, the site with intermediate affinity)
Useful work is performed by the binding energy released during the ATP binding step in the user molecule (e.g., the myosin-actin system of muscle)	The enthalpy change upon ATP hydrolysis is transduced to useful work (untilting of the myosin head and dragging of actin filament with it) in the user molecule

of the adenine ring. Considering the fact that the interactions of the adenine ring within the pocket are primarily hydrophobic in nature, critical atoms among these were identified. These atoms were taken to be the constituents of the adenine binding sub-domain. To compare the differences among the three conformations of the sub-domain, during the loose, tight and open states of the β-subunits, the effective space within the sub-domain was estimated in the following way: the coordinates of the centroid in each conformation were determined and then the root mean square deviations of the constituent atoms of the sub-domain from the centroid were calculated. The r.m.s. values of the tight and loose conformations were close to each other (18.05 Å and 18.93 Å, respectively), but the r.m.s. value of the sub-domain for the open conformation was significantly higher at 22.06 Å. This implies that the adenine-binding sub-domain in the open conformation contains 22.2% more space than in the tight conformation. This provides quantitative evidence that it would not be possible for the adenine ring to bind properly to the sub-domain in the open conformation. Figs. 1a–c depict the adenine binding sub-domain in the tight, loose and open conformations (observed at the same magnification) and provide visual evidence for the above conclusions.

Similar calculations performed for the phosphate-binding sub-domain showed that there exists 35.8% and 34.8% more space in the open conformation as compared to the tight and loose conformations, respectively. For the Mg^{2+} binding sub-domain, there was 24.4% and 37.1% more space in the open conformation over the tight and loose conformations, respectively. This shows that the Mg^{2+} coordination with its ligands is *different* in each of the three conformations, indicating that changes in the Mg^{2+} binding to its ligands are crucial for catalysis, as conceived by the torsional mechanism from the very inception.

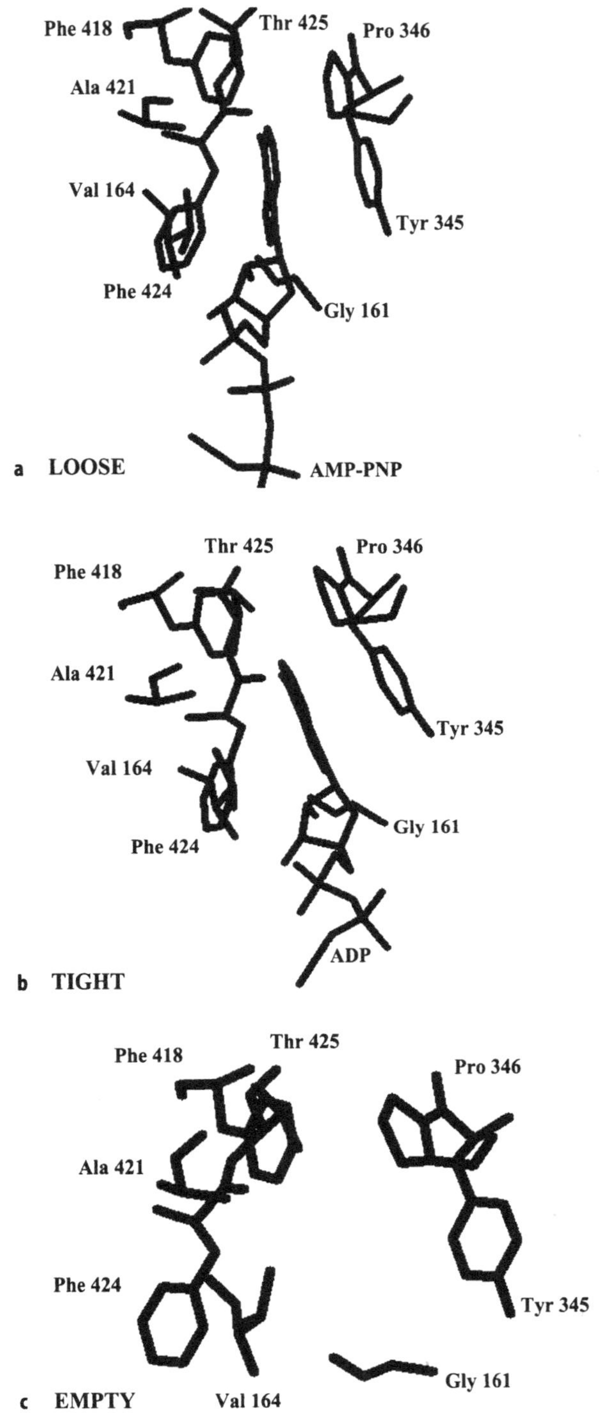

Fig. 1a–c. The adenine-binding sub-domain in the (**a**) loose, (**b**) tight, and (**c**) open conformations viewed using RasMol

2.3
Catalytic Site Occupancies During ATP Hydrolysis by F_1-ATPase

A breakthrough on the experimental front was made by Weber and Senior through the design of optical probes by insertion of tryptophan residues at appropriate locations in the catalytic sites of F_1 [42,43]. This permitted the first *direct* monitoring of nucleotide occupancy of the catalytic sites in the hydrolysis mode by a true equilibrium technique. Their results showed that the steady state hydrolysis activity by F_1-ATPase was due to enzyme molecules with all three catalytic sites occupied by nucleotides ("tri-site" catalysis). They even proposed that a mode of catalysis with two substrate-filled catalytic sites ("bi-site" catalysis) may not exist [44]. Boyer has recently proposed that bi-site activation continues even at high substrate ATP concentrations when three catalytic sites are filled [45]. In his opinion, showing V_{max} hydrolysis activity only when three sites are filled means nothing: one is still seeing bi-site catalysis. In this reviewer's view, he is now implying very subtly that "bi-site" does not mean "two catalytic sites filled" and is attempting to change the very definition of "bi-site" accepted for the last 30 years: it hasn't anymore to do with *physical occupancy* of the sites but with "activation" (e.g., changes at catalytic sites). In other words, at any time, one catalytic site, although filled, is not working, i.e., not undergoing any changes. There may be no scientific way to ever prove or disprove such an assertion (in the Popperian way), because whatever is happening, by default, is bi-site! It should be pointed out that the binding change mechanism has had its chances for three decades; several modifications have already been made to it over the years, and very recently, major changes have been postulated. Unfortunately, none of the changes has offered a true mechanistic understanding and has made the situation harder to resolve. Perhaps the time has come to give alternative mechanisms a chance. Finally, if in future it is postulated that bi-site activation operating under tri-site conditions is different from bi-site activation under bi-site conditions, we would be in great danger of scientific anarchy. This will also affect other fields, for example, those dealing with myosin and hemoglobin research. One way to maintain harmony is to continue with the definition of n-site based on physical occupancies. Moreover, if rapid enzyme turnover is obtained with two (or three) sites filled, it should be referred to properly as bi-site (tri-site) catalysis.

2.3.1
Other Specific Difficulties with the Binding Change Mechanism

Numerous other difficulties arise. After championing bi-site mechanisms for decades, we are suddenly informed that "the important consideration should be, however, not the number of catalytic sites that may be occupied, but what sites must be occupied for rapid enzyme turnover to occur" [45]. The proposal is that site 1 (highest affinity or T) and site 2 (intermediate affinity or L) are occupied in synthesis mode, but site 1 and site 3 (lowest affinity or O) are occupied in the hydrolysis mode. Thus, a *different* second site (site 2 or site 3) is conceived to be occupied during steady state synthesis and hydrolysis, respectively.

Why this should be is not clear. If only two catalytic sites are occupied out of three, then (whether it is steady state synthesis or hydrolysis) one would expect them to be the site with the highest affinity (site 1) and the site with intermediate affinity (site 2), but not site 3 in any case. In bi-site synthesis ADP+P_i enter and bind in site 2, ATP is made reversibly in site 1 and is released from site 3, while in bi-site hydrolysis, ATP enters and binds in site 3, ADP and P_i form in site 1 and are released from site 2 (Fig. 1 of ref. [45]). Thus, in the hydrolysis mode, site 3 is occupied by ATP but site 2 of higher affinity remains empty, which is not logical, as pointed out earlier [1]. On the other hand, if site 2 were also occupied, then as discussed above, it should be termed tri-site hydrolysis, not bi-site hydrolysis. Moreover, it is difficult to understand how a site (in this case site 2 during ATP synthesis) has "greater affinity for ADP than ATP" [45]. The catalytic site binding pocket is for the adenine moiety (Fig. 1) which is the same for both ADP and ATP. Even if the nucleotide phosphates contribute, how the triphosphate has a lower affinity for the catalytic site than the diphosphate is hard to conceive. Further, in the recent X-ray structure of Menz et al. [4], the ADP binds to the catalytic site that remained *unoccupied* in the 1994 Walker structure, i.e. it binds to β_E (site 3), and not to site 2 (which is site 1 in Boyer's nomenclature in Fig. 1 of ref. [45]). Hence, this fact cannot be taken as supporting the binding change mechanism; in fact, it supports tri-site catalysis.

2.3.2 Possible Resolution of Some Specific Difficulties in the Binding Change Mechanism: The Importance of the Transport Steps

High ATP concentrations are not expected to be present during rapid ATP synthesis in the physiological mode of functioning. ATP will only be produced on demand. So there will exist a cut-off, which is a problem of regulation. Significantly, elementary transport steps in the ADP-ATP translocator and the P_i-OH^- antiporter are critical: if the ATP produced is immediately transported out and exchanged for an ADP, as in the physiological situation, ATP synthesis will not proceed with "high" ATP concentrations present. In fact, if ATP leaves from site 3 during synthesis in bi-site catalysis and ATP enters site 3 during bi-site hydrolysis, and if ATP synthesis were to take place with high ATP concentrations prevailing, then it is difficult to conceive what prevents ATP from re-binding to site 3 and causing its own hydrolysis. We have repeatedly emphasized that it is important to study not just the reaction but also a whole series of transport steps. Kinetic schemes incorporating transport steps and chemical reaction for ATP synthesis under true steady-state conditions have been presented and quantitatively analyzed for the first time [32, 33]. The occurrence of competitive inhibition of ATP synthase by ATP as the inhibitor in the synthesis mode has also been suggested. In a population of ATP synthase molecules, a fraction of the population can carry out synthesis and another fraction can work in the hydrolysis, but according to the torsional mechanism, a single ATP synthase molecule can either be working in the synthesis mode or in the hydrolysis mode at an instant of time, i.e., synthesis and hydrolysis can be carried out *simultaneously* only by different enzyme molecules.

2.3.3
Discriminating Experimental Test of Proposed Molecular Mechanisms and Biological Implications

The basic issue can be stated as follows: if bi-site conditions (110, 101, 011 individually or together, where 1 refers to occupation and 0 to non-occupation of sites 1, 2 and 3, respectively) do not contribute significantly to the rate of steady turnover by themselves (as % of V_{max}, say), then one should not postulate them to contribute when three catalytic sites are occupied. In other words, if there exists no "bi-site activation" during bi-site catalysis, then it is not reasonable to postulate bi-site activation to have a "predominant role" under tri-site conditions. Since the filling of the third site should cause little (if any) rate enhancement according to the binding change mechanism, the fraction of V_{max} attained due to the 111 enzyme species should remain more or less the same as in bi-site conditions (110, according to the binding change mechanism, but even stretching it to the extreme, 110+101+011 occupied enzyme species). This prediction can be tested. Moreover, "bi-site activation" can be considered to remain at the same level as in bi-site catalysis (and not "stop") by comparing the rate due to 111 species, various bi-site species, and the sum of 111+various possible bi-site species among themselves and with the experimentally measured hydrolysis rate. Selected results are shown in Fig. 2. It is found that the theoretically predicted rate due to species 111 alone accounts perfectly for the experimentally observed rate data [44] over four decades of substrate MgATP concentration, providing *unequivocal* evidence for tri-site catalysis as the only mode of catalysis (Fig. 2). This has profound biological implications for any proposed mechanism. It should also be emphasized that the values of dissociation constants of the sites treated as independent from each other are *sufficient* to match the calculated rates with the experimental data over the entire range of substrate concentration. Experimental evidence supporting the torsional mechanism in the F_1 portion of ATP synthase has recently been reviewed in consummate detail [1].

2.4
The Torsional Mechanism of ATP Hydrolysis

The primary intention behind the development of the torsional mechanism was to understand the functioning of ATP synthase in the *synthesis* mode. However, in order to clarify and fully appreciate the aspects raised above, the torsional mechanism has been developed for the hydrolysis mode (Fig. 3). In steady-state hydrolysis, ATP binds to enzyme that has 1 ATP (in β_{DP}) and 1 ADP (in β_{TP}) already bound; in the tri-site state, the enzyme has 2 ATP (in β_{DP} and β_C) and 1 ADP (in β_{TP}) bound to the catalytic sites. The conformations of the catalytic sites are depicted in Fig. 3. Details of the ATP hydrolysis cycle are as follows: the ε-subunit is located close to (and interacts with) the O site (β_E). To start the cycle, first Mg^{2+} and ATP enter the nucleotide-free "T" site (β_{DP}) (which, in the absence of Mg nucleotides has an open conformation; see ref. [9]). Mg^{2+} and ATP enter "L", bind, change its conformation to L and hydrolyze to ADP and P_i;

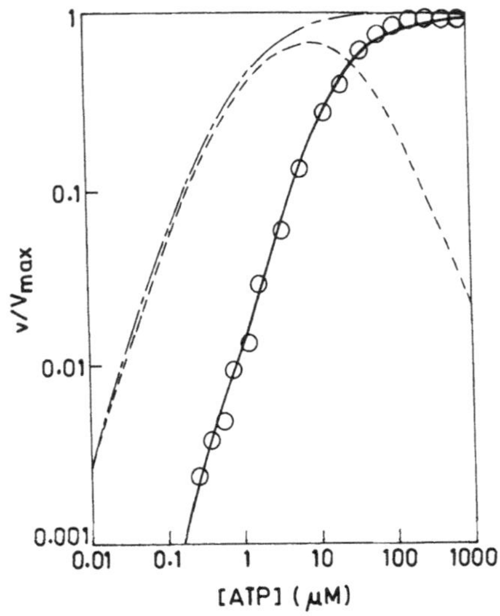

Fig. 2. Relative rates of ATP hydrolysis by F_1-ATPase as a function of substrate concentration for 2.5 mM Mg^{2+} excess over ATP. ○ denotes experimentally measured relative ATPase activity [44], — represents the calculated relative activity due to enzyme species with all three catalytic sites filled (111) as predicted by the torsional mechanism, - - - that due to all three possible bi-site species (110+101+011), and —-— that given by the sum of tri-site and all possible bi-site species. The sum is obtained assuming the species to possess the same specific activity. K_d values of sites 1, 2, 3 are 0.02, 1.4 and 23 μM, respectively [44]

P_i leaves L. Due to the hydrolysis event in β and the resulting change in electrostatic potential, torque is generated at the β-γ interface causing the top of the γ-subunit to rotate by 120°. Due to the load of the c subunits and the membrane itself, the bottom of γ does not rotate immediately; hence there is torsional strain in the γ-subunit. This torsion strains the ε-β_E interaction. The C-terminus of β_E sterically hinders movement of γ. The MgATP binds and its binding energy can break the strained ε-β_E interaction and the β_E (O or site 3) site changes its conformation to β_C (C), as described before in detail [9] and we have state 5. The change in conformation of β_E to β_C relieves the steric hindrance and the ε and bottom of γ now move in steps of 15°/30°. The conformations of β change: C (β_C)→T (β_{DP}), T→L (β_{TP}) and L→O (β_E) and we reach a state of the enzyme 6 in Fig. 3. The ε-subunit has now rotated from O to L and has converted the L site to O and helped release product ADP and the steady-state cycle now repeats (7–9) (Fig. 3). ATP hydrolysis in L (site 2) drives the rotation, but unless ATP binds in O and changes its conformation to C, the ε-subunit and the middle and bottom of the γ-subunit cannot rotate due to steric clash between γ and β_E. Moreover, unless ATP hydrolyzes in L and the torsion in γ strains the ε-β_E interaction, the ATP cannot bind and change the conformation of β_E to β_C. Finally, note that in the absence of the ε-subunit, ADP cannot be released and eventually all three catalytic sites will contain bound MgADP (the "ADP-inhibited state") and the enzyme will stop working as there exists no way by which ATP can enter and bind to the catalytic site.

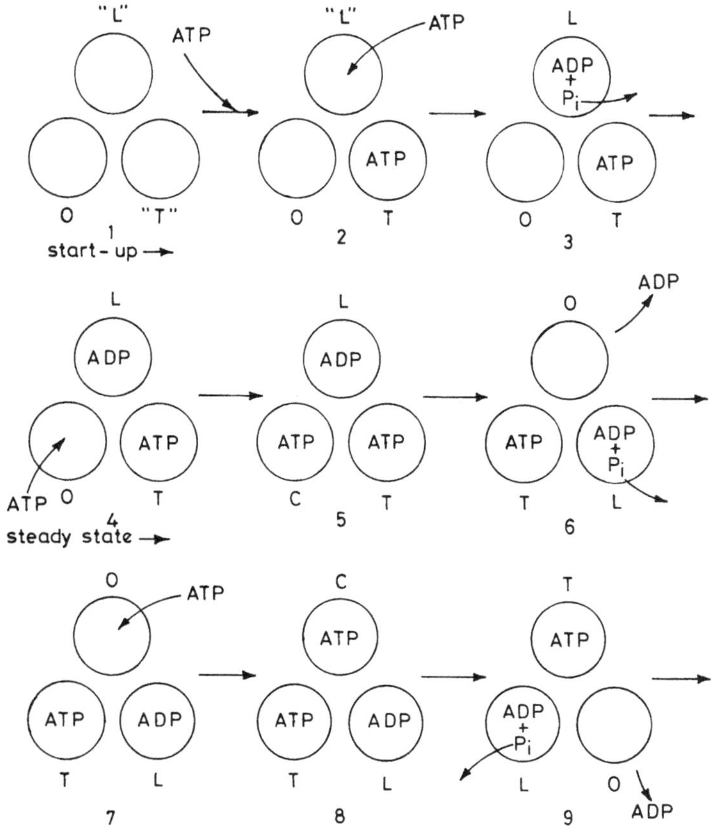

Fig. 3. The torsional mechanism of ATP hydrolysis

3
Molecular Mechanisms of Energy Transduction in the F_0 Portion of ATP Synthase

The inventive chemiosmotic hypothesis of oxidative phosphorylation was first proposed by P. Mitchell in 1961 [46, 47] and generated a great deal of controversy in the bioenergetics community for two decades. That era failed to provide any challenging alternatives, and the chemiosmotic hypothesis was accepted "for the time being" as "the best available hypothesis" of ATP synthesis. According to chemiosmotic postulates, the rate of ATP synthesis (J_{ATP}) is solely determined by the electrochemical potential difference of protons between two bulk aqueous phases, $\Delta\tilde{\mu}_H = F\Delta\psi - 2.303RT\Delta pH$, consisting of a linear addition of the pH difference and a delocalized electrical potential difference across the membrane created by the uncompensated, electrogenic translocation of protons themselves on the redox side. Thus, according to chemiosmosis, a unique correlation should exist between $\Delta\tilde{\mu}_H$ and J_{ATP}. Complete consensus could not be reached because several lines of biochemical evidence did not support the fundamental tenets or the implications of the hypothesis. Two of the major experimental anomalies [48] are taken up in this article: (i) the relation between the flux (J_{OX} or J_{ATP}) depends on how $\Delta\tilde{\mu}_H$ is varied, i.e., there is no unique dependence between flux and driving force, and (ii) inhibition of the enzymes on either the redox or the ATPase side does not lead to compensation of the rate of ATP synthesis by the remaining non-inhibited enzymes. These anomalies go against the fundamental tenets of chemiosmosis and cannot be explained by it.

The torsional mechanism of ion translocation, energy transduction and storage, and ATP synthesis explains the cornucopia of experimental observations on ATP synthesis without exception. The torsional mechanism itself has been reviewed and covered in great detail in the original publications, as well as in several inaugural and plenary lectures at various conferences. In order to understand the mode of ion translocation, the spatial and temporal *pattern* of elementary transport processes, and energy coupling, it is important to analyze the source of the electrical potential, $\Delta\psi$. Electrogenic ion transport has often been proposed to explain ion transport in the F_0 portion of ATP synthase [46, 47]. The chemiosmotic theory considers the uncompensated, electrogenic transport of protons by redox complexes as the source of $\Delta\psi$, i.e., a single source results in the creation of both a delocalized ΔpH and a delocalized $\Delta\psi$. However, various experimental observations obtained over the past several decades do not satisfy the electrogenic mode of ion transport. Experiments with ATP synthase reconstituted into liposomes demonstrate ATP synthesis at physiological rates even though no redox complexes are present in the system [49–51]. Similar experimental observations were first reported on submitochondrial particles and it was concluded that "an electrochemical gradient of protons can drive the synthesis of ATP independent of electron transport" [52]. According to the chemiosmotic hypothesis, an electrical potential difference of 180 mV exists across the membrane in state 4. Considering the fact that, in state 4, no proton translocation is mediated by the redox complexes, and proton leak through the membrane is extremely small [27–29, 47, 53], it is difficult to account for such a high $\Delta\psi$ across the membrane. In addition, the experimentally observed variation in the K^+/ATP ratio from 0 to 4 [54, 55] with K^+ as well as valinomycin concentrations cannot be satisfactorily explained by an electrogenic mode of ion transport. Lastly, a laborious, decade-long program of experimental studies aimed at directly measuring the presumed delocalized $\Delta\psi$ in giant mitochondria using microelectrodes did not detect any significant electrical potential [56, 57]. These observations, obtained using a variety of techniques over a period of more than 30 years, pointed to the absolute need to perform a reappraisal of the mode of ion transport across the membrane in the F_0 portion of ATP synthase. After a systematic reappraisal, we concluded that either no $\Delta\psi$ is created, or that $\Delta\psi$ is created in the vicinity of the ATP synthase complex by an *independent* source other than protons, and that the overall driving force for ATP synthesis are the ion gradients due to protons and counter-ions (anions transported through symsequenceport or cations transported through antisequenceport), and in this context, we proposed a *dynamically electrogenic but overall electroneutral mode of ion transport* [35, 38]. This mode of ion transport involves a membrane-permeable anion (e.g., chloride in chloroplasts, succinate/fumarate in mitochondria) moving in the same direction as the proton, or a cation being transported in a direction opposite to the direction of proton movement (e.g., valinomycin-K^+ in vitro) (Fig. 4). Thus, the energy-transducing complexes in mitochondria function as anion pumps [38]. However, both proton and anion (or counter-cation) do not move together or simultaneously (as proposed in ion-exchange mechanisms, in electroneutral ion transport mechanisms, or electroneutral pump-leak mechanisms) (Fig. 4) but sequentially. Hence the ion transport is step-wise or dynamically electrogenic, but overall electroneutral. However, in order to extract energy from the anion/counter-cation, it is critical to understand the temporal sequence of events.

The possibilities of simultaneous transport of proton and anion (or counter-cation) or proton transport preceding anion (or counter-cation) translocation

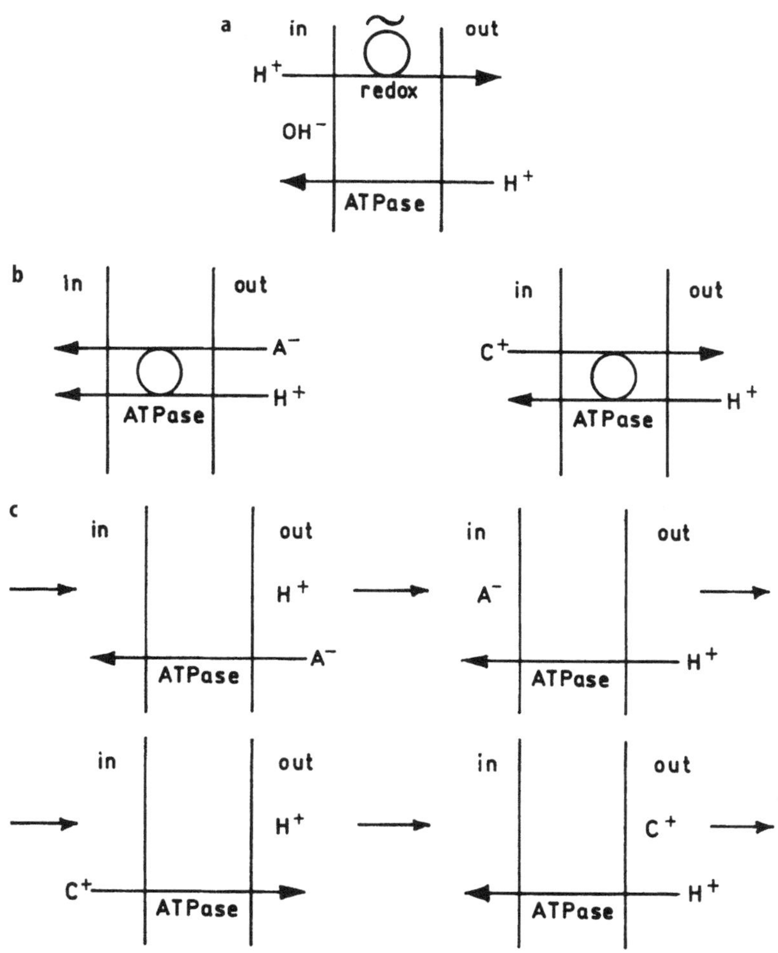

Fig. 4a–c. Schematic representation of a) electrogenic, b) electroneutral and c) dynamically electrogenic but overall electroneutral modes of ion transport

are ruled out because in either case, the energy stored in the anion (or counter-cation) gradient is not made available to the proton; therefore in the absence of sufficient *quanta* of energy, complete rotation of the c-rotor in the F_0 portion of ATP synthase (by 15°) cannot take place. Thus, anion transport or counter-cation transport (K^+ transport from inside to outside in the presence of valinomycin) must precede proton transport through the proton half-channels. In this mechanism, the energy of oxidative phosphorylation is stored in the overall sense as the proton and the anion/counter-cation gradients. The counter-ion gradients are converted to a diffusion potential, $\Delta\psi$, so that the true driving forces for ATP synthesis are ΔpH and $\Delta\psi$. The ion-protein interactions due to proton binding/unbinding in the presence of a $\Delta\psi$ involve the creation of a $\Delta(\Delta\psi)$ as an intermediate step for rotation of the c-rotor and subsequent storage of torsional energy in the γ-subunit to be used thereafter for synthesizing ATP [1, 9, 30–40]. Hence, the energy transiently stored in ΔpH and $\Delta\psi$ is converted to torsional energy through the mechanoelectrochemical process of ion-protein interactions. The *localized* nature of $\Delta\psi$ created by ion permeation events in the vicinity of the ATP synthase, and the strictly ordered temporal sequence of the permeation processes generate a complex pattern in which the overall fraction of energized spatial domains/regions (for a constant stimulus) remains more or less constant at each time, but the region involved in the elementary processes fluctuates with time, so that different spatial domains/regions or sites in the vicinity of the enzyme molecules are brought into play with the passage of time. We believe that the dynamically electrogenic but overall electroneutral mode of ion transport via symsequenceport or antisequenceport may prove to be a general principle governing ion transport and temporal and spatial pattern formation in biological systems.

3.1 Resolution of the Experimental Anomalies by the Torsional Mechanism

It will now be shown how the mechanism of ion translocation discussed in Sect. 3 [38, 39] resolves the apparent experimental anomalies in a natural, almost self-evident way. Suppose that the proton and anion gradients (i.e., the total energy available to the system through that ion I, as measured by the commonly employed expression $RTF^{-1}\ln[I_{out}/I_{in}]$) are distributed (through ion permeation) among n ATP synthase enzyme complexes ($n<n_{total}$) such that the $\Delta\psi$ contribution per ATP synthase complex is 60 mV, and that the ΔpH contribution also measures 60 mV per enzyme molecule. Let a rate of ATP synthesis J_{ATP} be measured under these conditions. Increasing the proton gradient such that ΔpH (and hence $\Delta\tilde{\mu}_H$) increases (to>60 mV per complex), keeping $\Delta\psi$ the same will not increase J_{ATP}, because the $\Delta\psi$ component (the anion) which is not in excess will limit the rate; the excess ΔpH alone cannot lead to increased rates of ATP synthesis by itself, according to the torsional mechanism of ion translocation. Hence, although $\Delta\tilde{\mu}_H$ increases, J_{ATP} remains unchanged in such a situation. Similarly, increasing the $\Delta\psi$ component will increase $\Delta\tilde{\mu}_H$ (as calculated by the chemiosmotic equation) but cause no increase in J_{ATP}. Similarly, a decrease in the individual driving forces from >60 mV to 60 mV (keeping the other driving force clamped to 60 mV) will cause no decrease in J_{ATP}, even though the presumed driving force ($\Delta\tilde{\mu}_H$) has decreased. Now consider the case when the total $\Delta\tilde{\mu}_H$ is kept constant at 120 mV. If, starting from a proton gradient equivalent to 60 mV and an anion gradient equivalent to 60 mV, the ΔpH component (or $\Delta\psi$ component) is increased to say 90 mV and the $\Delta\psi$ component (or ΔpH component) is decreased to 30 mV, J_{ATP} will decrease. The reverse transition will enhance J_{ATP} at constant $\Delta\tilde{\mu}_H$ because in the final state, the energy provided by both the components can be fully utilized by the active enzyme complexes. In fact, an increase in $\Delta\tilde{\mu}_H$ will cause an increase in phosphorylation rate if the increase leads closer to a 1:1 optimal balance in the energy provision capacity of the anions and the protons in the final state with respect to the operating levels of the enzyme complexes, as compared to the initial state. An increase in $\Delta\tilde{\mu}_H$ resulting in further imbalance of the $\Delta\psi$:ΔpH ratio from the initial ratio will not lead to any increase in the flux. In such a situation, either the excess energy of the ion gradients cannot be utilized and will remain stored, or a greater fraction of enzyme complexes will be "energized" by permeant anions/countercations creating a $\Delta\psi$ but there will be insufficient energy to synthesize ATP, or the energy of the excess ΔpH will be transduced to a rotation of half the requisite amount, after which the enzyme complex will stop working. It should be emphasized that if the overall energy provided by both the proton as well as the anion is increased such that a greater fraction of the enzyme complexes can be recruited and made active, J_{ATP} will keep increasing with increases in the so-called $\Delta\tilde{\mu}_H$ until $n=n_{total}$ is reached, after which J_{ATP} will saturate. Thus, there exists no unique relationship between $\Delta\tilde{\mu}_H$ and J_{ATP}, as found experimentally, and the rate will depend on *how the so-called $\Delta\tilde{\mu}_H$ is varied*, as clearly seen from our molecular mechanism.

According to the torsional mechanism, J_{ATP} will depend upon the anion and proton concentrations on both sides of the membrane and the number (n) of active enzyme complexes. In chemiosmosis, inhibition of a small fraction of the ATP synthase enzyme complexes should not affect the phosphorylation rate because the value of $\Delta\tilde{\mu}_H$ remains the same before and after. In other words, in Mitchell's theory, the remaining, non-inhibited enzyme complexes should "see" a larger driving force and should compensate for the inhibition by working at a faster rate and thus keep J_{ATP} unchanged. In the framework of the torsional mechanism, on the other hand, in the presence of sufficiently high anion and proton concentrations (i.e., under experimental conditions when *the anion and proton concentrations do not limit the rate*), the number of ATP synthase complexes (n) participating in ATP synthesis decreases due to addition of the inhibitor; hence J_{ATP} should decrease in proportion to the fraction of ATP synthase complexes inhibited. This is in harmony with experimental observations (ii) stated at the beginning of this section, which till now had been considered as "anomalous". We now see that these so-called anomalies are perfectly correct experimental observations that should not be ignored in the development of any theory. In fact, a real molecular mechanism and theoretical framework should be able to explain them, and not merely regard them as artifacts, or as inconvenient observations to be swept under the carpet. A novel prediction of the torsional mechanism is that under the above conditions, the relative inhibition of J_{ATP} is equal to the fraction of inhibited ATP synthase enzyme complexes (measured, say with DCCD or oligomycin as inhibitor) as well as the fraction of inhibited redox enzyme complexes (measured with rotenone or antimycin as inhibitor). Thus, both the redox as well as the ATPase enzymes are completely rate-limiting. We therefore find that the torsional mechanism can unambiguously explain all the apparently contradictory experimental observations of the past fifty years without exception. Moreover, it provides us with a true *mechanistic* understanding of the elementary events underlying ATP synthesis.

3.2
In vitro and in vivo Situations

It should be noted that in the above in vitro experiments there are two *independent* agents to vary $\Delta\psi$ and ΔpH, e.g., K^+-valinomycin and H^+, respectively, and varying one does not affect the other. Similarly, in experiments on mitochondria/chloroplasts with anions, if sodium succinate (where Na^+ is a non-permeant ion) is used, as opposed to succinic acid, we again have succinate monoanion and H^+ as separate agents that can be used to vary $\Delta\psi$ and ΔpH, respectively. Thus, in the above experiments, it is a requirement that changing K^+ (or succinate$^-$) concentration shall *not* affect H^+ concentration, and vice-versa. Under physiological conditions in mitochondria/chloroplasts, we may have H^+-succinate$^-$ (and not Na^+-succinate$^-$), or, in general, H^+A^- as permeant ions, and no valinomycin is present, i.e., in vivo, both permeant ions, H^+ and A^- are adducts of H^+A^- and are present as an ion pair. In such a situation, we cannot vary one independently of the other. In this way, the energy provision capacity of anions and protons will always be in a 1:1 ratio. Thus, a self-regulation of the *distribution* of energy quanta takes place and no excess of quanta is unnecessarily generated. These predictions of the torsional mechanism are nicely supported by recent measurements of the steady state and kinetics of the light-induced electrochromic shift in isolated thylakoids which estimate that ~50% of the total energy of the "protonmotive force" in vivo is stored as $\Delta\psi$ [58].

3.3
Biological Implications

The mechanism has profound biological implications [1, 33, 35, 37–39, 69, 70]. In Mitchell's chemiosmotic theory, energy flow is confined to concentration and electrical gradients associated with protons, and a *macroscopic, delocalized* driving force (the protonmotive force, $\Delta p = \Delta\psi - RT\Delta pH/F$, conceived as a linear addition of the two gradients) between two energized aqueous media separated by an *inert, rigid and insulating* membrane is envisaged. In the chemiosmotic framework, no force acts on membrane constituents, and no energy is stored in the membrane. This is also the essence of Mitchell's protonmotive osmotic energy storage equation. Thus, in chemiosmosis, *two* protons flow from the aqueous medium through a channel to the ADP site, and ATP is synthesized directly *without any changes taking place in the membrane.* Our detailed molecular mechanism shows that the ion-protein interaction energy is transiently stored as a twist in the α-helices of the c-subunits of F_0 and that membrane conformational changes are intimately connected to energy transduction, and emphasizes the dynamic cyclical changes in protein structure in the membrane-bound F_0 portion of ATP synthase. Hence there is an imperative need to understand not only what happens across the membrane but also what happens within it. Finally, there is nothing inherently osmotic about the mechanism of ATP synthesis, and osmotic energy is not directly converted to chemical energy, and our molecular mechanism implies that energy transduction and transient storage cannot be understood using osmotic principles alone. Energy can indeed be stored as ion gradients across a membrane in two bulk aqueous phases; however, the membrane is not just an insulator, and according to the torsional mechanism, *molecular interactions* between ion and protein-in-the-membrane are critical for elementary steps involving transduction, storage and utilization of the energy of the ion gradients. Thus, the fundamental process of energy coupling in ATP synthesis is not chemiosmotic, but *mechano(electro)chemical* [1, 9, 37, 38, 69].

Several related issues emerge. In chemiosmosis, for each pair of electrons transferred in mitochondrial respiration, up to a maximum of six protons may be produced ($H^+/O=6$) and the number of H^+ ions transported per O consumed cannot exceed the number of hydrogen carriers present in the respiratory chain. Thus, the number of H^+ transported per O atom=6 includes two transported over NAD, two over flavins and two over quinones, and two protons are required for each mole of ATP synthesized from ADP and P_i ($H^+/ATP=2$). Several experiments, the energy balance in the torsional mechanism, as well as a non-equilibrium thermodynamic analysis [27–29] show that these stoichiometries need to be doubled to account for the coupling protons [$H^+/O=12$, $H^+/ATP=4$]. These numbers have important thermodynamic consequences because smaller values of the stoichiometries require a larger protonmotive force to make the free energy change energetically competent for ATP synthesis. The moment experimental evidence and basic non-equilibrium thermodynamic computation that the active proton transport machinery on the redox side must be an ion pump that works with higher stoichiometries than that postulated in chemiosmosis is accepted, Mitchell's mechanism of redox loop transport along the respiratory chain breaks down, because there are simply not enough hydrogen carriers to transport 12 protons per oxygen atom. Where are the extra protons going to come from?

In the chemiosmotic theory, permeant ions lead to collapse of the membrane potential generated by the redox complexes. This leads to activation of respiration and to H^+ extrusion in mitochondria. In this framework, H^+ translocation is primary, while cation transport is secondary and passively compensates the primary electrogenic translocation of protons. Thus, K^+ ions distribute passively at electrochemical equilibrium in response to the delocalized $\Delta\psi$ created by respiration, i.e., the proton gradient drives the movement of cations. This has in a large measure contributed to the prevailing, so-called "well-established" view that $\Delta\psi$ is dissipated by counter-ion fluxes. According to Mitchell, valinomycin makes the inner mitochondrial membrane passively permeable to K^+ ions, the K^+ moves instead of H^+, and the $\Delta\psi$ collapses.

The observed large, valinomycin-induced uptake of K^+ is not consistent with chemiosmotic principles [35, 38]. Moreover, the measured K^+_{in}/K^+_{out} ratio and the K^+/ATP ratio are variable and *depend on the valinomycin concentration*, which is completely inconsistent with chemiosmotic theory because the valinomycin concentration should not affect the H^+/ATP stoichiometry of the primary electrogenic H^+ ion pump. Further, an increase in the Nernst diffusion potential (RT/F) ln $[K^+_{out}/K^+_{in}]$ due to increased external K^+ in the presence of valinomycin (keeping K^+_{in} constant) increased the rate of ATP synthesis in both reconstituted chloroplasts as well as *Escherichia coli* ATP synthase, a result contradictory to chemiosmosis. In the chemiosmotic framework, an increase in K^+ concentration can only dissipate $\Delta\psi$, i.e., an *increase* in external potassium concentrations would cause a *decrease* in the driving force $\Delta\psi$ but lead to *enhanced* ATP synthesis rates in the reconstitution experiments, which cannot be explained by chemiosmotic theory. Finally, the addition of valinomycin can cause either net influx or net efflux of K^+ depending on the experimental conditions, which is difficult to explain by a permeability effect alone, as postulated by the chemiosmotic theory. Thus, the role of the anion/counter-cation in ATP synthesis has never been satisfactorily explained by any version of the chemiosmotic theory.

It is difficult to rationalize the stoichiometry of potassium accumulation with chemiosmotic theory. The uptake of potassium in the presence of valinomycin and the concomitant extrusion of protons is found to be dependent on the permeability of mitochondria to anions. In the presence of permeant anions, lesser K^+-H^+ exchange occurs than in the presence of impermeant ions. Anions enter along with K^+ and water movement into mitochondria and swelling of mitochondria takes place. If H^+ transport were the primary process, entry of anions should not take place, and cation entry would then be an exchange reaction imposed by the electrical potential generated by H^+ ion extrusion. But this electrical force would not be operating on anions since OH^- is created within the organelle for each H^+ ion pumped out. A possibility is that an electrical potential is created by K^+-valinomycin transport into the mitochondrion, and proton extrusion as well as anion entry both operate to maintain electrical neutrality. This hypothesis can readily explain the associated rise in intramitochondrial pH, the reciprocity in H^+-K^+ movement, the anion movement with K^+ and the concomitant water entry due to the need for osmotic equilibration, and the swelling of mitochondria.

A *delocalized* $\Delta\Psi$ of 180 mV (4×10^5 V/cm) will apply very large electrical forces on membrane components. It is difficult to see how the enzyme will sense this $\Delta\Psi$ and how field-driven chemistry can take place, as opposed to concentration gradient-driven reactions in the torsional mechanism. If instead of supplying substrate to an enzyme, we supply an equivalent energy of a $\Delta\Psi$, will it make the product? It is hard to conceive how the $\Delta\Psi$ is a driving force that can be *directly* utilized in ATP synthesis. Note that if the extruded proton immediately returns through the ATPase H^+ channel, then only a negligible delocalized $\Delta\Psi$ will be created; if a separation exists between the creation of Δp and its utilization, as conceived in chemiosmosis, then first the Δp will have to be built up solely by proton translocation before it is utilized, and the principle of electroneutrality in the bulk will be violated. In fact, the presence of valinomycin should prevent the generation of a delocalized $\Delta\Psi$. It is difficult to conceive why the K^+ will wait till the Δp is created and only move in thereafter, and not earlier. These difficulties do not exist in our mechanism. Furthermore, important experimental evidence that energy coupling occurs in membranes that are too permeable to maintain an electrochemical potential gradient has been documented by the group of Sitaramam [59, 60].

3.4 Variation in K^+/ATP Ratio with K^+-Valinomycin Concentration According to the Torsional Mechanism

The K^+/ATP ratio can be taken as

$$\frac{K^+}{ATP} = \frac{v_{K+}}{v_{syn}} \quad (1)$$

where v_{K+} is the rate of K^+ efflux and v_{syn} is the rate of ATP synthesis. According to the dynamically electrogenic but overall electroneutral ion transport, ATP synthesis will occur due to proton transport in response to membrane-permeable anion as well as in response to K^+-valinomycin. Hence, for our mechanism,

$$\frac{K^+}{ATP} = \frac{v_{K+}}{v_{syn,an} + v_{syn,K+}} \quad (2)$$

where, $v_{syn,an}$ is the rate of ATP synthesis due to anion transport and v_{syn,K^+} is the rate of ATP synthesis due to K^+-valinomycin transport. Since the stoichiometry of H^+:anion (for the symsequenceport, i.e., sequential H^+ and anion transport in the same direction) and H^+:K^+ is 1:1 (for the antisequenceport, i.e., sequential H^+ and cation transport in opposite directions), and H^+:ATP is 4:1 [27–29, 31], we have

$$\frac{K^+}{ATP} = \frac{4v_{K+}}{v_{an} + v_{K+}} \quad (3)$$

with v_{an} as the rate of anion influx. The rate of K^+ transport is proportional to the concentration gradient of K^+-valinomycin across the membrane. The adsorption of valinomycin itself to the membrane sites can be described by a Langmuir adsorption isotherm, i.e.,

$$V_m = \frac{aV_t}{1 + bV_t} \quad (4)$$

where V_m is the valinomycin concentration on the membrane sites, V_t the total valinomycin concentration in the medium, and a and b are constants for a given system. The K^+-valinomycin complex formation at the membrane surface can be described by

$$V_m + K^+ \rightleftharpoons VK_m \quad (5)$$

with equilibrium constant K_v. Therefore,

$$VK_m = \frac{V_m K^+}{K_v + K^+} \tag{6}$$

and from Eq. (4),

$$VK_m = \frac{aV_t K^+}{(K_v + K^+)(1 + bV_t)} = \frac{CK^+}{K_v + K^+} \tag{7}$$

where $C = V_m$ for a constant valinomycin concentration, V_t.

Hence, based on our analysis,

$$v_{K+} = k_{K+}(VK_{mi} - VK_{mo}) = k_{K+}\left[\frac{CK^+_{in}}{K_v + K^+_{in}} - \frac{CK^+_{out}}{K_v + K^+_{out}}\right] \tag{8}$$

where k_{K^+} is the constant of proportionality between the rate of K^+ efflux and the concentration gradient of the K^+-valinomycin complex, VK_{mi} and VK_{mo} are the K^+-valinomycin complex concentration inside and outside, respectively, and K^+_{in} and K^+_{out} are the K^+ concentrations inside and outside, respectively. The rate of K^+ efflux may be altered either by changing the K^+ concentration gradient across the membrane (which changes the rate per molecule) or by changing the fraction of ATP synthase molecules involving antisequenceport between K^+ and H^+ (f_{anti}) (which changes the number of ATP synthase molecules), keeping the total fraction of ATP synthase molecules carrying out ATP synthesis (f_{total}) constant. f_{total} itself is a function of the proton concentrations on either side of the membrane [33]. The fraction f_{anti} may be changed by adding another counter-cation or counter-anion to the system and is a parameter that controls k_{K^+} in Eq. (8).

Further,

$$\frac{K^+}{ATP} = \frac{4Ck_{K+}\left[\frac{K^+_{in}}{K_v + K^+_{in}} - \frac{K^+_{out}}{K_v + K^+_{out}}\right]}{Ck_{K+}\left[\frac{K^+_{in}}{K_v + K^+_{in}} - \frac{K^+_{out}}{K_v + K^+_{out}}\right] + v_{an}} \tag{9}$$

which is Michaellian in nature with respect to $Ck_{K+}\left[\frac{K^+_{in}}{K_v + K^+_{in}} - \frac{K^+_{out}}{K_v + K^+_{out}}\right]$ which itself increases hyperbolically with respect to K^+_{in}, or decreases hyperbolically with respect to K^+_{out}. Based on the above analysis, we have explained all the relevant experimental observations on K^+ efflux on mitochondria [54, 55] and have tabulated them in Table 2. It should be noted that H^+_{out} is the concentration of protons outside.

Table 2. Explanation on the basis of the torsional mechanism of diverse classical experimental observations related to ATP synthesis that have never been satisfactorily explained by any other mechanism

Figure	Experimental observation	Proposed explanation based on our analysis
Massari and Azzone (1970) [54]		
Fig. 2	Outside K^+ concentration increases and rate of K^+ efflux correspondingly decreases.	As the K^+_{out} increases, the K^+ concentration gradient decreases, and therefore the concentration gradient of K^+-valinomycin as well as rate of K^+ efflux decreases.
Fig. 3	For a fixed valinomycin concentration and pH, rate of K^+ efflux decreases with increase in K^+_{out}. K^+ efflux increases with valinomycin concentration as well as H^+_{out}.	Increase in K^+_{out} decreases the concentration gradient of K^+ thereby decreasing the rate of K^+ efflux. On increasing the valinomycin concentration, V_t, the value of C in Eq. (8) increases hyperbolically and the rate of K^+ efflux shows the same increasing trend. Increase in $H^+{}_{out}$ increases the fraction of ATP synthase complexes involved in synthesis in a population as well as the rate of H^+ transport due to increase in the H^+ concentration gradient across the proton half-channels. Thus rate of K^+ efflux increases due to antisequenceport.
Fig. 4	On addition of succinate, v_K+ decreases for a constant K^+_{out}.	On addition of succinate, overall electroneutrality with protons is maintained by K^+ as well as by succinate parallely and independently; therefore v_{K+} decreases to accommodate for succinate symsequenceport by decreasing f_{anti}.
Fig. 9	v_K+ increases with increase in H^+_{out} (for a constant K^+_{out}) and decrease in K^+_{out} (for a constant H^+_{out}).	On decreasing K^+_{out}, the K^+ concentration gradient increases and rate of K^+ efflux increases. On increasing H^+_{out}, the fraction of ATP synthase molecules involved in synthesis and rate of H^+_{out} transport increase and $v_{K^\cdot}$ correspondingly increases.
Fig. 11	v_K+ varies hyperbolically with valinomycin concentration.	$v_{K^\cdot}$ is directly proportional to C which varies hyperbolically with the valinomycin concentration, V_t. Therefore, with all other conditions remaining the same, a hyperbolic dependence of $v_{K^\cdot}$ on valinomycin concentration is found.
Azzone and Massari (1971) [55]:		
Fig. 4	A decrease in log (K^+_{in}/K^+_{out}) decreases the rate of ATP synthesis. For a constant log (K^+_{in}/K^+_{out}), the rate of ATP synthesis increases with increase in H^+_{out}.	On increasing K^+_{out}, the K^+ concentration gradient decreases and due to a decrease in coupled proton transport by antisequenceport, the rate of ATP synthesis decreases. On increasing H^+_{out}, the rate of ATP synthesis increases due to increase in the fraction of ATP synthase molecules in synthesis mode and the rate of H^+ translocation through proton half-channels.

Table 2 (continued)

Figure	Experimental observation	Proposed explanation based on our analysis
Fig. 6	On decreasing log (K^+_{in}/K^+_{out}), rate of ATP synthesis decreases. Presence of ATP in the medium reduces the rate of ATP synthesis.	Increase in K^+_{out} decreases the K^+ concentration gradient and therefore, the rate of H^+ transport in response to K^+ translocation decreases thereby decreasing v_{syn}. However, ATP in the medium causes competitive inhibition of ATP synthesis leading to an observed decrease in v_{syn}.
Fig. 11	K^+/ATP increases with valinomycin concentration in the medium and saturates to nearly 4 for high valinomycin concentrations.	K^+/ATP ratio increases hyperbolically with respect to C [Eq. (9)] which itself increases hyperbolically with V_t. Therefore, on increasing V_t, C increases and finally reaches a constant value. Increase in C increases K^+/ATP until it becomes constant for constant C. This maximum value of nearly 4 is observed for low K^+_{out} compared to K^+_{in}.
Figs. 8, 12	K^+/ATP increases with increase in K^+_{out} with 3-hydroxybutyrate as the anion in the medium.	As seen in Fig. 12, v_{K^+} and v_{syn} decrease with increase in K^+ as discussed above. However, the decrease in v_{syn} is steeper than that for v_{K^+} because of inhibition of the anion channels for hydroxybutyrate transport, by the presence of K^+ outside, which is in addition to the decrease in v_{syn} due to a decrease in v_{K^+}. Therefore, as K^+_{out} increases (above ~1 mM), the hydroxybutyrate transport becomes extremely small and can be neglected with respect to the rate K^+ transport; K^+/ATP ratio becomes 4 [Eq. (9)]. For low K^+_{out} concentration (less than 1 mM), v_{an} is comparable to v_{K^+} and therefore the observed K^+/ATP ratio varies from a low value (~1) to nearly 4 with an increase in K^+_{out}.

3.5 The Torsional Mechanism and the Laws of Energy Conservation, Electrical Neutrality and Thermodynamics and Their Biological Implications

We now show that the dynamically electrogenic but overall electroneutral mechanism of ion translocation of the torsional mechanism satisfies the laws of a) energy conservation, b) electrical neutrality and c) thermodynamics ($\Delta\tilde{\mu}_i = 0$). Let a chemical potential be μ_i^{I} and μ_i^{II} on either side of the membrane before the primary translocation; these are purely chemical potentials, because no electrical imbalance exists before the primary ion movement, and no external electrical potential has been applied. The difference between the chemical potentials is therefore $\Delta\mu_i$ initially. After the primary ion has moved through the specific, regulated ion channel, the corresponding difference between the two aqueous compartments on either side of the membrane is the electrochemical potential difference, $\Delta\tilde{\mu}_i$. The electrical part of the electrochemical potential difference can now be looked upon as a production due to the primary translocation. Since Input=Output+Accumulation−Production, we can write the energy conservation law for our novel situation as

$$\Delta\mu_i = \Delta\mu_f - nFE \tag{10}$$

where n is the valency, F the Faraday and E the emf (positive). Since no further primary ion movement occurs according to our mechanism, the thermodynamic condition that the electrochemical potential difference of the primary ion be zero must be satisfied, thermodynamically speaking. Thus,

$$\Delta\tilde{\mu}_f = 0 = \Delta\mu_f + nFE \tag{11}$$

Eqs. (11) and (10) give us $\Delta\mu_f = -nFE$, and $\Delta\mu_i = -2nFE$, i.e.,

$$\Delta\mu_i = 2\Delta\mu_f \tag{12}$$

Thus, physically speaking, the movement of the ion creates a diffusion potential that balances the final chemical potential difference existing across the channel; hence no further movement of that ion can take place. The movement of the secondary ion now takes place; overall electrical neutrality is maintained, and the two gradients are utilized as proposed in detail in the torsional mechanism. The implications of this self-regulatory mechanism are that charge imbalance can indeed be created, but it cannot be *sustained* for long; hence a discrete, step-by-step mechanism of transport is favored. Dynamically the transport mechanism creates a $\Delta\psi$ that prevents translocation of the next ion. In fact, the transfer of a counter-ion is favored over translocation of another co-ion, which implies that the requirement of electroneutrality is very stringent. In the overall sense, the whole transport process is initiated because of the concentration gradient, or, more precisely, the chemical potential difference of the species across the membrane. This has important biological implications and enables us to answer the fundamental chicken-and-egg question: which came first – electrical potential differences or concentration differences? If electrical potential differences arise first, then they would apply large electrical forces on membranes and their components even at locations where (and times when) they are not needed, which may be quite undesirable. According to the torsional mechanism, the concentration differences come first, and potential differences appear as a consequence of concentration differences. These concentration differences are of fundamental significance and are precisely what differentiate the internal and external compartments of the cell/organelle.

3.6 The Major Differences between the Torsional Mechanism and the Chemiosmotic Theory

The major differences between chemiosmotic theory and the mechanism of transport discussed above and their biological implications can now be outlined. In chemiosmosis, a large $\Delta\mu$ needs to be built up before useful work can be done; in the dynamically electrogenic but overall electroneutral ion translocation mechanism, we can do useful work with small $\Delta\mu$ values, and we do not need to work against large heads, a fact that should lead to far greater efficiencies. Furthermore, in our mechanism, the driving force acts in situ and produces useful work at the site where it is needed. In chemiosmosis, on the other hand, the driving force is produced by a site far away from the site where useful work is needed to be performed; hence the effect of the driving force has to be sensed far away. Finally, overall electroneutrality is satisfied by our mechanism but not by the electrogenic transport of chemiosmosis. It has already been pointed out by Green in an incisive critique that chemiosmosis has taken "impermissible liberties with the canons of chemistry, such as the necessity to observe electrical neutrality in chemical reactions. The postulate of uncompensated protons moving freely through membranes is one example of such a violation" [61]. The salient differences between the torsional mechanism and the chemiosmotic theory are summarized in Table 3. These may again be regarded as two *poles* vis-à-vis the molecular mechanism in the F_0 portion of ATP synthase. Once again, since the mechanisms deal with the most *fundamental* issues, it should be possible for a scientist, irrespective of his or her specialized discipline, to evaluate the merits of these alternatives (Table 3).

Table 3. The salient differences between chemiosmosis and the torsional mechanism

Chemiosmosis	Torsional mechanism
$\Delta\tilde{\mu}_H$ is the driving force for ATP synthesis	ΔpH and ΔpA are the overall driving forces for oxidative phosphorylation. The anion/counter-cation gradient is converted to a $\Delta\psi$; hence ΔpH and $\Delta\psi$ are the driving forces for ATP synthesis
$\Delta\psi$ and $\Delta\tilde{\mu}_H$ are delocalized	ΔpH and ΔpA are delocalized but $\Delta\psi$ is localized
ΔpH and $\Delta\psi$ are equivalent and additive	ΔpH and $\Delta\psi$ are kinetically inequivalent driving forces that each affect the rate of ATP synthesis independently of the other
A decrease in ΔpH is compensated exactly by an increase in $\Delta\psi$ and vice-versa	Need not be so because each is a separate entity created by two independent sources
Ion-well; $\Delta\psi$ is converted to ΔpH	Not so; Δp(anion/counter-cation) is converted to $\Delta\psi$ and then both $\Delta\psi$ and ΔpH create a $\Delta(\Delta\psi)$ by ion-protein interactions
H^+ is primary and generates $\Delta\psi$	Anion/counter-cation generates $\Delta\psi$ and precedes H^+ translocation and is primary in that sense. Both proton as well as anion/counter-cation contribute half the energy required for ATP synthesis
Energy flow is confined to protons; no role of anions/counter-cations in energy coupling	Role of anions/counter-cations in energy coupling explained
Counter-ion gradients always dissipate $\Delta\psi$	Not necessarily so; counter-ion gradients may even generate $\Delta\psi$

Table 3 (continued)

Chemiosmosis	Torsional mechanism
K^+ distributes passively in response to $\Delta\psi$ created by H^+ transport	K^+-valinomycin creates a transient $\Delta\psi$ that is utilized by H^+ antisequenceport
Electrogenic and violates electro-neutrality in the bulk aqueous phases	Dynamically electrogenic but overall electro-neutral; does not violate overall electroneutrality
$\Delta\psi$ is ~180 mV in state 4	No substantial $\Delta\psi$ in state 4
Membrane is just an insulator	Cyclical dynamic changes take place in membrane constituents during energy transduction; the membrane plays a key mechanical, electrical and chemical role and participates in ion-protein interactions
Chemiosmotic	Mechano(electro)chemical
Macroscopic	Energy is stored as macroscopic ion gradients, but molecular interactions between ion and protein-in-the-membrane are key to energy transduction and utilization. Torque generation in the c-rotor of F_0 is a result of change in electrostatic potential, $\Delta(\Delta\psi)$ brought about by the ion gradients
Redox loop; H^+/O per site=2; H^+/ATP=2	Ion pumps; H^+/O per site~4; H^+/ATP=4 (if coupling protons alone are considered) or 5 (if the overall oxidative phosphorylation process is considered and the proton needed to neutralize the OH^- exchanged via the P_i-OH^- antiporter is taken into account; note that this fifth proton comes from the external medium and is not pumped out by the redox enzymes)
Role of various uncouplers explained only as dissipaters of $\Delta\tilde{\mu}_H$	As in chemiosmosis + explained as interfering with conformational transitions in F_0 or F_1
$\Delta\psi=[(RT/F)\ln(K^+_{in}/K^+_{out})]$	The equation is only a measure of macroscopic energy; increase in $\Delta\psi$ does not mean greater driving force per molecule. The $\Delta\psi$ per ATP synthase molecule still remains the same. At higher $\Delta\psi$, more enzyme molecules are capable of synthesis and diffusion potential is created in the vicinity of more enzyme molecules that can then be utilized by proton translocation
Protons participate directly in ATP synthesis	Conformational; protons do not participate directly at the F_1 catalytic site in synthesis
No real molecular mechanism coupling $\Delta\tilde{\mu}_H$ and ATP synthesis presented	Detailed molecular mechanism coupling ion gradients to ATP synthesis proposed
Analogy with a fuel cell	Analogy with an enthalpic non-equilibrium molecular machine

4 Thermodynamics of Oxidative Phosphorylation

4.1 Non-Equilibrium Thermodynamic Analysis and Comparison with Experimental P/O Ratios

A non-equilibrium thermodynamic analysis of the coupled processes of oxidative phosphorylation by rat liver mitochondria was carried out for the steady state as described by Nath [29] based on the principles laid by the important work of Caplan [62], Stucki [63], Westerhoff and van Dam [64]. The results are shown in Fig. 5. The values of the redox and ATPase pump stoichiometries n_O, n_P were varied from the Mitchellian (6, 2) to (9, 3) and (12, 4), keeping the ratio of these numbers constant and all other conditions the same for 3-hydroxybutyrate as substrate. The value of the flux ratio, J_P/J_O was plotted as a function of the affinity ratio A_P/A_O (Fig. 5). The experimental P/O of ~2.1–2.2 for long times and >2.5 for short time (<1 min) pulse mode experiments at operating affinity ratios between −0.25 to −0.3 [38] can be predicted by the stoichiometries of the torsional mechanism but cannot be predicted by the Mitchellian stoichiometries, as revealed by this most basic non-equilibrium thermodynamic computation. Further thermodynamic calculations were made for the overall oxidative phosphorylation process with n_O=12 and n_P=5 for 3-hydroxybutyrate as substrate, which implies an ideal mechanistic stoichiometry (Z) of 12/5=2.4 (without proton leak) and with n_O=8 and n_P=5 for succinate as substrate, which implies an ideal mechanistic P/O ratio of 8/5=1.6 (without proton leak). The results are tabulated in Table 4. The tabulated values have been determined for the stationary steady state of H^+ flux (J_H=0) using experimental data [64] for L_{OO}, L_{PP} and C_H. Pump stoichiometries of 12 H^+/O (n_O) and 5 H^+/ATP (n_P) were em-

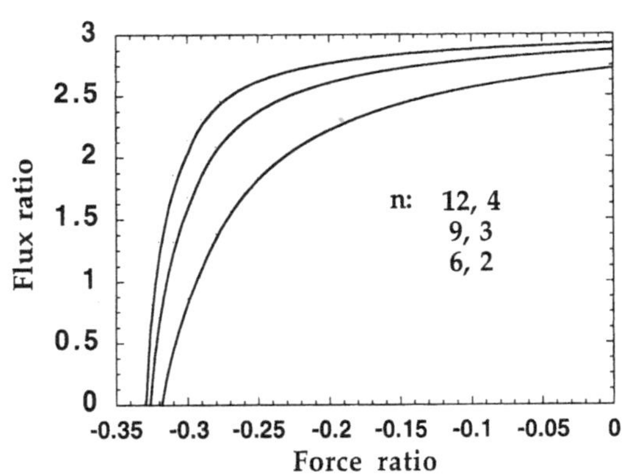

Fig. 5. P/O (flux) ratios as a function of the affinity (thermodynamic force) ratio for oxidative phosphorylation in rat liver mitochondria with 3-hydroxybutyrate as substrate computed using the basic phenomenological coefficients (L_{OO}, L_{PP} and C_H) given in Table 4 for three sets of redox and ATPase pump stoichiometry values (n_O, n_P): 12, 4 (*top curve*); 9, 3 (*middle curve*) and the Mitchellian 6, 2 (*bottom curve*)

ployed for 3-hydroxybutyrate and succinate as substrates. In this table, L_{00}, L_{11} and L_{01} stand for the coefficients $L_{PP}-L_{PH}L_{HP}/L_{HH}$, $L_{OO}-L_{OH}L_{HO}/L_{HH}$ and $L_{PO}-L_{PH}L_{HO}/L_{HH}$, respectively. $Z=(L_{00}/L_{11})^{1/2}$ represents the mechanistic stoichiometry and $q=L_{01}/(L_{00}/L_{11})^{1/2}$ the degree of coupling; mg refers to mg of mitochondrial protein. It is a true test of consistency that the thermodynamic analysis based on *experimentally measured values* [65] of the conductances L_{OO}, L_{PP} and the proton leak C_H (that were used to compute Z values) matches mechanistically expected values in both cases (Table 4). The computed values of the degree of coupling, q using the experimental conductances and the stoichiometries based on the torsional mechanism were 0.986 and 0.979 for 3-hydroxybutyrate and succinate, respectively.

The computed q values in Table 4 are of fundamental significance. For steady-state operation (or for long incubation times) and 3-hydroxybutyrate as substrate, q=0.986, or extending Stucki's terminology, n=6, which implies that the system optimizes output power, efficiency η (defined as $-[J_PA_P/(J_OA_O)]$), and the developed phosphorylation affinity, A_P of both protons as well as anions, i.e., it optimizes the function $(J_PA_P)(\eta)(A_P)(A_P')$. This can be interpreted physically as follows: both affinities (i.e., both species concentrations) play an important role in energy transduction and we need two *independent* processes that are coupled; both are essential for energy coupling. Thus, A_P would correspond to protons and A_P' to anions, according to the torsional mechanism. Thus, complex I–IV in mitochondria are anion pumps performing active transport [1, 38]. If succinate is taken as substrate, we expect a reduction in dimensionality, i.e., since succinate is present in excess, A_P' should not appear in the expression. Hence the system should optimize $(J_PA_P)(\eta)(A_P)$ which implies n=5. This should lead to a degree of coupling based on non-equilibrium thermodynamic theory of ~0.98 [29, 63], which agrees with our results in Table 4. The computed value of q is also in *perfect* agreement with the experimentally determined value of the degree of coupling under these conditions [66, 67].

Table 4. Phenomenological coefficients (conductances) for oxidative phosphorylation in rat liver mitochondria

Coefficient	Units	Value for 3-hydroxybutyrate ($L_{PO}=0$)	Value for succinate ($L_{PO}=0$)
L_{OO}	natom O/(mg min mV)	1.9	1.9
L_{PP}	nmol ATP/(mg min mV)	7.9	7.9
C_H	nmol H^+/(mg min mV)	3.2	3.2
L_{OH}	natom O/(mg min mV)	–22.8	–15.2
L_{PH}	nmol ATP/(mg min mV)	39.5	39.5
L_{HH}	nmol H^+/(mg min mV)	474.3	322.3
L_{00}	nmol ATP/(mg min mV)	4.610	3.059
L_{11}	natom O/(mg min mV)	0.804	1.183
L_{01}	nmol ATP/(mg min mV)	1.899	1.863
Z		2.395	1.608
q		0.986	0.979

4.2 Consistency Between Mechanism and Thermodynamics and Agreement with Experimental Data

We can now compute the actual P/O ratio and the operating efficiency for rat liver mitochondria with 3-hydroxybutyrate as substrate. Thus, we have [29, 62–64]

$$J_P/J_O = Z(q + Zx)/(1 + qZx) \tag{13}$$

and

$$\eta = -J_P A_P/(J_O A_O) = -Zx(q + Zx)/(1 + qZx) \tag{14}$$

where $x=A_P/A_O$.

Using Eqs. (13), (14) and Table 4, the results are tabulated in Table 5. Since 3 ATP molecules are formed from a supply of energy of A_O (~220 kJ/mol theoretically, but measured values were 208 kJ/mol) [66, 67], the value of x should be −1/3 for 3-hydroxybutyrate [38]. This is in perfect agreement with the experimental measurements of Stucki who found Zx at the operating point to measure −0.792 [63]. With our mechanistic stoichiometry, Z of 2.4, we obtain x= −0.792/2.4=−0.33. From Table 5 we find the operating efficiency at x=−0.33 to be 0.702, i.e., 70.2%. This value of efficiency can also be derived from the torsional mechanism. Since A_P includes the energy stored in ATP, the energy to bind P_i and the energy required to torsionally strain a bond so that ADP can bind (which then occurs without energy input), we only have to consider the losses due to uphill pumping of ions on the redox side by complexes I, II, III and IV and the proton leak. Assuming a similar type of translocation operating on the redox side, and that the entire machinery is regulated as one whole (redox+ATPase sides) rather than separately, the average efficiency of each redox complex

Table 5. Calculated P/O ratios and efficiencies of energy coupling as a function of the affinity ratios in oxidative phosphorylation by rat liver mitochondria for 3-hydroxybutyrate as substrate using the parameters obtained in Table 4

A_P/A_O	J_P/J_O	η (%)
−0.412	0	0
−0.400	1.228	49.14
−0.375	1.848	69.29
−0.350	2.045	71.58
−0.333	2.127	70.20
−0.325	2.142	69.63
−0.320	2.156	69.00
−0.300	2.200	66.01
−0.280	2.232	62.50
−0.250	2.266	56.65
−0.200	2.302	46.04
−0.100	2.341	23.41

can be estimated to be 0.7060. Incorporating the small loss due to the proton leak (Table 4) at the operating point of $x=-0.33$ yields a mechanistic efficiency of $0.7060\times(1-3.2/474.3)=0.7013$, which is in perfect agreement with the thermodynamics.

The torsional mechanism of ATP synthesis is consistent with thermodynamics as well as with the excellent experimental measurements of the P/O and A_P/A_O ratios (and hence the efficiencies of energy conversion) in ATP synthesis by Lemasters [66, 67] as already demonstrated by Nath [29]. The macroscopic behavior is a consequence of the proposed molecular mechanism and can be accurately predicted from the molecular mechanism. Thus, molecular and macroscopic approaches, each independent of the other, and each suggestive by itself, stand unified, and lend our molecular mechanism of ATP synthesis a cumulative force.

4.3 Thermodynamic Principle for Oxidative Phosphorylation and Differences from Prigogine's Principle

The innovative thermodynamic principle for the coupled process of oxidative phosphorylation formulated by the author [27–29] can be stated as follows: "The physical system/mechanism of coupling selected by an energy-transducing biological system at stationary steady state (from all possible localized and delocalized systems/mechanisms of coupling) is one that corresponds to *minimum rate of entropy production*. Further, the distance from equilibrium (which depends on the species concentration/thermodynamic affinity) at which the system operates is selected to maximize the product of the efficiency of energy transfer, the output power, and the operating affinities (or equivalently, to minimize, once again, the entropy production when all non-linear processes operating in the system have been taken into account). This is derived, strictly speaking, for linearly coupled systems close to equilibrium satisfying Onsager symmetry; however, it appears to have a validity beyond these restrictions." It *differs* from the principle formulated by Prigogine, which compares the minimum entropy produced by equilibrium or stationary steady states (e.g., those of zero H^+ flux) with other unsteady states for a *single* physical system, while the principle formulated by Nath is applicable to *different* physical systems all of which operate at steady state. This double optimization, i.e., the optimization with respect to conductances (the L values in Table 4) as well as species concentration, is in accordance with the physical interpretations of the torsional mechanism of ion translocation presented in Sect. 3.5 and suggests that the concentrations of various chemical species as well as the process of distribution of energy of the metabolism of glucose among an appropriate number of ATP molecules have been remarkably *tuned* by evolution for optimal performance as proposed earlier by us [29]. We now see with great clarity that the cell is even more highly coordinated and perfectly organized than what has been suspected to date.

4.4 Overall Energy Balance of Cellular Bioenergetics and its Biological Implications

If the harmony between the torsional mechanism and the thermodynamics of oxidative phosphorylation is as good as described above, the mechanism should be able to withstand the ultimate challenge of satisfying the overall, macroscopic energetic constraints of metabolism (keeping the constraints imposed by the oxidative phosphorylation process intact). Thus, we should look at the overall energy balance for the complete oxidation of glucose and the cytoplasmic ATP yield in the cell. Since, according to the torsional mechanism, the fifth proton is not pumped out by the redox side, the number of ATP per mole of glucose remains 38, with 28 arising from oxidative phosphorylation ($2\times2+8\times3$) and the remaining 10 from glycolysis and succinyl CoA ($2+2\times3+2$). Each ATP molecule is taken to be identical in every respect, including in terms of the energy required to make it. Then, for a basis of 1 mole glucose, the energy available from the supply side for ATP synthesis is $672\times0.7020 = 471.7$ kcal/mol. On the user side the energy production is $220/3\times0.7020\times(1/4.18)\times38 = 468$ kcal/mol. Hence the overall energy balance is satisfied perfectly. Note that if a P/O ratio of 10/4=2.5 [68] (where 3 protons are translocated through F_0 and 1 proton through the substrate translocator) were used literally, then only 31 ATP molecules would have been produced per mole of glucose. This has important implications from the point of view of cellular bioenergetics in general and also in particular because it illustrates how a mechanistic P/O=12/5=2.4 at the level of the overall oxidative phosphorylation process may prevail under steady-state conditions, yet only four protons may be pumped out by the redox side. The role of the fifth neutralization proton thereby acquires a special significance. The agreement of the torsional mechanism with the overall energy balance lends further strength to it. Our recent bioinformatic work and experimental studies on the chloroplast enzyme [69, 70] support the predictions of the torsional mechanism. These give us complete confidence that the mechanism is correct in every respect.

5 Muscle Contraction

5.1 Molecular Mechanisms of Muscle Contraction

More than 130 years have elapsed since the actomyosin complex was isolated from muscle [71], yet the molecular origin of the force produced during muscle contraction is unknown and remains one of the most outstanding enigmas in biology. Various models have been proposed to explain muscle contraction: the swinging crossbridge model [72–75], the swinging lever arm model [20, 76–80], and the recent rotation-twist-tilt (RTT) energy storage mechanism [22] are the important ones. Here we summarize the chief features and carry out a critical evaluation of these models/mechanisms. Some specific difficulties associated with prevalent models of muscle contraction are delineated and novel proposals to overcome these difficulties are suggested.

The fundamental problem of how force is generated by the actomyosin complex in the muscle sarcomere has proved very difficult to solve. A crucial step towards understanding the molecular basis of muscle contraction was taken in the middle of the 20th century through the formulation of the swinging crossbridge model of muscle contraction by H. Huxley and A. F. Huxley, which occupies a prominent place in most textbooks of cell biology. Several decades were spent trying to experimentally verify the model; however, despite the use of extremely sophisticated spectroscopic tools, the conformational changes predicted by the model have simply not been observed. This led to modification of the swinging crossbridge model into the swinging lever arm model in the 1980s. Recently, another novel molecular mechanism of muscle contraction, the RTT energy storage mechanism, has also been proposed in the literature.

Here, we review the above-mentioned three models/mechanisms of muscle contraction. We summarize the major tenets of each model/mechanism, describe what aspects of the problem are addressed by each of them and how, what facets of the puzzle they are unable to satisfactorily explain and what are the specific shortcomings associated with them. The comparisons and the probable way out of the present impasse provides deep insight into the molecular mechanism, and a wealth of new and original ideas for experimentalists to resolve with finality the outstanding enigmas in the field of motility, and thereby elucidate the molecular mechanism of muscle contraction.

5.1.1
The Swinging Crossbridge Model

The first model for muscle contraction, the swinging crossbridge model, was postulated in 1954 by the pioneering effort of H. Huxley [72, 73] and A. F. Huxley [74, 75]. Based on tryptic digestion studies, the myosin molecule was characterized as being composed of a heavy part (heavy meromyosin or HMM) constituting the globular head and the helix region (S-2), and a light part (light meromyosin or LMM), making up the thick filament. The model postulated the formation of crossbridges between the HMM and the actin filament. Movement was proposed to occur due to sliding of actin and myosin filaments past each other. However, the details of the exact nature of the movements and force generation were not specified. In this model, the origin of force generation is the globular head and its attachment to actin filament. The head is presumed to be attached to the backbone of the myosin filament by a 400 Å long linkage behaving as an inextensible thread having flexible couplings at each end [73]. This flexible linkage allows the myosin head to attach to actin in a constant configuration and undergo the same structural changes in each cycle over a wide variation of interfilament spacing. The motion of actin takes place when hydrolysis of ATP causes a change in the effective angle of attachment of the globular head to actin (tilting). This tilting can take place either by relative movement (sliding) between two interacting subunits of myosin, or by an independent change in the angle of attachment of each subunit. This tilt pushes the actin filament in one direction. The tilting transmits a force through the S-2 linkage which is under tension during the power stroke. This transmitted force moves the myosin filament in a direction opposite to the movement of actin. In this mode, the myosin molecule does not detach from actin during the cycle, and thus, to relieve this tension, the thick filament is pulled in the forward direction. As a result, the net movement is that of myosin and actin filaments in opposite directions.

The swinging crossbridge model was not clear about the detailed mechanism of motion and force generation. In the original version of the swinging crossbridge model, actin stays bound to myosin throughout the cycle. However, the kinetic studies carried out by Lymn and Taylor [81] and other groups conclusively show that ATP hydrolysis takes place when myosin is detached from actin. Further, the model necessitates pulling of myosin filaments in each cycle so as to increase the overlap between thick and thin filaments, and thereby release the tension in the S-2 linkage. This is highly unlikely, in our view, since the structure and arrangement of myosin filaments and the M-line do not permit such a movement. The arrangement of myosin filaments is such that to increase the overlap with actin, the same filament would have to be pulled in opposite directions on the two sides of the M-line. This will lead to tearing of the filament. More recent versions of the swinging crossbridge model incorporate the detachment of myosin from actin prior to ATP hydrolysis in a Lymn-Taylor cycle. However, these modified mechanisms still do not address the details of the movement and force production. For instance, how exactly the myosin reattaches to actin after ATP hydrolysis is not specified; further, how exactly the power stroke takes place and how nucleotide release is coupled to it are not mentioned. Finally, the large magnitude of changes in the angle of attachment of myosin to actin (from 45° during rigor to 90° after hydrolysis) have not been observed despite several decades of experimental effort.

5.1.2
The Swinging Lever Arm Model

The vagueness of the swinging crossbridge model regarding the details of the molecular mechanism and the motion led to a new hypothesis in the 1980s: the swinging lever arm model [20, 76–80]. Since the motion postulated by the swinging crossbridge model could not be experimentally detected, a new kind of motion was envisaged in which the regulatory domain (lever arm) moved about a fulcrum at the joint of the catalytic and the regulatory domains, instead of at the site of attachment of actin to myosin, as proposed in the swinging crossbridge model. According to the lever arm model, the seat of force generation is the globular part of the head, which binds to actin in a fixed orientation. Movement is caused by a change in the angle of attachment of the lever arm (i.e., the distal part of the myosin head) with respect to the catalytic domain as well as the S-2 helix. ATP hydrolysis in the catalytic domain swings the lever arm about the fulcrum site, changing its orientation with respect to both catalytic domain and the S-2 region. The swing is postulated to be of the order of

90°, sweeping a distance of over 10 nm [20, 76–80]. The reverse movement of the lever arm causes the power stroke.

The swinging lever arm model can be considered as a gross statement of the kind of motion myosin is envisaged to undergo during the power stroke. It does not attempt to address the complete contractile cycle of muscle. The model proposes a swing of the regulatory domain of the myosin molecule by ~90° after hydrolysis spanning a distance of ~10 nm, with the reverse stroke causing the motion [20]. According to the crystal structure of Rayment and colleagues [82–85], the top of the myosin C-terminal 20 kDa region rotates by almost 20° due to ATP hydrolysis. This motion is of a different kind from the swing of the lever arm, one being rotation, and the other akin to tilting. However, in the lever arm terminology, both these motions are referred to as rotation. Moreover, how the former is converted to the latter is not specified. While one can conceive of amplification in terms of length, it is difficult to imagine how a ~20° rotation can amplify into a change of ~90° (unless one wishes to revise the principles of geometry). Once the lever arm swings, the cause of the reverse stroke and the mechanism by which the reverse stroke is converted to the power stroke is not mentioned in the model. Moreover, such large motions spanning an angle of ~90° and a distance of almost ~10 nm have not been detected experimentally to date. The lever arm model postulates that only a small fraction of myosin heads (~15%) actively participate in the cycle. The function of the remaining ~85% of the heads is ambiguous. The model also does not specify a mechanism for reattachment of actin to myosin before the power stroke. How nucleotide release when myosin is bound to actin is coupled to movement is not addressed by the model. To summarize, the questions which the lever arm model does not address or does not provide even a rudimentary explanation are:

1 How does myosin bind to actin? Or, in particular, how does the envisaged "rotation" of the lever arm help in myosin-actin binding? And, if this lever arm "rotation" does not cause the binding of myosin to actin, then what agent causes it?
2 How does the change in orientation of the lever arm come about?
3 How does the change in lever arm orientation cause the power stroke?
4 How is release of ADP and P_i coupled to the power stroke and by what mechanism?

5.1.3 *The Rotation-Twist-Tilt (RTT) Energy Storage Mechanism*

Recently, a novel mechanism for the contractile cycle of muscle has been proposed [22]. The mechanism, called the rotation-twist-tilt (RTT) energy storage mechanism of muscle contraction, besides describing the exact nature and details of motion, also sheds light on the process of storage of energy of ATP hydrolysis, its subsequent conversion to useful work, and the generation of force in the actomyosin system. A central tenet of this mechanism is the storage of energy of ATP hydrolysis as an increase in twist between the coiled coils of the S-2 region and its subsequent untwisting causing the power stroke. According to the mechanism, ATP hydrolysis in the catalytic site rotates the top of the regulatory domain, which, being connected to the coiled-coils of the S-2 region, causes twist between them to increase, and leads to tilt in the myosin head. Since, at the time of hydrolysis, the myosin molecule is not bound to the actin filament, the head is free to rotate and tilt. The increase in twist is instrumental in storing the energy of ATP hydrolysis, while the rotation and tilt of the myosin head bring it sufficiently close to actin so as to form the actomyosin complex [22]. Untwisting of the coiled coils and subsequent untilting and constrained reversal of rotation of the head cause the power stroke. The untwisting releases energy, and the untilting of the head drags the actin filament. Since the myosin is bound tightly to actin, the reversal of rotation is restricted, generating a strain in actomyosin bonds during the power stroke. This strain decreases the energy of interaction between actin and myosin and thus enables ATP (which, by itself, has a lower binding energy to myosin than that of the actomyosin complex) to dissociate myosin from actin. The system is now in such a state that after the next ATP hydrolysis event, the myosin head can bind to actin, and thus, a new contractile cycle can be initiated.

5.2 Attempts to Address the Difficulties Associated with Other Models by the RTT Energy Storage Mechanism

The RTT energy storage mechanism appears to be novel in terms of explaining the details of energy storage and force generation. As described in Sect. 5.1.3, the mechanism envisages the myosin molecule to store energy through increase in twist between coiled coils of the S-2 region and then move the actin filament by force generated by the untwisting of these coils. While the large-scale conformational changes required by the lever arm model have not been experimentally verified, no need for such large changes arises in the RTT energy storage mechanism. The twist and the tilt predicted by the mechanism are in accordance with recent experiments [86]. The other mechanisms proposed for muscle contraction do not feel the need to tackle the problem of energy storage, which according to the RTT mechanism is a central one. The need for energy storage arises since, during ATP hydrolysis, the myosin head is detached from actin, and hence a non-equilibrium conformational state in which energy can be stored (as internal energy) until the time myosin can again bind to actin and execute the power stroke is essential (Fig. 6). Note that in this Figure, d signifies the initial distance of the end of the actin filament from the M line and x measures the distance between two adjacent myosin-binding sites on actin filament. Only the catalytic (thick bold line) and the regulatory (thin bold line) domains of myosin molecule are depicted; the S-2 region is not shown.

The mechanism of force generation is not elucidated in previous models (Sects. 5.1.1 and 5.1.2), i.e., how the reverse stroke of the lever arm transmits force to the tip of the myosin head or the actomyosin interface is very hard to envisage. Such difficulties do not arise with the RTT energy storage mechanism, which explicitly explains how the force is generated. Furthermore, since the myosin head returns to a position after the motion from which it can bind to

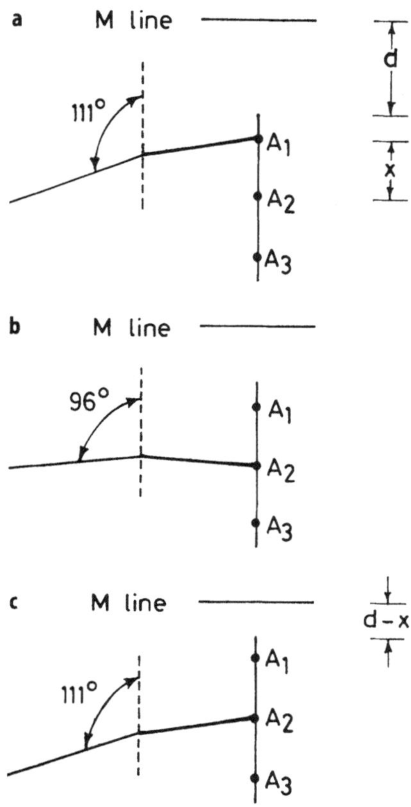

Fig. 6a–c. Schematic representation of the rotation-twist-tilt (RTT) energy storage mechanism of muscle contraction showing the changes in the actomyosin complex during the contractile cycle with reference to the M line. (**a**) The rigor state. Binding of ATP to this state will cause the myosin to get detached from actin in a post-power-stroke and pre-hydrolysis state (not shown, but see ref. 22). (**b**) The state of the actomyosin complex in the post-hydrolysis and pre-power-stroke state. (**c**) The post-power-stroke: the myosin has dragged the actin filament towards the M line, reducing the distance between the end of the actin filament and the M line by x

actin in the next cycle, there is no need to drag the myosin filament. As a result, the RTT mechanism will not cause any tearing of the thick filament as in the original version of the swinging crossbridge model.

According to currently accepted mechanisms, only a small fraction of myosin heads actively participate in the cycle. Neither the function nor the importance of dormant heads, nor the mechanism determining the active fraction is specified. No such difficulties arise in the RTT energy storage mechanism since ideally (at high load), all heads are taken to be actively participating in the contractile cycle, although only a few of them may be executing the power-stroke at any one instant of time. The strength of the load will determine the number of ATP molecules released by the regulatory mechanism and the number of myosin heads that will be recruited. The RTT mechanism also clearly specifies the conformational changes taking place after ATP hydrolysis, the mechanism by which these store energy, how their reversal causes the power stroke, and how strain in the actomyosin complex assists ATP binding to release myosin from actin.

5.3
A Distinguishing Feature of the RTT Energy Storage Mechanism and its Validation

As in lever arm models, velocity is proportional to length of the lever arm, l. However, a key prediction and distinguishing feature of the RTT energy storage mechanism lies in the fact that the force, F, that drags the actin filament is independent of l, or, at least, it has no direct relationship with l, unlike in the lever arm model (where $F \propto l^{-1}$) or the modified lever arm model (in which $F \propto l^{-2}$). This prediction is experimentally supported by force measurements carried out on short- and long-necked lever arm constructs for the first time [87]. It is also validated by the principle of energy conservation. The distance moved by the actomyosin system during the power stroke is constant (~5.3 nm) [88, 89]. Further, the effective energy (E) available for the power stroke arising from the hydrolysis of an ATP molecule is also constant. Hence, from the energy conservation relation,

$$F_z \cdot \boldsymbol{d} = \text{Work done} = \text{E} \tag{15}$$

As F_z and d are in the same direction (along z), we have $F_z\, d = \text{E}$, where d and E are constant. Therefore,

$$F_z = E/d = \text{constant} \tag{16}$$

Hence, the force to produce the power stroke is independent of the length of the lever arm. The novel and original approach and insights offered by the RTT energy storage mechanism should greatly accelerate the attainment of a thorough understanding of the molecular mechanism of muscle contraction.

The swinging crossbridge model of the 1950s was based on physical (X-ray diffraction and electron microscopy) observations. The lack of experimental verification of the major conformational changes predicted by the swinging crossbridge model led to the formulation of the swinging lever arm model in the 1980s. Unfortunately, to date, the large-scale motions predicted by the swinging lever arm model have also not been directly observed experimentally. This is simply and logically explained within the framework of the RTT mechanism by the fact that such large amplitude motions do not exist, and the mechanism shows that there is no need for such motions. Hence, a critical reassessment of the fundamental assumptions on which current mechanisms are based is sorely needed, which may lead to new ways of looking at the problem of the molecular origin of motility. The RTT energy storage mechanism of muscle contraction is a crucial first step in this direction. The aspects dealt with in this mechanism may constitute the *key* elements whose lack of detailed consideration has held back the progress of research in the important field of motility.

5.4 Engineering Analysis of the RTT Model

As discussed in Sect. 5.1.3, according to the rotation-twist-tilt energy storage molecular mechanism of muscle contraction [22], hydrolysis of ATP to ADP and P_i causes rotation of the top of the regulatory domain which results in twist motion in the S-2 region of the muscle fiber, rotation of the myosin head about an axis (x-axis) that passes through the S-1–S-2 hinge and the center of the arc of the circle swept by the rotation motion, and the tilt motion of the myosin head about the S-1–S-2 hinge (Fig. 6). The twist motion stores the energy of the enthalpy change upon ATP hydrolysis in the S-2 region, and the rotation and tilt motion of the myosin head lead to binding of the myosin head to the actin fiber. Untwisting of the S-2 region leads to release of the stored energy and generation of force due to constraint in the untilt motion of the myosin head independently without actin (because the myosin head is bound to actin fiber) [22]. However, the combined actin-myosin system can untilt about the S-1–S-2 hinge; this drags the actin filament upward (Fig. 6) along with the myosin head [22]. A detailed mechanical analysis of this process is presented in this section.

5.4.1 *Storage of Energy and Concomitant Motions*

Figure 7 depicts a simplified representation of the myosin twisting process and the Cartesian coordinate axes employed in our analysis. The twisting moment, τ, is in a direction tangential to the length of the tail (T) of the myosin fiber. This results in energy storage in the two α-helices forming the coiled coil of the S-2 region as an increase in twist. The joint between S-1 and S-2 has been shown to possess flexibility by electron microscopy studies.

If the S-1–S-2 joint had been completely rigid, then the whole myosin fiber (S–1+S–2) could have been regarded as one unit, and only rotation of the S-1 subunit about an axis passing through the S-1–S-2 hinge and the center of the arc of the circle swept by rotation (the x-axis) and twist of the S-2 region would have been possible. The same rotation takes place in each of the constituent α-helices in the S-2 region, but as they are coiled around and interacting with each other so that each cannot rotate independently of the other, it manifests itself as twist motion in the S-2 region. However, no tilt motion about S-1–S-2 can take place in this case of a completely rigid joint. Therefore, no power stroke will be generated in the later part of the cycle because there will be no untilt motion. Hence, complete rigidity of the S-1–S-2 hinge can be ruled out.

If the S-1–S-2 hinge (joint B) had been completely flexible, then there would only be the rotation of the top of the regulatory domain and there would be no twist in the S-2 region because joint B (Fig. 7) would not provide any constraint to any type of motion. In effect, the enthalpy change of ATP hydrolysis to ADP+P_i would be dissipated as heat and no useful work would be performed. The actin-myosin system cannot be this type of machine. Hence joint B must possess some flexibility and some rigidity.

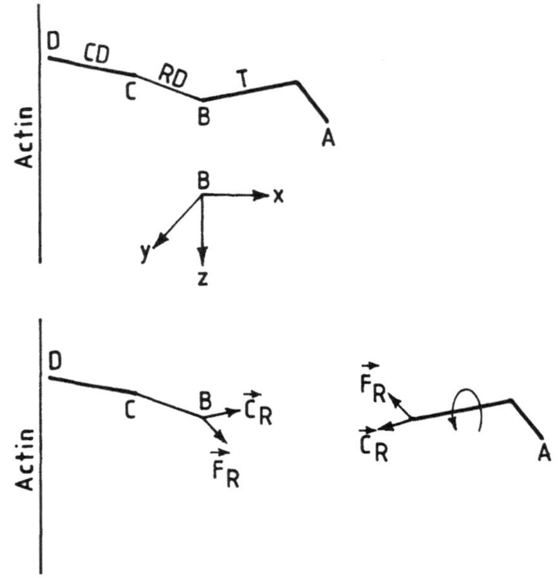

Fig. 7. *Top*: Simplified representation of myosin and actin system of muscle and the coordinate system employed. CD stands for the catalytic domain of myosin, RD for the regulatory domain, T for the myosin tail and B for the S-1–S-2 hinge. *Bottom*: Free body diagram for the generation of force F_R and couple C_R at the S-1–S-2 hinge upon rotation of the top of the regulatory domain of myosin head due to ATP hydrolysis

For such a joint, upon rotation of the top of the regulatory domain of myosin head, force C_R and corresponding couple F_R is generated at the joint. This occurs due to partial rigidity in the S-1–S-2 hinge and the three-dimensional structure of the myosin head and the S-2 region. In particular, due to the components of F_R in the z and y directions (Fig. 7), couples are generated in the y and x directions, respectively which are responsible for the tilt and rotation motions, respectively (Fig. 8). For a rigid body (S-1 region) with no additional forces and couples acting on the catalytic and regulatory domains, the force generated at the S-1–S-2 hinge can be taken to act anywhere in the S-1 region; hence the force generated will produce a couple (Fig. 8). Note that the axis of tilt motion is the axis that passes through the S-1–S-2 hinge and is perpendicular to both x- and z-axes, i.e., along y, while the z axis is taken along the actin fiber and passes through B (Fig. 7). Due to this tilt and rotation motion the myosin head gets attached to the actin fiber, as schematically shown in Fig. 9.

5.4.2
Release of Stored Energy and Upward Motion of Actin Fiber

After attachment of the myosin head to actin as discussed above, the energy stored in the tail of myosin fiber (S-2 region) is released by untwisting of the twisted myosin fiber tail. Again, various cases of the type of joint B can be considered. If it is fully flexible, no motion of the catalytic and regulatory domains

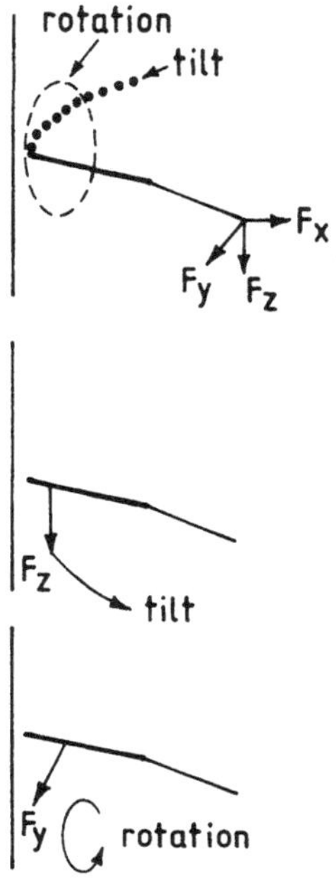

Fig. 8. Simplified diagram depicting the combined rotation and tilt motion of myosin head due to ATP hydrolysis (*top*). The individual motions of tilt (*middle*) and rotation (*bottom*) are also shown separately

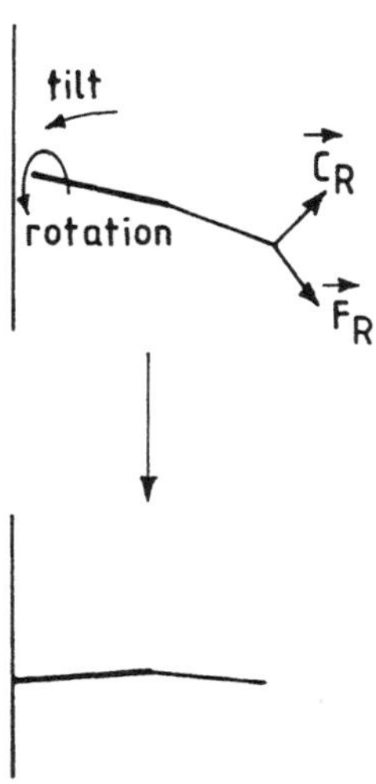

Fig. 9. Attachment of myosin head to actin fiber (*bottom*) due to the rotation and tilt motion of the head (*top*)

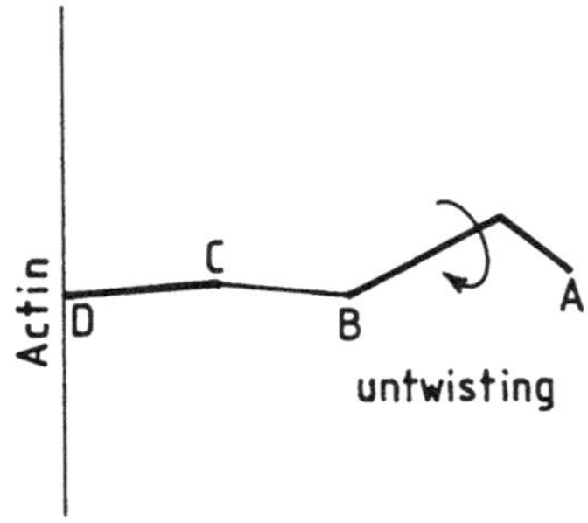

Fig. 10. Untwisting of the S-2 region of the muscle tail and release of the stored energy of ATP hydrolysis after the twisting process of myosin S-2 has been completed and myosin head has bound to actin

(CD and RD in Fig. 7) is possible and all the stored energy will be dissipated as heat. If joint B is fully rigid, there will be a tendency for CD and RD to rotate, but because the myosin head is bound to actin fiber, the interactions between myosin head and actin will strain, but no real free rotation of CD and RD is possible. Further, no untilting motion of the myosin head-actin system can occur due to the absence of any motive force in the z-direction. Hence, no muscle contraction can take place. Hence, again we are forced to consider joint B as a partially rigid and partially flexible joint.

For a partially rigid and partially flexible elastic S-1–S-2 hinge, untwisting of the S-2 region of the myosin fiber (the myosin tail) will lead to generation of forces and couples at joint B that, from the principles of energy conservation and microscopic reversibility, are equal in magnitude but opposite in direction to those generated during the energy storage process (Fig. 10). The generation of forces is due to the partial rigidity in the S-1–S-2 hinge and the three-dimensional conformation of the myosin head and the S-2 region. The axes of the possible rotation and untilt motions will remain the same as in Sect. 5.4.1 (i.e., x- and y-axes, respectively). However, in this case, the system consisting of myosin head and actin fiber cannot rotate freely about the x-axis as the myosin head is strongly bound to actin. Hence, the interactions between myosin head and actin will be strained, which is also of great importance as it will be easier (i.e., it will require less energy) to unbind the myosin head from its actin-binding site in subsequent elementary steps of the contractile cycle [22].

The component of the force F_R in the z-direction due to the untwisting process in the myosin S-2 region will tend to cause untilting of the actomyosin system (Fig. 11). However, the myosin head cannot untilt independently of the actin filament. The entire actomyosin system cannot untilt about the y-axis due to the absence of a degree of freedom in the actin fiber for the untilt motion owing to the physical structure and linkage of the actin filament and the Z-line. Hence only a linear motion of the actomyosin system along the z-direction is possible due to the force on the system in that direction [22] (Fig. 11).

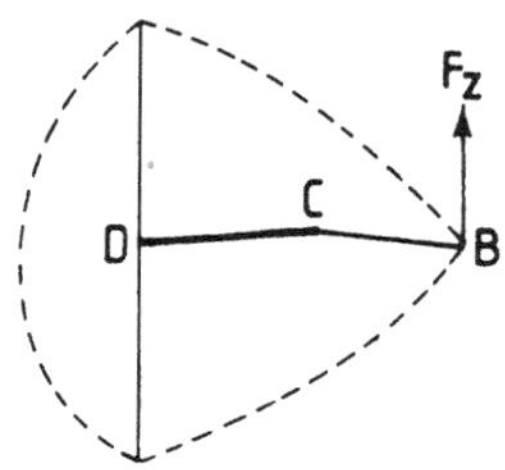

Fig. 11. Untilting of the actomyosin system of muscle because of the component of the force F_R in the z-direction during the untwisting process of myosin S-2 and dragging of the actin filament along with the myosin head in the z-direction. For details see text and ref. [22]

6 Engineering Applications

Our work has profound ramifications for the design of macroscopic and molecular machines in engineering and technology; it revolutionizes approaches to the design of machines. Till now, in mechanical, chemical and biochemical engineering, engines have been conceived as *thermal* machines based on the interaction of the machine system with the surroundings, and a heat exchange step lies at the heart of the design of each machine. This formalism is the root cause of low efficiencies of energy conversion (<35% for power plants, and only 8–9% for fuel cells). Our work shows that to escape the entropic doom imposed on all processes by the second law of thermodynamics and to increase these efficiencies phenomenally, the second generation machine needs to convert energy *directly* from one form to another, *without equilibration with the surroundings, without an intervening heat exchange step*. In our view, energy in equilibrium with the surroundings cannot be stored, and must be dissipated, and *to prevent wastage and dissipation as heat, the* ***energy must be stored within the system*** *of the macromolecule* ***in a non-equilibrium state*** *(as internal energy/enthalpy)*. The design of future machines will have to be ***enthalpic***, and not ***entropic***, as is the case today. This is in accord with the prophetic statement and vision of McClare and Blumenfeld [90, 91].

Second generation machines will need to transduce stored energy from one form to another directly, without intermediate thermal steps. Thus, they will have to be designed as *enthalpic* machines (i.e., their operation is governed by the ΔH part of the ΔG change) that carry out their motive step *faster* than heat flow and never equilibrate with the thermal degrees of freedom of the surrounding medium. This rapid mode of operation ensures a high efficiency of energy coupling (between donator and acceptor molecules, say) and prevents dissipation of the stored energy as heat. Hence, any entropy production (or high rate of entropy production for a process at steady state) by such an enthalpy (or internal energy)-driven macroscopic or molecular machine is a wasteful process. This notion of an enthalpic non-equilibrium machine is in harmony with Nath's minimum ϕ thermodynamic principle for coupled bioenergetic processes (Sect. 4.3) where the values of the coefficients (conductances), which are related to the different kinds of microscopic biological couplings, are varied, keeping the concentrations of the various chemical species constant and/or the concentrations (or thermodynamic affinities) are varied, for constant values of the coefficients [29].

Further, it should be stressed that in our molecular mechanism of energy transduction by enthalpic non-equilibrium machines, *conservative forces* have been used. In *dissipative structures* [92], the ordering of the system is maintained by an exchange of matter/heat with the surroundings beyond a certain level. Mathematically, in terms of the second law of thermodynamics,

$$dS_{system} = d_eS + d_iS, \text{ or } dS_{system}/dt = d_eS/dt + d_iS/dt \tag{17}$$

For dS_{system} to decrease (ordering of the system), for a particular value of the entropy internal to the system due to irreversible processes taking place within the system (d_iS, a positive definite quantity), the entropy exchanged by the system with the surroundings (d_eS) must be large and negative (i.e., heat must be given off by the system to the surroundings), as seen from Eq. (17). On the other hand, in the mechanical process of energy transduction discussed here, heat exchange has no relevance; the molecular energy transducer exhibits a non-equilibrium state that stores internal energy without allowing that energy to become heat, and entropic terms in Eq. (17) cannot be a major contributor to its action. Release of this stored energy is used to perform useful external work or is transduced into another form of stored energy without losing/dissipating that energy as heat in the process. Hence, we would term biological energy transducers as *conservative non-equilibrium structures*. Needless to say, the ordered non-equilibrium structures that result differ from equilibrium types of structures (e.g.. those that occur at phase transition points).

This research has paved the way for the development of the new field of **Molecular Engineering** [1, 34, 69], in which the engineering principles of thermodynamics, kinetics, transport, mechanics, dynamics, elasticity, machine design and electrical science are applied innovatively to biological systems at the *molecular* level to understand their functioning and to apply them to design, develop and fabricate novel macroscopic as well as molecular devices and machines. In our daily experience, we are familiar with macroscopic machines that convert mechanical energy to electrical energy and vice-versa (the generator), electrical energy to heat and vice-versa (the toaster), electrical energy to chemical energy and vice versa (the battery charger); however, it is difficult to think of machines that use a direct conversion of chemical energy to mechanical energy or vice-versa (without a heat intermediate, although an electrical intermediate is permissible). Thus, the most ubiquitous molecular energy conversion in the living cell has hardly been applied in our industrial technology. Just as the electrochemical works of Faraday, Galvani and Volta led to the development of a host of new devices (the lead storage battery, the dry cell) in the 19th century, similarly, mechanochemical and mechano(electro)chemical research has the potential to lead to novel energy conversion devices in the 21st century.

We have built a simple, macroscopic mechanical device assembled from readily available materials to show that a machine based on energy storage and release as envisaged by our molecular mechanism is, in principle, possible

(Fig. 12). Thus, an 8 mm mild steel torsional spring simulated the γ-subunit, and an audio-cassette rotor served as the pulley! A mechanical, anti-rotation mechanism was devised that allowed the bottom disc, a mechanical equivalent of the c-rotor, to rotate in one direction only. The lower portion of the spring was fixed to the disc, while the upper portion was fastened to a bolt with two nuts. Three vertical rods were arc-welded to a drilled-out mild steel piece (which represents the F_1) to make for three-fold symmetry. A metal strip/film was fixed to the drilled-out mild steel piece with strong adhesive. The strip physically interacted with the bolt and simulated the interactions of the top of γ with the catalytic sites of ATP synthase. The bottom disc was rotated in steps of approximately 30°. The strength of the bolt-strip interaction was adjusted such that the contact at the top could withstand the torsional strain generated in the spring due to three rotations of the bottom (Fig. 12b). Upon the fourth rotation step of the bottom, sufficient torsional energy was stored in the spring to break the strip-

a

Fig. 12a–c. A working macroscopic internal-energy based prototype engine/machine built by us to illustrate the principles of energy storage and release embodied in the torsional mechanism. **a** The resting state before start up. **b** The non-equilibrium energized state. The bottom has moved in three steps of 30°, but the top has remained stationary and there is torsional strain in the central shaft (representing γ; see text); note that the head of the bolt faces the right side of the drilled-out mild steel piece (representing F_1) at the top of the shaft and the load has not yet been lifted. **c** After the fourth step, the contact of the top of the central shaft with the F_1 has broken and the top of the shaft has rotated rapidly in a single 120° step, releasing the torsional strain (the bolt head now faces the left side of F_1) and lifting the load upwards. In steady-state operation of the machine, system configurations similar to the middle and bottom snapshots follow each other in rapid succession. (For **Fig. 12b, c** see next page)

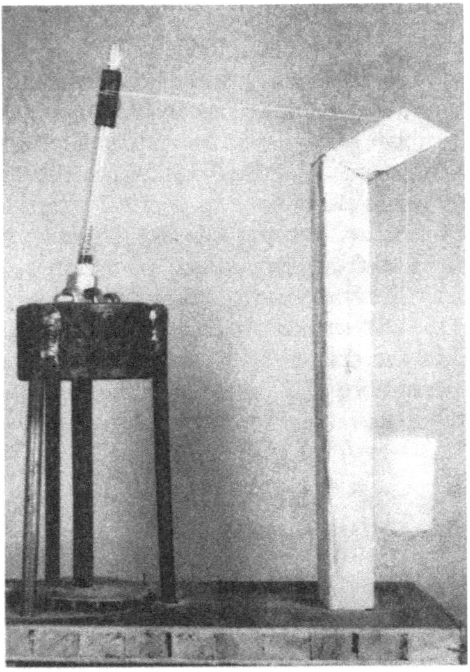

b

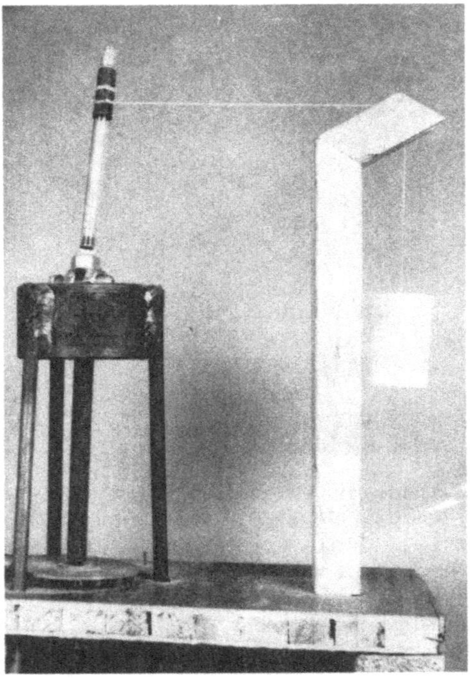

c

Fig. 12 (continued)

bolt (γ-β) interactions at the top, and now the top rotated very rapidly in one ~120° step. This rotation was communicated to a rod (built from the body of a ball point pen!), which rotated and lifted a load attached to it by a thread via the pulley (Fig. 12c).

Such a machine will not use our scarce resources of fossil fuels (oil and natural gas, shale oil and coal). It is perfectly conceivable that in a more sophisticatedly evolved version of the machine, ion movements could be induced by concentration gradients produced by light energy or the energy of redox reactions and be made to rotate the bottom disc (now of molecular dimensions), as proposed in detail in the torsional mechanism. It would then even be possible to utilize the 10^{17} W of energy radiated by the sun, our only real renewable energy source. Such mechanochemical devices offer us hope of a future solution to the energy crisis. The development of such devices will pose great challenges to our abilities in nanobiotechnology and molecular engineering in this century. The ultimate in molecular engineering has already been achieved by the molecular machines of the cell. We would do well to learn from it.

The energy storage and other mechanical aspects of the torsional mechanism of ATP synthesis and the RTT mechanism of muscle contraction draw attention to the solid state physical nature of biological systems. Thus, gas or liquid state theories are totally inadequate to understand the torsional energy storage in the γ-subunit and the c-subunit in the F_1F_0-ATP synthase and in the S-2 region of myosin during function. In fact, our recently performed sequence alignments of c-subunit molecules from over 50 sources show the presence of key hydrophobic residues towards the end of the C-terminal helix of subunit c, which are necessary in the torsional mechanism to ensure that the new incoming c-subunit does not untwist and untilt along with the remaining ten protonated c-subunits during energy transduction [1, 37, 69]. The location of the conserved Asp/Glu residue close to the middle of the C-terminal helix (and not towards either end) is again required in our mechanism to ensure that the proton gets exposed to the exit access half-channel only after the bottom of the γ-subunit has rotated by 15° and not at any other time (or never get exposed) during the energy transduction cycle (emphasizing the critical importance of the *timing* of the elementary steps). The presence of the Pro after the Asp/Glu may help the process by causing a bend in the vicinity of the C-terminal. The preponderance (as high as 40%) of a high number of small, uncharged amino acids (Gly and Ala) in the N-terminal helix of the c-subunit, and especially the presence of five conserved small residues [93] in the middle of the N-terminal helix directly across the Asp/Glu on the C-terminal helix is also essential to accommodate the large rotation-twist-tilt motions of the C-terminal helix proposed in the detailed torsional mechanism within F_0 with minimal structural perturbations so as to prevent possible disruption of the c-subunit oligomer during function. This shows the importance of *evolutionary* arguments and points to the *individual* role of each amino acid residue or groups of atoms/residues, which is different and not equivalent to the roles of other atoms/residues, in a *mosaic* structure of a solid-like nature. Fine-tuned by billions of years of evolution, this situation may be unique to biological systems and makes a strong case for the development of a *solid-state biology* and the attainment of a true understanding of certain aspects of biological energy transduction/storage processes that have a solid-state physical nature.

The electrochemical/mechanoelectrochemical basis of pattern formation discussed in Sect. 3 is highly relevant to other biological processes, e.g., morphogenesis and development processes. It is quite conceivable that genes will produce the required molecules in the correct cells/spatial domains, i.e., control *composition*. Thus, genomic research on gene products will supply *necessary* conditions for pattern formation and morphogenesis; however, this may prove to be *insufficient* by itself to understand how the system is organized structurally and dynamically. In other words, it alone cannot explain how characteristic spatial and temporal order arises in biology. The role of ion fluxes of the kind discussed here and their interaction with the cytoskeleton based on physico-chemical laws/principles may be key to the dynamic properties of the morphogenetic system that may indeed possess a mechanoelectrochemical basis.

Finally, what about the "magic molecule" ATP itself (Fig. 13), with which we began this article? On comparing the corresponding bond lengths of ADP and ATP [5], no significant change is observed in any bond length due to phosphate addition to ADP during ATP synthesis. However, when we compare the bond angles of (i) $P\beta$ in ADP with $P\beta$ in ATP, and of (ii) $P\gamma$ in ATP with the P_i, significant changes are observed as seen from Table 6. We find that the $O1'P\beta O2'$ bond angle decreases by as much as 7° in ATP compared to ADP (122.66° in ADP versus 115.73° in ATP). All other angles between the $P\alpha$ and $P\beta$ and the correspondingly attached oxygen atoms remain more or less the same with a variation of only ~±1°. On the other hand, the terminal phosphate in ATP has an almost tetrahedral structure as in inorganic phosphate with all the bond angles close to 109.5°, which implies that the conformation of the terminal phosphate remains almost unchanged even after binding to the enzyme-bound ADP. Hence, in our interpretation, the major conformational change is observed to occur through a change in the bond angle $O1'P\beta O2'$.

As per our torsional mechanism of ATP synthesis, during the transition from β_{TP} (loose conformation) to β_{DP} (tight conformation) due to conformation changes caused by rotation of the top of the γ-subunit, the positively charged atoms of the key catalytic residues move closer to and interact with the O1′ oxygen atom of the ADP [9, 32] (Fig. 13). For example, the distance of the O1′ oxygen to the N and NZ atoms of the critical catalytic residue Lys 162 (*Escherichia*

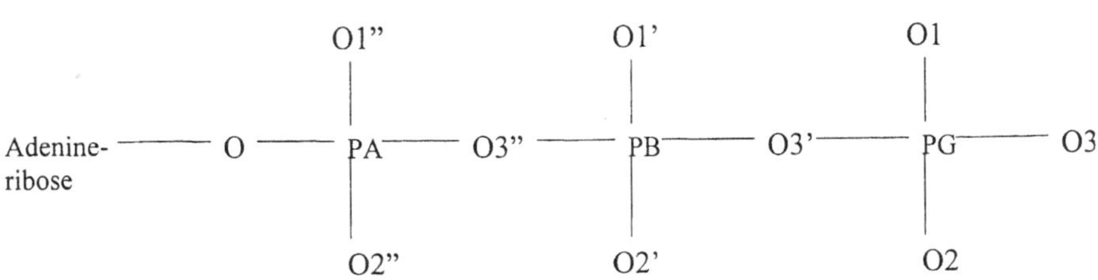

Fig. 13. Line diagram of ATP depicting the notation and the numbering of the atoms as used in our analysis

Table 6. Bond angles in ADP and ATP bound to the F_1 portion of ATP synthase

Bond	Bond angle in ADP	Bond angle in ATP
O1PγO2	-	111.29°
O2PγO3	-	105.58°
O1PγO3	-	111.58°
PγO3′Pβ	-	140.83°
O1′PβO2′	122.66°	115.73°
O2′PβO3′	106.78°	108.7°
O1′PβO3′	109.00°	110.03°
PβO3″Pα	127.46°	131.68°
O1″PαO2″	111.7°	109.94°
O2″PαO3″	116.01°	115.34°
O1″PαO3″	109.68°	109.41°

coli amino acid residue numbering) reduces from 2.81 Å and 3.28 Å, respectively, to 2.50 Å and 2.73 Å, respectively. Furthermore, the Mg^{2+} interacts with the O2′ oxygen of the substrate. These interactions lead to the development of a better and more effective ADP-O^- nucleophile. The increased nucleophilicity of ADP-O^- is a major contributor to the driving force for ATP synthesis as postulated by our torsional mechanism [1, 9] and validated by the computational results shown in Table 6 obtained from structural information [5]. Thus, the interactions of the oxygen atoms of the enzyme-bound ADP with Mg^{2+} and the critical catalytic residues (e.g., Lys 162) orient the substrate in the proper conformation for the nucleophilic attack and are hence key to the catalysis.

A schematic diagram showing the electrostatic interactions among the negatively charged oxygen atoms, which result in the conformational changes in the ATP is depicted in Fig. 14a. These interactions stabilize the structure so that the forces on each oxygen atom are balanced, i.e., the net force is zero. Upon hydrolysis of ATP due to the nucleophilic attack by H_2O, the terminal phosphate bond is broken and the ADP returns to the original conformation thereby releasing the stored energy. It should be noted that the terminal phosphate bond itself is not the means of storage of energy but its presence forces the O1′PβO2′ to attain the conformation which stores this energy, and its removal causes the same bond angle to attain the original conformation and consequently release the energy.

Based on our analysis, ATP can be modeled as two like-charged spheres attached to the ends of hinged bars and connected by an inextensible string forcing the spheres to remain close to each other, i.e., in a high energy conformation relative to its resting state in ADP [Fig. 14b]. ATP hydrolysis is equivalent to cutting the string thereby freeing the spheres in terms of their movement away from each other as a result of mutual repulsion. This movement of charges may be used to carry out useful work like rotation of the γ-shaft during ATP hydrolysis in F_1-ATPase or transduced into and transiently stored as an increase in twist in the S-2 coiled coil of myosin during muscle contraction [22]. All the forces used in our proposed mechanism, which we refer to as the "*locally strained but overall at equilibrium mechanism*" of energy storage in ATP, are

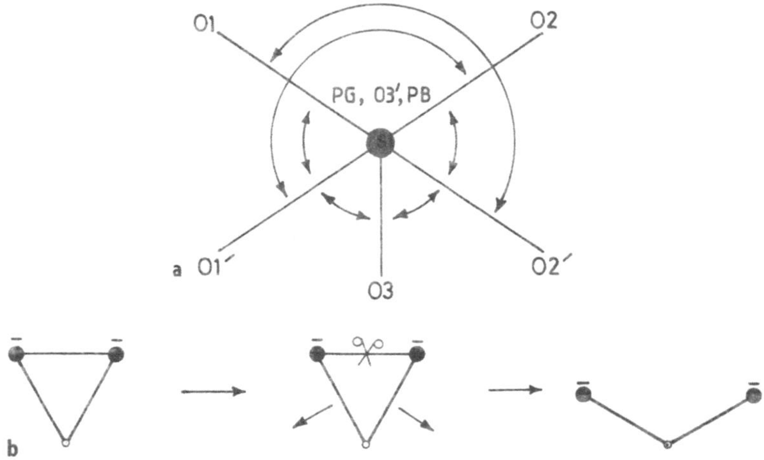

Fig. 14a, b. Schematic diagram for locally strained but overall at equilibrium mechanism for energy storage in ATP. (**a**) Electrostatic interactions among the oxygen atoms resulting in the conformational changes in the ATP molecule (**b**) Mechanical model idealizing energy storage in ATP. The bold lines represent the bars, the thin line the inextensible string, the filled circles the negatively charged spheres (oxygen atoms) and the open circle the hinge. The scissors denote ATP hydrolysis

conservative in nature; hence, our mechanism provides a way to transduce energy without causing any wastage or dissipation. This alternative is very different from other proposals in the literature [94] to effect dissipation-free energy transduction, and, in our view, it has great merit. Structures that are locally strained but nonetheless are overall at equilibrium are important in a variety of engineering applications. The technology of the future will have to deal with the exorbitant cost (and unavailability) of fossil fuel energy for all large-scale manufacturing activities [95]; hence future machines will have no alternative but to use a "high energy" (in terms of energy storage by a locally strained but overall at equilibrium molecule, as discussed above) compound such as ATP.

In conclusion, we see the recurrence of the very principles proposed and detailed in the biochemical theory consisting of the torsional mechanism of ion translocation, energy transduction and storage and ATP synthesis and the rotation-twist-tilt energy storage mechanism of muscle contraction, and hence we believe that these principles are of a very general and universal nature in biological systems. The developed theory is accurate [as far as the problems of experiments on complex *biological* systems (as opposed to simpler physical systems), with their inherent assumptions, errors, and difficulties in interpretation permit], consistent within itself and with all the known laws of science, detailed in each part yet broad in overall scope, reasonably simple and making no unnecessary assumptions, quantitative and possessing the ability to make novel predictions that are experimentally testable, and, finally, fruitful and pregnant with possibilities as a guide to further experimentation and for future new discoveries and inventions. It meets all the criteria laid out by Kuhn for a good scientific theory [96]. It offers *unifying* principles of energy transduction in biological systems, and a unique opportunity for the unification of bioenergetics itself. In my view, the aspects dealt with in our work constitute the *key* elements whose lack of detailed consideration has held back the progress of research in this important field.

7 Conclusion

In this paper, the molecular mechanisms of energy transduction by some fascinating molecular machines of the cell have been described. In particular, two of the most fundamental processes in biology - ATP synthesis and muscle contraction - have been dealt with. The molecular mechanisms of energy transduction by the F_1 and F_0 portions of ATP synthase have been systematically addressed in consummate detail. Emphasis has been laid on our novel torsional mechanism of ion translocation, energy transduction, energy storage and ATP synthesis, a result of dedicated research over the past twelve years. The differences between the torsional mechanism and other mechanisms have been interpreted and presented in great detail. The recent pioneering experimental research of key groups has been pointed out and a comparison of the mechanisms with the new data has been made and their biological implications have been discussed at length. The resolution of experimental anomalies by the torsional mechanism and a mathematical analysis of its transport aspects have been carried out. The consistency of the mechanism with the laws of electrical neutrality and thermodynamics of the oxidative phosphorylation process has been scrutinized. The various mechanisms of muscle contraction have been reviewed and the distinguishing and original features of our rotation-twist-tilt energy storage mechanism have been delineated. An engineering analysis of the mechanism has been summarized. Finally, the engineering applications and ramifications of our work have been addressed. The design of new machines based on these novel concepts and insights has been explained and a brief account of a working prototype of such a machine has been provided, verbally and pictorially. The leading role of the new field of molecular engineering for further progress has been accentuated; in particular, how molecular engineering can offer us a future (but concrete) solution to the energy crisis has been suggested. Finally, some ideas on the generality and universality of the proposed principles and the possible unification of energy transduction in seemingly disparate biological processes have been presented.

Acknowledgement. My research program on the mechanism and thermodynamics of molecular machines has been generously funded over the decade by the Department of Science and Technology (1993–1995) (Grant No. SR/OY/GB-26/93), the All-India Council for Technical Education (1996–1999) (Grant No. 1-52/CD/CA/95-96) and by the Swarnajayanti Research Project under the Swarnajayanti Fellowships (2001–2006) (Grant No. DST/SF/Life-102/99-2000) specially instituted on the occasion of the Golden Jubilee of India's independence by the Ministry of Science and Technology, Department of Science and Technology, Government of India.

8
References

1. Nath S (2002) Adv Biochem Eng Biotechnol 74:65
2. Senior AE, Nadanaciva S, Weber J (2002) Biochim Biophys Acta, in press
3. Abrahams JP, Leslie AGW, Lutter R, Walker JE (1994) Nature 370:621
4. Menz RI, Walker JE, Leslie AGW (2001) Cell 106:331
5. Bianchet MA, Hullihen J, Pedersen PL, Amzel LM (1998) Proc Natl Acad Sci USA 95: 11065
6. Zhou Y, Duncan TM, Cross RL (1997) Proc Natl Acad Sci USA 94:10583
7. Allison WS (1998) Acc Chem Res 31:819
8. Hausrath AC, Grüber G, Matthews BW, Capaldi RA (1999) Proc Natl Acad Sci USA 96:13697
9. Nath S, Rohatgi H, Saha A (2000) Curr Sci 78:23
10. Cox GB, Fimmel AL, Gibson F, Hatch L (1986) Biochim Biophys Acta 849:62
11. Fillingame RH, Jiang W, Dmitriev OY, Jones PC (2000) Biochim Biophys Acta 1458: 387
12. Nakamoto RK, Ketchum CJ, Al-Shawi MK (1999) Annu Rev Biophys Biomol Struct 28: 205
13. Sorgen PL, Caviston TL, Perry RC, Cain BD (1998) J Biol Chem 273:27873
14. Sabbert D, Engelbrecht S, Junge W (1996) Nature 381:623
15. Noji H, Yasuda R, Yoshida M, Kinosita K (1997) Nature 386:299
16. Yasuda R, Noji H, Kinosita K, Yoshida M (1998) Cell 93:1117
17. Rayment I (1996) J Biol Chem 271:15850
18. Block SM (1996) Cell 87:151
19. Goldman YE (1998) Cell 93:1
20. Geeves MA, Holmes KC (1999) Annu Rev Biochem 68:687
21. Mitsui T (1999) Adv Biophys 36:107
22. Nath S, Khurana D (2001) Curr Sci 81:78
23. Boyer PD, Cross RL, Momsen W (1973) Proc Natl Acad Sci USA 70:2837
24. Boyer PD (1993) Biochim Biophys Acta 1140:215
25. Boyer PD (1997) Annu Rev Biochem 66:717
26. Boyer PD (2000) Biochim Biophys Acta 1458:252
27. Nath S (1994) 16th Int Congr Biochemistry and Molecular Biology, New Delhi, India, vol. II, p 390
28. Nath S (1997) ISBC X, Int Soc Biological Calorimetry, Monte Verità, Switzerland, p i9
29. Nath S (1998) Pure Appl Chem 70:639
30. Rohatgi H, Saha A, Nath S (1998) Curr Sci 75:716; erratum (2000) 78:201
31. Nath S, Rohatgi H, Saha A (1999) Curr Sci 77:167
32. Nath S, Jain S (2000) Biochem Biophys Res Commun 272:629
33. Jain S, Nath S (2000) FEBS Lett 476:113
34. Nath S (2000) 41st Annual Conference of the Association of Microbiologists of India, Jaipur, India p 3
35. Nath S (2001) National Conference on Recent Advances in Thermodynamics of Chemical and Biological Systems, Amritsar, India, p PL-2
36. Nath S (2001) Trends in Biochemistry, 8th Meeting, Pune, India, p 13
37. Nath S (2001) ISBC XII, Int Soc for Biological Calorimetry, Santiago de Compostela, Spain, p 82
38. Jain S, Nath S (2001) Thermochim Acta 378:35
39. Nath S (2001) Application of Chemistry for the Advancement of Life Sciences, Nagpur, India, p 3
40. Nath S (2002) 89th Session of The Indian Science Congress, Part III, Section of Biochemistry, Biophysics and Molecular Biology, Lucknow, India p 12
41. Vinogradov AD (2000) J Exp Biol 203:41

42. Weber J, Wilke-Mounts S, Lee RSF, Grell E, Senior AE (1993) J Biol Chem 268:20126
43. Löbau S, Weber J, Senior AE (1998) Biochemistry 37:10846
44. Weber J, Senior AE (2001) J Biol Chem 276:35422
45. Boyer PD (2002) FEBS Lett 512:29
46. Mitchell P (1961) Nature 191:144
47. Mitchell P (1966) Biol Rev 41:445
48. Westerhoff HV, Melandri BA, Venturoli G, Azzone GF, Kell DB (1984) Biochim Biophys Acta 768:257
49. Kaim G, Dimroth P (1999) EMBO J 18:4118
50. Possmayer FP, Gräber P (1994) J Biol Chem 269:1896
51. Fischer S, Gräber P (1999) FEBS Lett 457:327
52. Thayer WS, Hinkle PC (1975) J Biol Chem 250:5330
53. Nicholls DG (1977) Eur J Biochem 77:349
54. Massari S, Azzone GF (1970) Eur J Biochem 12:301
55. Azzone GF, Massari S (1971) Eur J Biochem 19:97
56. Tupper JT, Tedeschi H (1969) Science 166:1539
57. Tedeschi H (1975) FEBS Lett 59:1
58. Cruz JA, Sacksteder CA, Kanazawa A, Kramer DM (2001) Biochemistry 40:1226
59. Sambasivarao D, Sitaramam V (1985) Biochim Biophys Acta 806:195
60. Mathai JC, Sitaramam V (1989) Biochim Biophys Acta 976:214
61. Green DE (1981) Proc Natl Acad Sci USA 78:2240
62. Kedem O, Caplan SR (1965) Trans Faraday Soc 61:1897
63. Stucki JW (1980) Eur J Biochem 109:269
64. Van Dam K, Westerhoff HV, Krab K, Van der Meer R, Arents JC (1980) Biochim Biophys Acta 591:240
65. Jou D, Ferrer F (1985) J Theor Biol 117:471
66. Lemasters JJ (1984) J Biol Chem 259:13123
67. Lemasters JJ, Grunwald R, Emaus RK (1984) J Biol Chem 259:3058
68. Hinkle PC, Arun Kumar M, Resetar A, Harris DL (1991) Biochemistry 30:3576
69. Nath S, Jain S (2002) Thermochim Acta 394:89
70. Jha, RK, Nath S (2002) submitted for publication
71. Kühne W (1864) Untersuchungen über das Protoplasma und die Kontraktilität, Engelmann, Leipzig
72. Huxley HE, Hanson J (1954) Nature 173:973
73. Huxley HE (1969) Science 164:1356
74. Huxley AF, Niedergerke R (1954) Nature 173:971
75. Huxley AF (1957) Prog Biophys Biophys Chem 7:255
76. Cooke R (1999) Curr Biol 9:R773
77. Holmes KC (1997) Curr Biol 7:R112
78. Vale RD, Milligan RA (2000) Science 288:88
79. Uyeda TQP, Abramson PD, Spudich JA (1996) Proc Natl Acad Sci USA 93:4459
80. Highsmith S (1999) Biochemistry 38:9791
81. Lymn RW, Taylor EW (1971) Biochemistry 10:4617
82. Fisher AJ, Smith CA, Thoden JB, Smith R, Sutoh K, Holden HM, Rayment I (1995) Biochemistry 34:8960
83. Smith CA, Rayment I (1995) Biochemistry 34:8973
84. Smith CA, Rayment I (1996) Biochemistry 35:5404
85. Gulick AM, Bauer CB, Thoden JB, Rayment I (1997) Biochemistry 36:11619
86. Corrie JET et al (1999) 400:425
87. Warshaw DM et al (2000) J Biol Chem 275:37167
88. Ishijima A, Kojima H, Funatsu T, Tokunaga M, Higuchi H, Tanaka H, Yanagida T (1998) Cell 92:161
89. Kitamura K, Tokunaga M, Iwane AH, Yanagida T (1999) Nature 397:129
90. McClare CWF (1971) J Theor Biol 30:1
91. Blumenfeld LA (1971) Biophysics (USSR) 16:724

92. Glansdorff P, Prigogine I (1971) Thermodynamic Theory of Structure, Stability and Fluctuations, Wiley, New York
93. Hong S, Ko YH, Pedersen PL (2001) Arch Biochem Biophys 394:275
94. Oster G, Wang H, Grabe M (2000) Phil Trans R Soc Lond B 355:523
95. Lanyi JK, Pohorille A (2001) Trends Biotechnol 19:140
96. Kuhn TS (1977) The Essential Tension, Univ. Chicago Press, Chicago

Received: May 2002

Adv Biochem Engin/Biotechnol (2003) 85: 181–214
DOI 10.1007/b11048

Status of Immunodiagnosis and Immunocontraceptive Vaccines in India

S. K. Gupta

Gamete Antigen Laboratory, National Institute of Immunology, Aruna Asaf Ali Marg, New Delhi – 110067, India. *E-mail: skgupta@nii.res.in*

The article focuses on the Indian initiative of making kits for diagnosis of various infectious and non-infectious diseases as well as reproductive hormones and hormones in various other endocrine disorders. Indigenous diagnostic kits for the detection of various infections such as filariasis, typhoid, amebiasis, Japanese encephalitis, hepatitis, HIV, dengue, leishmaniasis, malaria, rabies, toxoplasmosis, rotavirus, and group A streptococci have been developed. Agreements to transfer the know-how of some of these leads to industries have been signed. The know-how of enzyme-linked immunosorbent assay (ELISA) for detection of hepatitis C has been successfully transferred to industry and is being commercially produced. For detection of HIV-1 and HIV-2, indigenous diagnostic kits based on three different formats, namely ELISA, Western blot and rapid test have been developed and are being commercially produced by Indian industries. The factors influencing the successful transfer of laboratory-scale diagnostic assays from academia to industry and their commercial exploitation have been discussed. Indian scientists have made seminal contributions in exploring the possibility to develop an effective and safe contraceptive vaccine to control the increasing human population of India. Achieving contraception by means of vaccine is a novel approach, which entails generation of a specific antibody response against antigens critically involved in the process of mammalian reproduction. In India, three major programs on contraceptive vaccines based on the beta-subunit of human chorionic gonadotrophin (βhCG) for women, ovine follicle stimulating hormone (oFSH) for men, and riboflavin carrier protein for both males and females have been initiated. The work at the National Institute of Immunology, New Delhi on contraceptive vaccine for women, based on βhCG, has demonstrated, for the first time, that it is feasible to regulate fertility by such an approach. Basic research being carried out to achieve immunocontraception by interfering at sperm-oocyte interaction level has been briefly discussed. These developments are still at the research stage. In addition to advances in the area of contraceptive vaccines, a non-steroidal contraceptive oral pill has been developed by Central Drug Research Institute, Lucknow, commercially produced by two Indian pharmaceutical companies and has been incorporated in the National Family Welfare Program. Another interesting approach for fertility regulation in male has been developed in India, which involves vas occlusion with styrene maleic anhydride (SMA) and is currently undergoing clinical trials in human subjects.

Keywords. Diagnostic kits, Academia-industry interaction, Immunocontraception

Abbreviations

aa	Amino acid
AFB	Acid fast bacilli
AIDS	Acquired immunodeficiency syndrome
AIIMS	All India Institute of Medical Sciences
APS	Streptococcus A polysaccharide
βhCG	Beta-subunit of human chorionic gonadotrophin
BCIL	Biotech Consortium of India Limited
bmZPA	Bonnet monkey zona pellucida glycoprotein A
bmZPB	Bonnet monkey zona pellucida glycoprotein B
bmZPC	Bonnet monkey zona pellucida glycoprotein C
bp	Base pair
BSA	Bovine serum albumin
cDNA	Complementary deoxyribonucleic acid
CDRI	Central Drug Research Institute
CMC	Christian Medical College
CRI	Cancer Research Institute
CSF	Cerebrospinal fluid
CSIR	Council of Scientific and Industrial Research
CTL	Cytotoxic T lymphocyte
CTP	Carboxy terminal peptide
CTP-DT	Carboxy terminal peptide conjugated to diphtheria toxoid
DBT	Department of Biotechnology
DCI	Drug Controller of India
DMSO	Dimethyl sulfoxide
DST	Department of Science and Technology
DT	Diphtheria toxoid
ELISA	Enzyme-linked immunosorbent assay
FSH	Follicle stimulating hormone
GAS	Group A streptococcal pyogenes
GnRH	Gonadotrophin releasing hormone
HAV	Hepatitis A virus
HBV	Hepatitis B virus
HCV	Hepatitis C virus
hCG	Human chorionic gonadotrophin
HDV	Hepatitis D virus
HEV	Hepatitis E virus
hFSH	Human follicle stimulating hormone
HGP	Hepatitis G virus
HIV-1	Human immunodeficiency virus-1
HIV-2	Human immunodeficiency virus-2
HSD	Heterospecies dimer
HSD-DT	Heterospecies dimer diphtheria toxoid conjugate
HSD-TT	Heterospecies dimer tetanus toxoid conjugate
ICGEB	International Centre for Genetic Engineering and Biotechnology
ICMR	Indian Council of Medical Research

IgM	Immunoglobulin of M isotype
IICB	Indian Institute of Chemical Biology
IISc	Indian Institute of Science
IIT	Indian Institute of Technology
K_a	Association constant
LH	Luteinizing hormone
LPS	Lipopolysaccharide
MAb	Monoclonal antibody
MOU	Memorandum of understanding
NARI	National AIDS Research Institute
NICD	National Institute of Communicable Diseases
NICED	National Institute of Cholera and Enteric Diseases
NII	National Institute of Immunology
NIRRH	National Institute for Research in Reproductive Health
NTP	N-terminal peptide
NTP-DT	N-terminal peptide conjugated to diphtheria toxoid
oFSH	Ovine follicle stimulating hormone
PCR	Polymerase chain reaction
PGIMER	Post Graduate Institute of Medical Education and Research
r	Recombinant
RCM-RCP	Reduced and carboxymethylated riboflavin carrier protein
RCP	Riboflavin carrier protein
RF	Rheumatic fever
RHD	Rheumatic heart disease
SDS	Sodium dodecyl sulfate
SDS-PAGE	Sodium dodecyl sulfate polyacrylamide gel electrophoresis
SMA	Styrene maleic anhydride
TSH	Thyroid stimulating hormone
TT	Tetanus toxoid
UDSC	University of Delhi, South Campus
UNDP	United Nations Development Programme
WHO	World Health Organization
ZP	Zona pellucida
ZPA	Zona pellucida glycoprotein A
ZPB	Zona pellucida glycoprotein B
ZPC	Zona pellucida glycoprotein C

1 Introduction

As per the census, the human population of India grew from 318.7 million in the year 1941 to more than one billion in the year 2001 [1]. The ever-increasing human population poses a major challenge for Government to provide basic amenities, education, employment and adequate healthcare. In the 1950s, to control the human population, Government of India adopted the National Family Planning Program. The development of newer technologies for contraception was identified as one of the objectives under this program. Achieving contraception by means of an immunocontraceptive vaccine emerged as a novel idea through the research carried out in Indian laboratories. With the establishment of the Department of Biotechnology (DBT) in 1986, this initiative was opted in a "mission mode" with adequate funding. In the early 1990s, DBT also initiated another major program on the development of diagnostic kits pertaining to various diseases prevalent in India. This program was initiated on the premise that early detection of the disease in question will enable institution of an effective therapeutic modality at the initial stages of the disease leading to better management and, in the absence of correct diagnosis, avoid empirical treatment. In addition, DBT also supported the development of diagnostic kits for the detection of reproductive hormones and those associated with various endocrine disorders. The program on diagnostics was also supported by other major national funding agencies such as Indian Council of Medical Research (ICMR), Department of Science and Technology (DST) and Council of Scientific and Industrial Research (CSIR). Table 1 enlists the diagnostic kits developed in India with financial support from national funding agencies. The know-how for some of these kits has been transferred to industry for their commercial production (Table 2).

This review summarizes the developmental efforts of some of the indigenous diagnostic kits. The current status of India's pursuits in the development of immunocontraceptive vaccines for fertility regulation has also been discussed.

2 Development of Diagnostic Kits for Infectious and Non-Infectious Diseases and Hormones

The turnover of the diagnostic industry in India constitutes approximately 2% of the turnover of the pharmaceutical industry as compared to approximately 7% in developed countries such as USA. To cater for the requirements of well-being of more than one billion Indians, it is likely that the consumption of diagnostic kits will increase. The increased demand for diagnostics will pertain to

i) Increase in consumption of already established diagnostic systems due to increase in the size of population and awareness of health needs.
ii) With the change in lifestyle and re-emergence of infectious diseases, there is a need to develop new diagnostic kits.

The various government funding agencies recognized these issues and initiated funding of various projects to develop indigenous diagnostic kits. Diagnosis at an early stage of the onset of a disease provides the basis for its successful therapy. To achieve wider applicability of diagnostic kits, these should be cost effective, simple and easily be performed by semiskilled para-auxillary personnel, as well as have the requisite sensitivity, specificity and unambiguous readability. Judicious use of technological developments based on hybridoma technology, DNA recombinant technology, enzymology, fermentation, and amplification techniques have made it possible to develop sensitive and specific assays for a variety of diseases and proteins of interest. Some of these efforts are briefly described below.

Table 1. List of some of the diagnostic kits developed or under development at various institutions in India

Diagnostic kit	Developed by
ELISA and dot blot for pregnancy detection	National Institute of Immunology (NII), New Delhi; National Institute for Research in Reproductive Health (NIRRH), Mumbai
ELISA for filariasis	Mahatma Gandhi Institute of Medical Sciences (MGIMS), Wardha
Agglutination, ELISA and Western blot for HIV-1 and HIV-2	University of Delhi, South Campus (UDSC), New Delhi; NII, New Delhi; Cancer Research Institute (CRI), Mumbai; Indian Institute of Science (IISc), Bangalore; International Center for Genetic Engineering and Biotechnology (ICGEB), New Delhi; All India Institute of Medical Sciences (AIIMS), New Delhi
ELISA for hepatitis A	National Institute of Virology (NIV), Pune
ELISA for hepatitis C and hepatitis E	AIIMS, New Delhi; ICGEB, New Delhi
ELISA for detection of dengue and West Nile	NIV, Pune
ELISA and PCR for Japanese encephalitis	NII, New Delhi; National Institute of Mental Health and Neuro Sciences (NIMHANS), Bangalore; King George's Medical College (KGMC), Lucknow; NIV, Pune
Hemagglutination for leishmaniasis	Indian Institute of Chemical Biology (IICB), Calcutta
ELISA for typhoid	NII, New Delhi
PCR for tuberculosis	AIIMS, New Delhi; Central Drug Research Institute (CDRI), Lucknow; NII, New Delhi
Lactate dehydrogenase (LDH)-based test for malaria	CDRI, Lucknow
Immunofluorescence assay for rabies	AIIMS, New Delhi
Agglutination assay for toxoplasmosis	AIIMS, New Delhi
ELISA for rotavirus	AIIMS, New Delhi
Latex agglutination and antibody capture assay for detection of group A streptococci	NII, New Delhi
ELISA for *Aspergillus*	Institute of Genomics and Integrative Biology (IGIB), New Delhi
ELISA for urinary reproductive hormones	NIRRH, Mumbai
Steroid and thyroid hormones	IICB, Calcutta; AIIMS, New Delhi
ELISA for alpha-fetoprotein	IICB, Calcutta

Table 2. List of the diagnostic assays/reagents for which a technology transfer agreement was signed with industry

Diagnostic assay	Transferred to	Status
Pregnancy detection	Ranbaxy Laboratories Ltd. (Diagnostic Division), New Delhi	Commercially produced from 1990–1993
ELISA for filariasis	Hindustan Antibiotics Ltd., Pune	Failure in technology adsorption
ELISA for typhoid	Lupin Biotech, Bhopal	Failure in technology adsorption
ELISA for amebiasis	Cadila Pharmaceuticals, Ahmedabad	Produced commercially for limited period
Blood grouping reagents	Cadila Pharmaceuticals, Ahmedabad	Produced commercially for brief period
ELISA for dengue, Japanese encephalitis, West Nile	Zydus Cadila Health Care, Ahmedabad	MOU signed recently, technology being transferred to industry
ELISA for alpha-fetoprotein	Shantha Biotechnics, Hyderabad	Technology being transferred to industry
ELISA for hepatitis A	Bharat Biotech Ltd., Hyderabad	Technology being transferred to industry
ELISA for hepatitis B	Ranbaxy Laboratories Ltd. (Diagnostic Division), New Delhi	Failure in technology transfer
ELISA for hepatitis C	XCyton Diagnostic Ltd., Bangalore	Kit being commercially produced
ELISA for pregnenediol glucuronide, estrone glucuronide, FSH, and LH	Zydus Cadila Health Care, Ahmedabad	Technology being transferred
ELISA for HIV-1 and HIV-2	XCyton Diagnostic Ltd., Bangalore; Lupin Biotechnology, Bhopal; Ace Diagnostic, Gurgaon	XCyton Diagnostic Ltd. is producing the kit Commercially. Others failed to manufacture this kit
Western blot for HIV-1 and HIV-2	J. Mitra & Co., New Delhi	Being produced commercially
Agglutination test for HIV-1 and HIV-2	Cadila Pharmaceuticals, Ahmedabad	Technology successfully transferred. Being produced commercially

2.1 Pregnancy Detection

Human chorionic gonadotrophin (hCG) has traditionally been used as an index of pregnancy and is the basis of all available bioassays and immunoassays. Its synthesis starts as early as 170 h after fertilization and can be detected in the circulation within 8–10 days post-fertilization [2, 3]. In an effort directed towards the development of sensitive and specific immunoassays, at the Department of Biochemistry, All India Institute of Medical Sciences (AIIMS), New Delhi, a mouse hybrid cell clone, P_3W_{80}, secreting an antibody against hCG was developed [4]. It was the first report from India of generating a mouse hybrid cell clone secreting an antibody of predefined specificity. The monoclonal antibody (MAb), P_3W_{80}, had high affinity for hCG: association constant $(K_a)=3.03\times10^{10}$ L/mol, reacted equally well with the β-subunit of hCG and had low cross-reactivity with the α-subunit of hCG (0.28%). It also had low cross-reactivity with human luteinizing hormone (hLH), human follicle stimulating hormone (hFSH) and human thyroid stimulating hormone (hTSH) [4, 5]. Another hybrid cell clone, P_22_3, secreting an MAb that recognized α-hCG, hCG and pituitary gonadotrophins was also developed [6]. It was devoid of cross-reactivity with β-hCG and had K_a of 1.3×10^9 L/mol with α-hCG.

Using a combination of these two MAbs, a highly sensitive and specific sandwich enzyme immunoassay was developed [7]. The test employed a P_3W_{80} coated microtitration plate, which was incubated with test urine samples followed by detection of antibody-bound hCG by a P_22_3 MAb-horse radish peroxidase conjugate. The detection limit of the assay was 1 ng hCG/mL (10 mIU hCG/mL). Keeping the domestic conditions in view, an interesting prototype kit for pregnancy detection was developed by using a "dip-stick" that has two nitrocellulose pads (0.45 µm pore size; 3×5 mm) pasted at one end of the plastic strip [8]. The technical know-how was transferred to Ranbaxy Laboratories Ltd., New Delhi and the kit was commercially produced, albeit for a limited period (2 years) only. National Institute for Research in Reproductive Health (NIRRH), Mumbai has also developed a prototype pregnancy detection kit based on polyclonal antibodies that is being used in some of the primary healthcare centers in Mumbai.

2.2 Typhoid

Routinely, bacteriological culture and Widal's agglutination test are used for the diagnosis of typhoid fever. Bacterial culture, though most definitive, is time consuming. Widal's agglutination test involves detection of circulating antibodies against *S. typhi*. Human subjects not suffering from active typhoid fever may have antibodies in their circulation due to prior vaccination against typhoid or the endemicity of the disease, which may influence the interpretation of the results. An ideal test for the diagnosis of the disease would be based on the identification of circulating *S. typhi* antigens or bacteria during the early phase of pyrexia. These efforts are hampered due to the very low levels of bacteremia during the initial stages of typhoid fever. Nonetheless, ELISA tests for the detection of *S. typhi* 09 antigen and flagellar antigen have been described [9, 10]. Another possible approach may be to develop a polymerase chain reaction (PCR)-based assay.

In early 1990s, taking advantage of the fast rate of multiplication of *S. typhi*, scientists at NII, New Delhi designed a method for the diagnosis of typhoid that involved culturing of the blood sample for 6–8 h followed by a highly specific immunoassay. In order to develop the specific immunoassay, epitope specific MAbs were generated against O, H, and Vi antigens of *S. typhi* [11–13]. The anti-O MAbs recognized different determinants on the 09 and 012 antigens of *S. typhi* lipopolysaccharide (LPS). The anti-Vi antibodies were directed against two determinants, one constituted by the *O*-acetyl group and the other by *N*-acetyl and carboxyl groups. The sandwich assay based on these MAbs was highly sensitive and specific [14]. The O ELISA, H ELISA and Vi ELISA detected 1 ng LPS/mL, 10 ng H antigen/mL and 0.1 ng Vi antigen/mL, respectively. The sensitivity and specificity of these assays were evaluated in a double-blind study with the bacterial culture as gold standard. The ELISA format was transformed into a "dot-enzyme immunoassay" using a "dipstick" to make it user friendly for routine clinical diagnosis. It is based on the simultaneous detection of O and Vi antigens. After enrichment of bacteria by short-term culture for 8 h, the assay can be completed in 30 min. A Memorandum of Understanding (MOU) was signed between NII, New Delhi and Lupin Biotechnology Pvt. Ltd., Bhopal for transfer of the technology and its commercial production.

2.3 Group A Streptococcal Pharyngitis

Group A *Streptococcal pyogenes* (GAS), a common human pathogen, is responsible for uncomplicated pharyngitis to debilitating and potentially fatal sequelae like rheumatic fever (RF), rheumatic heart disease (RHD) and toxic shock syndrome [15]. A conservative estimate suggests that 2–3 million children suffer every year from RF in India. A recent resurgence in the incidence of RF from developed countries has also been reported. One of the major issues involved in the effective management and prevention of RF and RHD is the rapid diagnosis of GAS infection followed by treatment with antibiotics. Traditionally, the presence of GAS infection is confirmed by culturing the organisms on a blood-agar plate followed by assaying the sensitivity of the β-hemolytic plaques with bacitracin or serological identification using a group specific antiserum. It takes 24–48 h. Commercial kits employing polyclonal or MAbs and based on the principle of agglutination or sandwich enzyme immunoassay can detect the GAS in a much shorter time.

At NII, New Delhi a latex agglutination assay, employing polyclonal antibodies raised against streptococcus A polysaccharide (APS) linked to bovine serum albumin (BSA), has been developed. The antibodies have been rendered monospecific by adsorption on Sepharose-4B-BSA and a bacterial cocktail column. The sensitivity of the assay is 12.5 ng APS/mL [16]. In order to have a better end-point read-out, another one-step "antibody-capture assay" was developed.

The assay employed nitrocellulose membrane strips dotted with 5 μg of MAb, MA-107, reacting specifically to GAS. The strip is dipped in a solution of standard APS or polysaccharide extracted from throat swabs (collected from children) using a nitrous acid extraction procedure along with colloidal gold conjugated to monospecific polyvalent antibodies against GAS. As the immune complexes of APS with antibody colloidal gold rise due to capillary action, these are captured by MA-107 streaked at a fixed point where a purple dot appears in the case of a sample positive for GAS. The sensitivity of the assay is 15.6 ng APS/mL [17]. The assay is highly specific as no colored dot appears when polysaccharides from streptococci B, C, G and *S. aureas* are tested. The efficacy of these two rapid tests has been evaluated in comparison with the traditional culture method in both asymptomatic children (1500) and symptomatic patients (233). Both latex agglutination and antibody capture assays showed a sensitivity of 100% and specificity of >98% as compared to culture and isolation in symptomatic patients [18]. However, among asymptomatic carriers, sensitivities of 100 and 87.5% and a specificity >95% were observed for latex agglutination and antibody capture assay, respectively.

2.4
Tuberculosis

Tuberculosis has been endemic to India for centuries and continues to pose a major public health problem. It remains one of the most infectious diseases in the world. The global incidence of tuberculosis is estimated to be 8–10 million cases approximately, causing 3 million deaths per year, more than that from any other single infectious disease [19]. In India, the number of deaths due to tuberculosis is estimated to be approximately 5,00,000 every year [20]. The worldwide incidence of tuberculosis is on the rise as a consequence of spread of the AIDS. According to WHO survey, in the year 1998, the estimated number of tuberculosis cases in India was more than 2 million. In addition to pulmonary tuberculosis, there is also an increase in the incidence of extrapulmonary tuberculosis, which presents a greater challenge for diagnosis of the disease. Conventionally, the diagnosis is confirmed by the presence of acid fast bacilli (AFB) and by isolation of *Mycobacterium tuberculosis* on culture. However, culture methods are time-consuming. Serological tests aiming to measure either circulating antibodies or *M. tuberculosis* antigens have not shown the requisite sensitivity and specificity due to extensive immunological cross-reactivity between *M. tuberculosis* and environmental mycobacteria. Furthermore, there will be variation in the antibody repertoire generated against *M. tuberculosis* antigens from individual to individual, making measurement of anti-*M. tuberculosis* antibodies a less reliable index for diagnosis of the disease. Nonetheless, scientists at Sree Chitra Tirumal Institute for Medical Sciences and Technology, Thiruvananthapuram have reported the development of a reverse passive hemagglutination assay, using rabbit polyclonal antibodies for the detection of circulating mycobacterial antigens in cerebrospinal fluid (CSF). The assay had a sensitivity of 94.1% and a specificity of 99.0% [21]. At the same institute, a dot-immunobinding assay for the detection of circulating anti-mycobacterial antibodies in CSF for the rapid laboratory diagnosis of tuberculosis meningitis has also been developed [22]. National Institute of Communicable Diseases (NICD), New Delhi has developed an ELISA assay based on mycobacterial antigen 60 for the detection of circulating antibodies. The assay showed a specificity of 92% and a sensitivity of 75.5% [23].

In a DBT-sponsored effort, three different institutions in India, namely NII, New Delhi, AIIMS, New Delhi and CDRI, Lucknow have developed PCR-based assays. One of the assays involved PCR amplification of a 169 bp fragment of the *M. tuberculosis* genome. The PCR-amplified fragment was detected by an *M. tuberculosis* complex-specific probe [24]. The other assay involved PCR amplification of a 513 bp product of the devR gene using dev Rf (5′-GGTGAG-GCGGGTTCGGTCGC-3′) and dev Rr (5′-CGCGGCTTGCGTCCGACG TTC-3′) primers. The PCR products were electrophoresed on a 1% agarose gel, transferred to a positively charged nylon membrane, hybridized with the γ-^{32}P labeled internal oligonucleotide der R1 (5′-CCGTCCAGCGCCCACATCTTT-3′) probe and exposed to X-ray film [25, 26].

The utility of these assays for the diagnosis of tuberculosis was evaluated in a systematic manner on sputum specimens under a multicentric DBT-funded program. The results showed reasonable promise. The false negatives observed by PCR may be due to several factors such as i) the presence of PCR inhibitors in the clinical samples, ii) a low bacterial load and iii) variable efficiency of DNA extraction due to sample diversity. The occurrence of false positives may also be due to cross-contamination. Recently, a rapid, single-tube method for the isolation of DNA suitable for PCR from a variety of clinical materials has been described [27]. Briefly, the procedure is comprised of concentration of the bacilli by high-speed centrifugation, removal of PCR inhibitors by buffer containing guanidinium isothiocyanate and release of mycobacterial DNA by heating in the presence of detergents and Chelex-100 resin. Being a single-tube method, the risk of cross-contamination while handling, at the same time, a large number of samples is minimized, further reducing the number of false positives. Removal of inhibitors allowed the use of more DNA equivalents in the PCR reaction thereby increasing the sensitivity of the assay procedure. The assay has been validated on 780 clinical samples of diverse nature such as sputum, CSF, pulmonary fluids, pus, fine needle aspirates, tissues, blood and milk.

In addition to the above, efforts have also been made at several institutes to develop PCR-based detection systems for *M. tuberculosis*. Scientists at the Department of Pathology, Postgraduate Institute of Medical Education and Research (PGIMER), Chandigarh have developed a PCR method for its detection. The assay has been found useful in cases of granulomatous uveitis and pleural fluids [28, 29]. A highly specific PCR assay, based on amplification of a 340 bp sequence, has also been developed at Radiation Medicine Centre, Bhabha Atomic Research Centre, Mumbai [30]. Molecular Biology Unit, National Dairy Research Institute (NDRI), Karnal have developed a PCR test based on insertion sequence IS1081 to detect *M. tuberculosis* complex organisms in peripheral blood. The method was applied to blood samples from immunocompetent individuals with localized pulmonary tuberculosis. Blood samples from 43.75% subjects were found to be positive for the circulating DNA copies of the *M. tu-*

berculosis complex [31]. Department of Microbiology, AIIMS, New Delhi has also developed and evaluated a PCR assay from the *M. tuberculosis* complex-specific MPB64 gene for the diagnosis of pulmonary tuberculosis [32]. In a double-blind study, 182 clinical samples (sputum, bronchioalveolar lavage and pleural fluid) from patients with a clinical diagnosis of pulmonary tuberculosis and 72 samples from patients with non-tubercular pulmonary lesions and normal healthy individuals were included. PCR was positive in 59% of the single sputum samples from clinically diagnosed pulmonary tuberculosis. The test could identify *M. tuberculosis* in 81.8% of the culture positive sputum samples. From clinically suspected cases, 71.4% of bronchioalveolar lavage and 60.7% pleural fluid samples were PCR positive [32].

2.5 Hepatitis

Viral hepatitis, a major public health problem, is caused by hepatitis A virus (HAV), hepatitis B virus (HBV), hepatitis C virus (HCV), hepatitis D virus (HDV) and hepatitis E virus (HEV) in the majority of cases. HAV and HEV are transmitted by the fecal-oral route and cause only acute hepatitis. HBV and HCV are transmitted parentally and are associated with acute and chronic hepatitis, liver cirrhosis and hepatocellular carcinoma. HDV uses the same route of infection as HBV and depends on it for entry into hepatocytes. HDV causes both co- and superinfection, modifying the course of HBV-associated acute and chronic liver disease. Hepatitis G virus (HGV) has also been described recently. It is transmitted parentally and has been detected in both acute and chronic liver diseases.

2.5.1 *HAV*

In order to detect HAV, at NIV, Pune, scientists have successfully infected rhesus monkeys with HAV. Liver from the infected animals has been used as a source of viral antigen. An ELISA has been developed for the detection of circulating IgM antibodies against the HAV. The know-how has been transferred to Bharat Biotech Ltd., Hyderabad for the commercial production of the HAV detection kit.

2.5.2 *HBV*

Hepatitis B is a major health problem world-wide. During the course of infection, HBV releases antigens such as HBsAg and HBeAg that serve as important serological diagnostic markers for HBV infection. Antibodies generated against viral antigens such as anti-HBs, anti-HBe and anti-HbcIgM have also been exploited for the diagnosis of HBV. Among these, HBsAg is the most widely used for the detection of HBV infection because of its early appearance, high titers in both acute and chronic cases, and persistence until the virus is cleared. A number of ELISAs are available commercially for assay of HBsAg. J. Mitra & Co., New Delhi has been producing the ELISA kits for the detection of HBsAg as well as anti-HBs. Large-scale epidemiological studies of HBV infection in various clinical conditions including maternal-child transmission have been carried out in India using the available commercial diagnostic kits [33–35]. The detection of HBV infection by a DNA probe has also been reported from India [36]. DNA probe-based diagnostic assays are of special significance in identification of variants of HBV in serum that is negative for all serological markers of infection.

In addition to contributing to the epidemiology of HBV, in the particular context of the Indian scenario, a great deal of basic work is being done on other aspects of HBV infection. The mode of attachment of HBV to liver cells is still being actively pursued as this may help in devising strategies to prevent the virus from entering into the liver cells. Both direct attachment of the virus to liver cells and polymerized albumin-mediated attachment have been suggested as possible mechanisms. An Indian group at AIIMS, New Delhi has convincingly shown that polymerized albumin-mediated attachment is insignificant in HBV attachment to liver cells [37]. This group has also identified and isolated a 34 kDa protein, the putative receptor for the pre-SI region of the HBV envelope protein [38].

Scientists at AIIMS, New Delhi in collaboration with International Center for Genetic Engineering and Biotechnology (ICGEB), New Delhi have used synthetic peptides to pre-SI, pre-S2 and S regions of HBV envelope protein and investigated the dynamics of the appearance and disappearance of both humoral and cellular immune responses at various stages of HBV infection. These studies suggested that the acute liver damage is mediated by an immune response to the envelope protein whereas the chronic liver damage is due to a cytotoxic T lymphocyte (CTL) response against the core protein [39]. These scientists have also designed a self-oligomerizing synthetic envelope peptide (124–147 amino acid residues) as a candidate vaccine and analyzed in detail both B and T cell epitopes on this peptide [40–42].

2.5.3
HCV

HCV is the major cause of post-transfusion non-A and non-B hepatitis. HCV is transmitted parenterally, primarily through intravenous drug abuse and contaminated blood and blood products. It was identified in 1989 as a single positive-sense RNA genome with an approximate size of 9.7 kb belonging to the flaviviridae family [43]. Early diagnosis of HCV is necessary because about half of the acute cases become chronic, and many eventually develop hepatocellular carcinoma. In spite of significant progress in the HCV area, its diagnosis based on detection of viral antigens by conventional serological methods has not been very successful. The majority of the assays are focused on detecting antibodies against HCV for the diagnosis of the present and past infection. In this direction, three generations of ELISAs have been developed. The first generation assay used c100–3 recombinant fusion polypeptide representing part of the NS4 region of the HCV genome. The second generation ELISA included two additional antigens from the HCV core (c22-3) and NS3 (c33c) regions. The third generation anti-HCV assays included an additional antigen from the NS5 region. Use of an anti-HCV assay, in the developed countries, for routine screening of blood and blood products brought down the incidence of HCV infection to less than 0.05%. However, the performance of these assays varied in different geographical regions, mainly due to the wide strain variation and genetic diversity of HCV. So far at least 11 genotypes and 72 subtypes of HCV have been described from different parts of the world.

An HCV RNA detection test based on reverse transcriptase-nested PCR for the highly conserved 5′ non coding region was standardized at AIIMS, New Delhi. This assay was evaluated with a second generation commercial ELISA on samples from 115 patients with biopsy proven chronic active hepatitis and 140 cases of cirrhosis of the liver [44]. It was observed that the sensitivity of anti-HCV ELISA in the detection of HCV infection is only 88.2%. Also, all the anti-HCV positive chronic hepatitis C patients were not viremic. Furthermore, some HCV seronegative patients were found to have HCV RNA (22.2%). These results indicate that the diagnosis of HCV infection is not possible if it is based solely on the available serodiagnostic tests. To circumvent this problem, the same group has cloned the core (9–459 nt) and part of the non structural region (NS5, 7959–8207 nt) of the virus from several Indian isolates (EMBL Accession Number X91297–91307 and X91416–91423). During sequence analysis, this group identified a new subtype of HCV in Indian patients. This has been labeled as type 3 g [45]. The core region of this new subtype 3 g showed amino acid (aa) changes unique to it. Three of these mutations were concentrated in the N-terminal immunodominant portion of the core polypeptide, which may lead to altered charge and structure of the core protein. Such changes may prove important in the differential immunodiagnosis of subtype 3 g cases and its pathogenicity. These studies demonstrate that HCV type 3 variants including a new subtype 3 g account for majority of HCV infection in Northern India. Furthermore, from this information, a diagnostic assay system based on synthetic peptides from core (1b type, 3 g type), NS3, NS4 and NS5 regions, which is in routine use for screening blood donors and liver disease patients for HCV infection at AIIMS, New Delhi, has been developed. The test was compared with 7 commercial tests available in the Indian market [46]. The sensitivity of the in-house ELISA was 90% whereas it varied with the commercial assays. It has so far been evaluated on more than 150,000 samples and the know-how transferred to XCyton Diagnostic Ltd, Bangalore. It is now commercially available in the Indian market. The know-how from Indian industry has also been transferred to IRM Pharmaceutical, USA for its commercial production in USA.

2.5.4
HEV

The group at AIIMS, New Delhi has succeeded in growing HEV in rhesus monkeys instead of chimpanzees [47]. The transmission pattern and replication of the virus in liver cells using strand-specific PCR was investigated [48]. The epidemiological pattern of HEV infection in India revealed that it is a major cause of sporadic acute hepatitis in India and is associated with more than 27% of the acute liver failure seen in this country [49]. The group has also cloned and expressed ORF1, ORF2 and ORF3 of HEV and produced for the first time an infectious cDNA of HEV [50–53]. The in vitro produced RNA transcripts on transfection of HepG2 cells undergo replication, transcription and translation. All the viral proteins could be metabolically labeled and immunoprecipitated from RNA transfected cells. The culture supernatant of such cells is infectious to experimental animals [54]. The group has also developed a diagnostic test for acute HEV infection by detection of anti HEV-IgM in patient's serum and has demonstrated protracted viremia in patients [55].

2.6 Human Immunodeficiency Virus (HIV)

Human immunodeficiency virus (HIV), the causative agent of acquired immunodeficiency syndrome (AIDS) is mainly of two types, e.g., HIV-1 and HIV-2. HIV-1 accounts for most of the HIV infection world-wide, while HIV-2 is largely confined to West Africa and some parts of the Europe, Central Africa, Canada, Brazil, America and India. There are various subtypes of HIV-1, which are classified under two major groups; i) Major or M group containing subtypes A to K, and ii) Outgroup or O group, containing 9 subtypes. The O group subtypes are less related to M group subtypes, still these are more close to HIV-1 than HIV-2. HIV-2 has five subtypes (A to E). In USA and Western Europe the majority of the HIV-1 isolates have been of the B subtype. In Africa subtypes A, C and D are found more frequently than the other subtypes. In Thailand subtype E dominates [56]. In India subtype C is more prevalent in the major part of the country, except for the north-east, where subtype E is encountered more frequently [56–59]. It is estimated that more than 36 million people suffer from HIV infection around the world. Approximately 70% of the HIV-infected people live in sub-Saharan Africa. However, the incidence of HIV infection is on the rise in South Asia. In India, the first HIV positive case was reported from Chennai in April 1986 and later from Mumbai in the same year among commercial sex workers. Since then, according to joint United Nations Program on HIV/AIDS (UNAIDS) and World Health Organization (WHO) statistics, there are more than 3.7 million HIV-infected people in India in the year 2000.

All persons infected with HIV pass through a phase of symptomless infection, which may last for several months to years. Asymptomatic HIV-infected subjects, however, can transmit HIV infection to other human subjects. In some of the HIV-infected subjects, it may lead to full-blown AIDS. The patients ultimately die due to various opportunistic infections. The virus is transmitted though various routes such as homo- or heterosexual contact with an HIV-infected partner, using contaminated needles or transfusion of contaminated blood. HIV is also vertically transmitted from infected mother to fetus through the placenta. Hence, the detection of HIV infection at early stages is very important for containment of the infection.

HIV is a retrovirus whose genome is a single-stranded RNA comprising of 9,500 bases and flanked at both ends by long terminal repeats. There are nine open reading frames. The first recognizable phase of HIV infection is defined, in some cases, by the presence of HIV core antigen in blood and, in all cases, by a host humoral response to HIV proteins. These antibodies are generally directed against viral structural proteins, coded by the gag, pol and env genes and less frequently against the non-structural products of the HIV genes, i.e., vif, tat, rev, nef, vpu and vpx. The tests are mainly of three types i) screening tests, ii) confirmatory tests, and iii) monitoring tests. Depending on the source of the antigen, these tests have been further classified as: first generation assays, which use the whole virus lysates, second generation assays, which use recombinant antigens, and third generation assays that use recombinant and/or synthetic peptides as the antigen source. First generation assays have the disadvantage of giving false positive results due to contamination with components of the host cells in which the virus was grown. There are approximately 20 different companies marketing HIV detection kits in India. Some of these companies are also manufacturing these kits indigenously based on the technology transferred from academic institutes as discussed later in this section. The total consumption of the HIV kits in India is approximately US $ 6 million in the year 1999–2000 with an upward trend of consumption. DBT, Government of India initiated 3 independent projects in 1993–1994 to develop indigenous kits for the diagnosis of HIV. The basic aim of this activity was to be self-reliant and to design these kits based on the HIV subtypes prevalent in the Indian subcontinent.

2.6.1
ELISA for HIV-1 and HIV-2

At NII, New Delhi an ELISA based on synthetic peptides corresponding to gp41 and gp120 for the detection of antibodies against HIV-1 and a synthetic peptide corresponding to gp36 for detecting antibodies against HIV-2 has been developed. This group further PCR-amplified, cloned and sequenced a 637 bp fragment corresponding to the p24 core protein from an Indian HIV isolate. The deduced amino acid sequence of the cloned p24 fragment revealed 98.1% sequence homology with the consensus B subtype and 92.9% with consensus C subtype [60]. The 637 bp *Kpn* I-*Hind* III fragment was cloned downstream of the His_6 tag in the pQE-30 vector under the control of phage T5 promoter leading to the production of a His_6-p24 fusion protein in *Escherichia coli.* The purified recombinant p24 reacted with serum samples from HIV-infected subjects when tested by Western blot and ELISA [60]. Validation of the assay on coded serum samples provided by 5 different institutions under the supervision of government agencies, revealed a sensitivity of 99.6% and a specificity of 98.8%. The technology was transferred to industry (Ace Diagnostic Limited, Gurgaon, Haryana), but unfortunately, the company closed down before the product could reach the market. Biotech Consortium of India Limited (BCIL), a DBT body that helps in the commercialization of biotechnology based leads, is looking for another industrial partner.

In addition to DBT-sponsored efforts, another ELISA kit based on synthetic peptides was made with the help of scientists at Indian Institute of Science (IISc), Bangalore. It is being manufactured and marketed by a Bangalore-based industry, XCyton Diagnostic Limited under the brand-name HIVChex and has been available since 1997. At AIIMS, New Delhi another ELISA kit based on synthetic peptides has also been developed for screening of HIV-1- and HIV-2-infected human subjects. The know-how has not been transferred to industry as yet.

2.6.2
Western Blot for HIV-1 and HIV-2

An indigenous Western blot kit capable of detecting HIV-1 and HIV-2 has been developed at Cancer Research Institute (CRI), Mumbai. The group succeeded in isolating HIV-1 and HIV-2 from Indian patients and was able to culture the isolated viruses in vitro. The procedures for purifying the virus and to prepare the viral lysate were optimized. The various virus proteins present in the viral lysate were resolved by SDS-PAGE and electrophoretically transferred to nylon membranes. Using these membranes, a Western blot procedure was developed to detect antibodies present in HIV-infected human subjects. The know-how was transferred to a Delhi-based industry, J. Mitra. It is now available in the market.

2.6.3
Rapid Test for HIV-1 and HIV-2

An interesting prototype for a rapid assay to detect HIV-1 and HIV-2 infection based on the principal of hemagglutination has been developed by scientists at University of Delhi, South Campus. Basically, the technique involves generation by recombinant technology of the fusion protein comprised of monovalent anti-RBC antibody (Fab) along with the various immunodominant epitopes of HIV-1 (gp41) and HIV-2 (gp36) [61]. Ideally, the monovalent anti-RBC antibody should react with a determinant that is represented by RBCs present in all human subjects. When such a recombinant fusion protein is mixed with a drop of whole blood, it results in visible agglutination of RBCs only when antibodies against the HIV-1/HIV-2 are present in the blood sample. One end of the recombinant fusion protein will bind through the gp41/gp36 epitope with the anti-HIV-1/HIV-2 antibodies present in the blood and the other end will bind to RBC through the monovalent anti-RBC antibody leading to the formation of a lattice network and hence agglutination. No agglutination is observed in cases of blood samples negative for antibodies against HIV-1/HIV-2. The results by this assay can be obtained in less than 10 min and are readable by the naked eye. It can be used at the field level as it does not require any equipment. The technology has been transferred to Cadila Pharmaceuticals Limited, Ahmedabad. The kit is commercially available since August 2001 as the NEVA HIV (naked eye visible agglutination assay) kit.

3 Constraints in Successful Commercialization by Industries of Diagnostic Assays Developed by Academia

A variety of diagnostic assays against various infectious and non-infectious diseases, hormones and other relevant proteins have been developed at various academic institutions/universities (Table 1). However, the know-how of only few of these leads has been successfully transferred to industry (Table 2). There are various factors contributing to the low rate of successful transfer of know-how to industry. One of the constraints is the non-availability of ELISA plates and various raw materials such as antibody-enzyme conjugate, etc. from an indigenous source. In order to assemble the diagnostic kits, these have to be imported which draws a 30–45% import duty. There is no import duty on the finished life-saving diagnostic kits. Hence, the imported kit is cheaper since the local producer has to pay not only the higher import duties on raw materials but also needs to pay central excise duty (16%) on the finished product. Consequently, biotech companies have considerable apprehension in venturing into indigenous production of diagnostic kits and prefer trading with the imported kits. The government has recently realized this and exempted from import duty certain raw materials used in the ELISA kits.

Another reason for the reluctance of diagnostic companies in venturing into totally indigenous production could be due to lack of capabilities in the specialized area of production of immunological reagents, recombinant proteins, synthetic peptides, etc. Many of these kits are meant for diagnosis of the disease prevalent in restricted regions of the country such as leishmaniasis and hence have low market potential. Major diagnostic companies do not want to venture into such products merely due to economic considerations.

In order to promote indigenous production of various diagnostic kits, it is desirable that

i) DBT may consider the setting up of one or more facilities for the production and supply of immunological reagents. This would not only reduce the dependence on imports but also facilitate development of core capabilities in this specialized area. Such a center could also cater to custom production of the reagents.
ii) Diagnostic kits classified as life-saving drugs should attract custom duty equal to the excise duty levied on similar products locally manufactured in India. Alternatively, there should be exemption from excise duty on the indigenously manufactured kits.
iii) All the imported raw materials required for the local manufacture of diagnostic kits should be considered as essential and should attract custom duty at par with the custom duty levied on imported finished kits/products.
iv) Local innovations need to be encouraged and, therefore, technologies, kits and services produced by the application of local technologies should attract lower internal duties than the equivalent custom duties leviable on such products when imported. Even if marginal discrimination is made in this regard in favor of the local manufacturers, this would encourage indige-

nous manufacturers to innovate and become competitive. The idea is not to encourage indigenous uncompetitive technologies or to protect them by tariff barriers.

v) Technology is the basis of industrial development. The entrepreneurs conducting research and development (R & D) or teaming up for contract R & D at the public funded institutes, should be encouraged by provision of greater fiscal incentives than are being provided currently.

4 Contraceptive Vaccines

The contraceptive vaccines entail generation of a humoral and/or cell mediated-immune response against antigens that have critical roles in the reproductive processes. In the mammalian reproductive system, there are multiple sites where an immunological intervention can lead to a block of fertility [62]. Gonadotrophin releasing hormone (GnRH), which is secreted by the hypothalamus and acts on the pituitary, leading to the secretion of the pituitary gonadotrophins such as follicle stimulating hormone (FSH) and luteinizing hormone (LH), is one such target. GnRH is critical both in the male and female reproduction. Blocking of its action by antibodies will lead to a block in the secretion of the pituitary gonadotrophins and thereby gametogenesis and hence the block in fertility. However, GnRH is not being pursued as a candidate antigen for the design of immunocontraceptive vaccine as such an approach will lead to generalized disturbances in the physiology of reproductive processes with undesirable side effects. Neutralization of pituitary gonadotrophins will interfere in the process of gametogenesis. The feasibility of a male immunocontraceptive vaccine by neutralizing the action of FSH has been investigated in great detail by the scientists at IISc, Bangalore. Another interesting target for the design of an immunocontraceptive vaccine that has been investigated by several groups around the world, including Indian scientists, is to interfere at sperm-oocyte interaction level. The efforts by scientists at Indian Institute of Technology (IIT), New Delhi led to the development of an interesting approach, although not an immunocontraceptive vaccine, based on the premise to interfere in the passage of sperm through vas deferens that is currently undergoing clinical trials in human subjects. The immunocontraceptive potential of riboflavin-carrier protein (RCP), in male and female mammals, has also been investigated by scientists at IISc, Bangalore. Post-fertilization, hCG is adjudged crucial for the establishment and maintenance of pregnancy for at least during the first 7–9 weeks of gestation in humans. Pioneering work from India by Professor G. P. Talwar and his colleagues (initially at AIIMS, New Delhi and subsequently at NII, New Delhi) has shown the feasibility of developing a contraceptive vaccine for females based on the β-subunit of hCG. The prospects and current status of some of these Indian efforts, in the field of fertility regulation, will be briefly summarized in this section (Table 3).

Table 3. Clinical trials in human subjects of indigenous technologies for fertility regulation

Formulation	Target sex	Clinical trials
oFSH	Male	Completed phase I safety trials in India
βhCG-TT	Female	Completed phase I trials in Helsinki, Uppsala, Santiago, Bahia, multiple centers in India
βhCG-αoLH-TT, βhCG-αoLH-DT	Female	Completed phase I and phase II trials at multiple centers in India
Centchroman	Female	Completed multi-centric phase I, phase II and phase III trials. Being commercially produced by two Indian pharmaceutical companies. Introduced in National Family Welfare Program in 1995
Styrene maleic anhydride + dimethyl sulfoxide (RISUG)	Male	Completed phase I and phase II trials at multiple centers in India. Phase III clinical trials initiated

4.1 Follicle Stimulating Hormone (FSH)

In females, FSH helps in the growth of ovarian follicles whereas, in males it regulates the growth of seminiferous tubules and spermatogenesis. Scientists at IISc, Bangalore have immunized male bonnet monkeys (*Macaca radiata*) with ovine FSH (oFSH) that led to the generation of anti-oFSH antibodies in all the immunized animals (n=10). The antibodies against oFSH cross-reacted with human FSH (hFSH) and neutralized its bioactivity [63]. Immunization for 4.7–5.7 years did not affect the health and libido of the animals. No decrease in the serum levels of testosterone was observed in the immunized animals. Within 150 days of immunization, a marked decrease (75–100%) in the number of spermatozoa was observed in the seminal ejaculate. The immunized animals, apart from being acutely oligospermic, ejaculated sperm that were markedly deficient in key acrosomal enzymes, such as acrosin and hyaluronidase and had decreased motility. Periodic boosting with oFSH was required to maintain the azoospermic status. Histopathology of testicular biopsies from immunized animals revealed impairment of spermatogenesis. Immunization with oFSH rendered the animals infertile as demonstrated by their failure to impregnate females of proven fertility. The process is reversible as, concomitant to a reduction in the antibody titers, the sperm count increased in most animals with an ability to impregnate the females [64].

Based on these results, a pilot study was undertaken in which human male volunteers (n=5) were immunized with oFSH [65]. The antibodies generated against oFSH not only cross-reacted with hFSH but also had high binding affinity for it (0.65×10^9 L/M). The antibodies blocked the binding of hFSH to its receptor. No significant changes in the levels of LH, testosterone and thyroid stimulating hormone (TSH) were observed in the immunized males. Immunization with oFSH, however, led to a marked reduction (30–90%) in seminal plasma transferrin – a marker for Sertoli cell and seminiferous tubule functions. A reduction in the sperm count was also observed in some immunized volunteers.

4.2 Human Chorionic Gonadotrophin (hCG)

After fertilization, hCG is synthesized and secreted by the growing blastocyst (subsequently placenta) and is responsible for the "rescue" of the corpus luteum. Progesterone secreted by the corpus luteum is essential for implantation of the growing blastocyst. hCG is also adjudged crucial for the establishment and sustenance of pregnancy, at least during the first 7–9 weeks of gestation in humans, as lower levels of the hormone are associated with threatened abortions.

Pioneering work from India by Professor G. P. Talwar and his colleagues has shown the feasibility of developing a contraceptive vaccine for females based on the β-subunit of hCG (βhCG). βhCG being a self-antigen, was made immunogenic by conjugation with tetanus toxoid (TT). The initial prototype vaccine (βhCG-TT) generated high anti-hCG antibody titers in only a small percentage of immunized women [66]. Women having low anti-hCG antibody titers were not protected from conception [67]. Subsequently, the immunogenicity of the vaccine formulation was enhanced by

i) incorporation of a sodium phthalyl derivative of lipopolysaccharide (SPLPS) as an additional adjuvant in the first injection,
ii) associating non-covalently βhCG with the α-subunit of ovine luteinizing hormone (oLH) to form a heterospecies dimer (HSD), and
iii) conjugation of HSD with two different carriers, e.g., TT and diphtheria toxoid (DT).

During immunization, the HSD-TT and HSD-DT conjugates were used in an alternating sequence to avoid carrier-mediated immunosuppression that has been observed in some immunized women on repeated injections of the conjugate with the same carrier [68]. After extensive pre-clinical toxicology studies in animals, the ethical and drug regulatory agencies gave an approval for phase I clinical trials in sterilized women to evaluate the safety and immunogenicity of the vaccine formulation. It was carried out at 5 different centers in India. The vaccine was able to induce anti-hCG antibodies in all immunized women, although the antibody response was variable [69]. The anti-hCG antibody response was reversible. The immunized women continued to ovulate and exhibited normal menstrual cyclicity patterns [70]. A careful study of 36 hematological, biochemical, metabolic, and endocrinological parameters of the immunized women revealed no significant side effects [71]. The immune sera did not show any reactivity with human and primate somatic tissues.

Based on these results, phase II clinical trials of HSD-TT/HSD-DT vaccine were undertaken in fertile women (25–35 years) at AIIMS and Safdarjang Hospitals, New Delhi and PGIMER, Chandigarh. All women volunteers had at least two living children, active sexual life and regular menstrual cycles. Immunization led to the generation of anti-hCG antibodies and a block of fertility [72]. Fertile immunized women exposed to conception over 1224 cycles recorded only one pregnancy at an antibody titer of >50 ng/mL. The block in fertility was reversible, as immunized women became pregnant when antibody titers fell below <35 ng/mL. The progeny born to 4 women, previously immunized with the above vaccine formulation, were followed-up for anthropometric indices from 2 to 3.5 years and data compared with elder siblings. These investigations revealed that low titers of anti-hCG antibodies, below the protective threshold, have no apparent side effects on the progression of progeny and on the early development of the progeny [73].

One of the probable reasons for the low anti-hCG antibody response in some women may be related to carrier-mediated epitopic suppression as observed in other systems [74–77]. To further improve the immunogenicity of HSD-DT/TT vaccine, the scientists at NII, New Delhi have further modified the vaccine formulation. Instead of DT/TT as carrier proteins, promiscuous T non-B cell epitopes have been employed to provide T-cell help to generate an effective anti-hCG antibody response in the context of a diverse major histocompatibility complex (MHC) background. Three promiscuous T helper (Th) epitopes used as carriers corresponded to measles virus fusion protein (aa residues 288–302), influenza virus hemagglutinin (aa residues 307–319), and HIV-1 reverse transcriptase (aa residues 38–52). Conjugation of βhCG with each of the Th cell peptides improved the antibody response against hCG in mice of different haplotypes. Immunization of mice with a cocktail of all three βhCG-peptide conjugates led to a generation of anti-hCG antibodies higher than that achieved by βhCG-TT [78]. The antibodies had a high affinity for hCG and neutralized its bioactivity. It is also desirable to reduce the number of injections since a schedule of multiple injections would not be practical for general use. For this purpose, encapsulation of the vaccine in microspheres has been proposed. Such microspheres have been seen to generate a long-lasting antibody response in experimental animals after a single administration [79].

4.3
Riboflavin Carrier Protein (RCP)

Riboflavin-carrier (or binding) protein (RCP), an estrogen-inducible phosphoglycoprotein (37 kDa), is present in the circulation of egg-laying hens, but not in immature birds or adult males. The obligatory role of RCP in yolk deposition of the vitamins for the growth and survival of the avian embryo is well documented. RCP, an evolutionary conserved vitamin carrier, is induced during the reproductive phase for the deposition of adequate amounts of the vitamins in the developing oocytes in oviparous species and to transport the vitamins to the developing embryo across the placental barrier in mammals. Elucidation of the crystal structure of RCP at 2.5 Å resolution has helped in delineation of the riboflavin binding hydrophobic pocket and receptor recognition motif at the C-terminus that is comprised of two α-helices located on either side of the flexible phosphorylated sequence [80]. The RCP receptors on chicken oocyte membrane belong to the LDL-receptor related proteins. The binding of RCP to chicken oocyte membrane is dependent on calcium [81]. RCP binds to protein with molecular weight of 380, 260 and 110 kDa. The purified phosphopeptide, prepared from a tryptic digest of RCP, inhibits the binding of RCP to the receptor protein, with an affinity comparable to that of native RCP, indicating that the phosphopeptide sequence is critical for RCP-receptor interaction. An RCP homologue that exhibits physico-chemical and immunological properties similar to avian RCP has been demonstrated in rodents and non-human primates [82, 83].

4.3.1
Role of RCP in Female Mammalian Reproduction

The group at IISc, Bangalore headed by Prof. P. R. Adiga has carried out pioneering work demonstrating the role of RCP in male and female reproduction and the possibility of using RCP-based immunological approaches for contraception. Administration of polyclonal antibodies against chicken RCP to day 11 pregnant rats led to 100% embryonic resorption, suggesting the role of RCP in sustained embryonic growth [84]. Female rats immunized with chicken RCP elicited anti-RCP antibodies. Immunized animals were mated with fertile males. On day 8 of gestation, laprotomy revealed bloated uterine horns with or without implantation/resorption sites, indicative of early embryonic loss. None of the immunized animals delivered pups as long as the antibody titers remained high [85]. Subsequent to a decline in antibody titers, animals became pregnant and delivered normal progeny. No discernible adverse reactions due to immunization were encountered as reflected by the undisturbed riboflavin status, estrous cyclicity and gonadal hormonal profile. These studies suggested that RCP is not required for maternal well-being. Presumably, RCP has a pregnant-specific role in channeling vitamins to the conceptus.

Active immunization of female bonnet monkeys (*Macaca radiata*) with chicken RCP also led to early embryonic loss without interfering in menstrual cyclicity, riboflavin status or ovarian endocrine profile [86].

Subsequently, this group showed that SDS-denatured RCP is comparatively more efficacious vis-à-vis native chicken RCP in eliciting neutralizing antibodies despite being less immunogenic [87]. Denaturation of RCP evokes preferentially an antibody repertoire against sequential epitopes. In subsequent studies, reduced and carboxymethylated RCP (RCM-RCP) treated with SDS was used for immunization of female bonnet monkeys. None of the immunized animals became pregnant. A total of 121 ovulated cycles were protected (7–30 cycles/animal). In some immunized animals, the duration of a few cycles was lengthened by 7–8 days with concomitant marginal elevation of progesterone on day 26 as compared to control females. It is possible that these animals might have conceived, but experienced early embryonic loss.

4.3.2
Synthetic Peptides of Chicken RCP as Candidate Immunogens

The polyclonal antibodies generated against SDS-treated RCM-RCP in mice, rats, rabbits and bonnet monkeys were used to map the sequential epitopes using PEPSCAN analysis [88]. These studies have led to the identification of 5 epitopic sequences (aa residues 10–17, 42–49, 134–141, 172–179 and 200–207) shared by all the antisera from the 4 mammalian species [89]. Additionally, scientists at IISc, Bangalore have also used MAbs to map functionally relevant B cell epitopes. Administration of MAb $6B_2C_{12}$ to pregnant mice led to embryonic resorption [90]. The MAb binds to a tryptic peptide corresponding to the C-terminus of chicken RCP (aa residues 200–219). The synthetic C-terminal peptide (aa residues 200–219; CTP) was conjugated to DT. The CTP-DT conjugate is immunogenic and elicited antibodies that recognized CTP, RCM-RCP and RCP. Passive administration of anti-CTP antibodies during 9–13 days of gestation led to termination of pregnancy [91]. The female rats actively immunized with CTP failed to conceive [92]. In CTP, stabilization of the helix by introduction of appropriately spaced salt bridges led to the generation of antibodies with higher affinity, and a longer-lasting antibody response with more prolonged immunocontraceptive effect as compared to the original CTP [93].

This group has also made a synthetic peptide corresponding to the N-terminus of chicken RCP (aa residues 4–24, NTP), in which cysteine residues at positions 5 and 24 were replaced with Ala and Tyr, respectively, and coupled the same to DT. Anti-NTP antibodies recognized NTP, RCP and RCM-RCP in ELISA. Passive administration of anti-NTP antibodies to pregnant mice during 8–10 days of gestation resulted in embryonic resorption. The contraceptive efficacy in female rats was also verified by immunization with the NTP [92]. The female bonnet monkeys immunized with both CTP-DT and NTP-DT conjugates failed to become pregnant when mated with the males of proven fertility [94]. Additionally, active immunization of fertile female rats with synthetic peptides corresponding to aa residues 64–83 and 130–147 also conferred protection from pregnancy [95].

4.3.3
Role of RCP in Male Reproduction

Using antibodies as specific probes, RCP has been localized both in the somatic cells of the testis and various germ cell populations including mature spermatozoa of several mammalian species including humans [96]. RCP is localized on the acrosomal cap of the sperm. Preliminary experiments with RCM-RCP-immunized male bonnet monkeys showed that their voided spermatozoa exhibited impaired motility with no apparent agglutinating antibodies. The testosterone levels were within the normal range implying normal Leydig cell function. When these immunized males (n=4) were repeatedly mated with fertile females (n=10) during days 10–16 of ovulated cycles, the females appeared to be protected from pregnancy during a total of 29 fertile cycles [94].

4.4
Spermatozoa and Oocyte Specific Antigens

The generation of an immune response against antigens unique to spermatozoa and/or oocyte, aiming to interfere in the process of fertilization, is an interesting approach to develop immunocontraceptive vaccines. Several clinical studies of women have shown an association of antibodies against sperm antigens with otherwise unexplained infertility [97]. These nature's experiments, prompted several groups across the world to identify and characterize the unique spermatozoa-associated antigens that are critical for sperm-oocyte interaction. These investigators have either employed polyclonal antibodies from infertile human subjects or MAbs generated against sperm that inhibit fertilization. Using such approaches, scientists at NII, New Delhi and NIRRH, Mumbai have been able to characterize sperm-based candidate antigens that are promising for designing immunocontraceptive vaccines [98–101].

4.5
Zona Pellucida (ZP) Glycoproteins

ZP, an extracellular matrix that surrounds the mammalian oocytes, is composed of three distinct glycoproteins. Based on the size of their mRNA transcripts, these have been classified as ZPA, ZPB and ZPC. ZPA is encoded by the longest mRNA transcript and ZPC by the smallest. The ZP glycoproteins mediate the initial recognition and binding of the spermatozoa to the oocyte in a species-specific manner, induce the zona-bound sperm to undergo acrosome reaction, block polyspermy and protect the pre-implanted embryo. By virtue of the critical involvement of ZP glycoproteins at various stages during fertilization and their tissue specificity, these have been proposed as candidate antigens to block fertility at a pre-fertilization stage. Initial active immunization studies were carried out with pig ZP glycoproteins due to

i) Availability of a large number of pig ovaries from abattoirs,
ii) Immunological cross-reactivity of antibodies thus generated with ZP from various species including humans, and
iii) Ability of polyclonal antibodies against pig ZP glycoproteins to inhibit in vitro human sperm-oocyte interaction.

Active immunization of the female subjects, using porcine ZP glycoproteins led to infertility in several species [102–106]. The observed immunocontraceptive effect was invariably associated with either transient or irreversible changes in the cyclicity, alteration of the sex steroid hormonal profile and disrupted follicular development. The group at NII, New Delhi evaluated the immunocontraceptive potential of highly purified porcine ZP3 (mixture of ZPB and ZPC) in female bonnet monkeys. The animals were immunized either with porcine ZP3 or porcine ZP3 coupled to βhCG employing adjuvants that are permissible for human usage [107]. All the immunized animals generated good anti-ZP3 antibody titers. In spite of high circulating anti-ZP3 antibodies, immunized animals continue to have ovulatory cycles. Animals remained infertile in the presence of high anti-ZP3 antibody titers and showed no disturbance in cyclicity. Subsequent to a decline in the anti-ZP3 antibody titers, out of the 8 immunized animals, 4 became pregnant. The animals that failed to regain fertility, however, had normal ovarian histopathology.

4.5.1
Cloning and Expression of Non-Human Primate ZP Glycoproteins

To evaluate the immunocontraceptive potential of ZP glycoproteins, scientists at NII, New Delhi have developed a non-human primate model. For this purpose the cDNA corresponding to the three ZP glycoproteins from bonnet monkey was cloned and sequenced [108–110]. Comparison of the deduced aa sequence of bonnet monkey ZPA (bmZPA), bonnet monkey ZPB (bmZPB) and bonnet monkey ZPC (bmZPC) revealed 94.2, 92.6 and 93.9 percent sequence identity with respective human homologues. The cDNA corresponding to bmZPA, bmZPB and bmZPC, excluding the N-terminal signal sequence and the C-terminal transmembrane-like domain, were cloned either in pQE-30 or pRSET vectors under T5 and T7 promoters, respectively (Fig. 1). The respective protein was expressed as a polyhistidine fusion protein in *E. coli* [109–111]. The

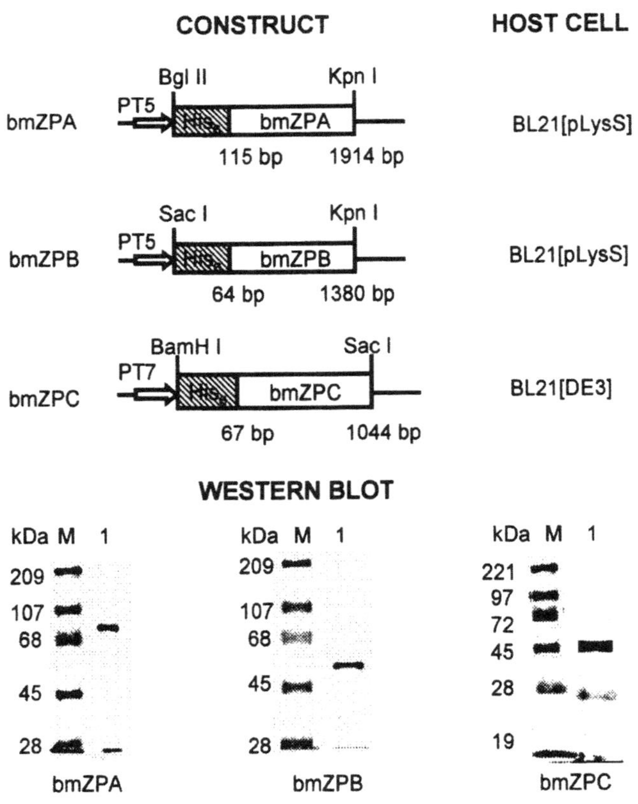

Fig. 1. Expression of bmZPA, bmZPB and bmZPC in *E. coli*. The upper panel is a schematic representation of the constructs of bmZPA and bmZPB in pRSET expression vector and bmZPC in pQE-30 expression vector. The host *E. coli* cell used for the expression of the respective recombinant protein is also shown. The lower panel represents the immunoblot of the respective purified recombinant protein. PT5, promoter of phage 5; PT7, promoter of phage T7; Lane M, molecular weight markers; Lane 1, purified protein

SDS-PAGE and Western blot analysis of the purified recombinant (r) proteins revealed a 70 kDa band for r-bmZPA, a 55 kDa band for r-bmZPB and a 43 kDa band for r-bmZPC [112–114] (Fig. 1). The r-bmZPA binds to the principal segment of the capacitated spermatozoa but the binding shifts to the equatorial segment, tip of the inner acrosomal membrane and mid-piece in acrosome-reacted spermatozoa [112]. The r-bmZPB binds to the principal segment of the acrosomal cap of capacitated spermatozoa. The binding of r-bmZPB shifts to the equator, post-acrosome and mid-piece in the acrosome-reacted spermatozoa [113]. The r-bmZPC binds to the head region of the capacitated spermatozoa but does not bind to acrosome-reacted spermatozoa [114]. These studies helped in elucidating the binding characteristics of the individual ZP proteins to spermatozoa and showed that the carbohydrate moieties are not critical for the binding of ZP glycoproteins to sperm receptor.

4.5.2
Immunocontraceptive Potential of Recombinant Non-Human Primate ZP Proteins

The rabbit polyclonal antibodies as well as murine MAbs generated against r-bmZPB not only recognized the native ZP but also inhibited, in vitro, the human sperm-oocyte interaction [115]. Based on these results, female baboons (*Papio anubis*) were immunized with r-bmZPB conjugated to DT using Arlacel-A and squalene as adjuvants (Table 4) [116]. All the immunized animals elicited a good antibody response against r-bmZPB and DT used as carrier. The animals exhibited normal ovulatory menstrual cycles without any disturbance in the cyclicity. In the presence of high anti-bmZPB antibody titers (>2×10^3 antibody units), the immunized animals failed to conceive following mating with the males of proven fertility. With the decline in anti-bmZPB antibody titers, the immunized females became pregnant, suggesting thereby that the block in

Table 4. Effect of immunization with r-bmZPB conjugated to DT on fertility in female baboons[a]

Experimental group	Antibody titer range in antibody unit (AU$\times10^3$) Against		Numer of protected ovulatory cycles	Remarks
	ZPB	DT		
Unimmunized group				
PAN57	–	–	0	Delivered
PAN84	–	–	0	Aborted
Immunized with r-bmZPB-DT				
PAN22	1.0–8.0	91.3–280	6	Died
PAN30	1.0–26.7	18.0–220	5	Delivered
PAN32	1.0–10.0	18.0–221	7	Delivered
PAN30	2.0–11.0	58.5–192	4	Delivered

[a] Adapted from ref. [116].

fertility by such an approach may be reversible [116]. Using a homologous animal model, immunization of female bonnet monkeys with r-bmZPA-DT and r-bmZPB-DT led to a block in fertility (unpublished observations).

4.5.3
Feasibility of ZP Based Synthetic Peptides as Immunogens for Fertility Regulation

The feasibility of using a synthetic peptide corresponding to the B cell epitope of mouse ZPC and devoid of the oophoritogenic T-cell epitope to block fertility without the concomitant autoimmune oophoritis has been demonstrated [117]. In such a situation, the T-cell help was provided by a "promiscuous" foreign T-cell epitope. At NII, New Delhi, scientists have used MAbs capable of inhibiting human sperm-oocyte binding to map the relevant B-cell epitopes. Alternatively, synthetic peptides have been made, based on computational information of the deduced aa sequence of the respective zona protein with respect to hydrophilicity, antigenic index, surface probability, probable glycosylation sites and aa sequence identity with the respective homologue from other species. The MAbs generated against r-bmZPB and capable of inhibiting the binding of sperm to zona were mapped to a common epitope corresponding to aa residues 136–147 [115]. The epitopic domain corresponding to aa 136–147 of bmZPB was completely conserved in human ZPB. Polyclonal antibodies generated against a 24-mer synthetic peptide corresponding to aa residues 321–347 of bmZPC inhibited the binding of human sperm to antibody-treated zona-encased oocytes [118]. Interestingly, the female bonnet monkeys immunized with the bmZPC peptide-DT conjugate showed a block in fertility [119].

5
Logistic Hurdles Associated with Contraceptive Vaccines

Vaccines traditionally have been used as "herd" immunization approach to combat infectious diseases. However, contraceptive vaccines meant for fertility regulation in humans demand:

1) Adequate protective antibody response in 100% of the recipients,
2) Sustenance of antibody titers above the protective threshold for a defined period.

In order to overcome these problems there is a need to develop more potent adjuvants admissible for human use, characterization of a "promiscuous" T-cell epitope and their use in the vaccine formulation instead of DT/TT and the development of new vaccine delivery strategies. Nonetheless, immunocontraceptive vaccines can be used successfully to control wildlife populations as demonstrated by the ZP-based immunogens to control the population of elephants [120], wild mares [121], dogs [122] and 27 species of captive zoo animals [123].

6 Indigenous Contraceptive Leads other than Contraceptive Vaccines

The development of contraceptive vaccines is still at the research stage. However, it is prudent to describe two other major contributions from India in the area of development of contraceptives, one of which is in practical use.

6.1 Oral Non-Steroidal Contraceptive Pill

Towards developing a non-steroidal oral contraceptive, CDRI, Lucknow synthesized DL-Centchroman ($C_{30}H_{35}O_3N{\cdot}HCl$; *trans*-1-[2-{4-(7-methoxy-2,2-dimethyl-3-phenyl-3,4-dihydro-2*H*-1-benzopyran-4-yl)-phenoxy}ethyl]pyrrolidine hydrochloride. After testing its contraceptive efficacy in various laboratory animals, the drug has undergone extensive phase I, phase II and phase III clinical trials employing a multiple-dose schedule in human subjects (for review see [124]). Centchroman has weak estrogen agonistic and potent antagonistic activities and prevents implantation. The drug is effective when taken immediately after coitus or routinely as a weekly pill. In phase II and phase III multicentric trials, children born of method-and-users failure pregnancies showed a normal growth pattern without any congenital abnormality. Since 1992, the drug is in the market and being sold in India with the tradenames, Saheli and Choice-7 by Hindustan Latex Ltd., Thiruvananthapuram. It is also being marketed under the tradename Centron by Torrent Pharmaceuticals India Ltd., Ahmedabad. The drug was included in the National Family Welfare Program in 1995.

6.2 Injectable Intravasal Contraceptive for the Male

A novel method to achieve long-term contraception in males, which is reversible, has been developed by Prof. Sujoy K. Guha and his colleagues at Indian Institute of Technology (IIT), New Delhi. Basically the technique involves injecting styrene maleic anhydride of a specific preparation dissolved in dimethyl sulfoxide (DMSO) into the lumen of the vas deferens [125–127]. This polymer alters the normal negative electrical charge distribution on the spermatozoa head thereby reducing their fertilization potential. In fact, sperm undergoes a premature acrosome reaction. Biochemical analysis revealed an enhanced release of acrosin and hyaluronidase from sperm that are necessary for their penetration into the ovum at the time of fertilization. Premature loss of the enzymes within the male tract renders the spermatozoa incapable of fertilization. Using this approach, the vas deferens remained in a non-occlusive mode. By flushing out the polymer with a single injection of the solvent DMSO or sodium bicarbonate, fertility can be restored [128].

Toxicity studies in rats and monkeys and teratogenicity in rats showed that this procedure has no significant untoward effects [129–132]. In view of the efficacy and safety demonstrated in the animal models, the Drug Controller of India (DCI) gave permission for a phase I clinical trials in male subjects whose wives had undergone sterilization operations. These studies showed that the injection of the drug into the lumen of the vas deferens is a safe procedure with no long-term adverse effects [133]. Subsequently, phase II clinical trials with males having wives in normal reproductive status indicated pregnancy protection efficacy [134, 135]. In phase I clinical trials the drug was tested on 14 male subjects and in phase II trials on 32 subjects. After assessment of the results, ICMR and DCI have permitted multicentric phase III clinical trials to further evaluate the contraceptive efficacy. ICMR is coordinating the trials and the Ministry of Health and Family Welfare is financially supporting the study. So far 150 subjects have been administered with SMA. The trials are being conducted in three different hospitals located in Delhi.

7 Concluding Comments

Diagnostic assays for various infectious diseases and hormones have been developed in various academic institutions with the financial support of DBT, CSIR, ICMR and DST. These kits are based on various formats such as latex agglutination assays, rapid agglutination assay, antibody capture assay, ELISA, Western blot and PCR etc. To a lesser extent, there is an initiative by some industries to develop on their own indigenous diagnostic kits. The know-how for some of the diagnostic assays has been successfully transferred from academia to industry for commencement of their commercial production. To harness the benefits of the developments in academia more optimally, there is a need to increase the interaction between academia and industries and provide more facilities and opportunities to young technical entrepreneurs. One of the possible solutions is to develop Science Parks as an interface, where people from industry or a young entrepreneur can work in close association with scientists from academic institutions. It is heartening to mention that several such Biotechnology/Science Parks have recently been established in various states of the country, to harness the maximum benefits from the advances in the field of biotechnology, for the welfare of mankind.

The initiative for the development of immunocontraceptive vaccines for fertility regulation in humans is still at the research stage. The demonstration that fertility can be regulated in women immunized with a βhCG-based immunocontraceptive vaccine is a landmark development. Due to logistic problems of inadequate antibody response in 100% of the recipients and the variable duration of the protective antibody response, such an approach needs refinement before it can be of practical utility. Nonetheless, scientific investigations with respect to oFSH, βhCG, RCP and ZP proteins have led to a better understanding of their functional relevance in reproductive physiology. Although more research input is needed before such technologies can be used in humans, some of these can be used for the control of wildlife populations with the present inadequacies. The development of vas occlusion technology with SMA for blocking fertility in males has shown very promising results in clinical trials and has a good potential for inclusion in the National Family Planning Programme.

Acknowledgement. The help in providing the source material to write this review article from Drs. V. K. Vinayak and Bindu Dey, DBT, New Delhi, Dr. S. K. Panda, Department of Pathology, AIIMS, New Delhi, Dr. Jaya Tyagi, Department of Biotechnology, AIIMS, New Delhi, Dr. Sujoy K. Guha, Centre for Biomedical Engineering, IIT, New Delhi, Dr. V. K. Chaudhary, UDSC, New Delhi, and Dr. N. C. Saxena, ICMR, New Delhi is gratefully acknowledged. I thank Dr. Kakoli Ghoshal, Dr. Ayub Qadri, Ms. G. K. Gahlay, NII, New Delhi for critically reviewing the article.

8
References

1. Gupta MD, Ramachandran V, Mutatkar RK (2001) J Biosci (Suppl) 26:437
2. Catt KJ, Dafau ML, Vaitukaitis JL (1975) J Clin Endocrinol Metab 40:537
3. Fishel SB, Edwards RG, Evans CJ (1984) Science 223:816
4. Gupta SK, Talwar GP (1980) Indian J Exp Biol 18:1361
5. Gupta SK, Ramakrishnan S, Talwar GP (1982) J Biosci 4:105
6. Gupta SK, Singh O, Kaur I, Talwar GP (1985) Indian J Med Res 81:281
7. Gupta SK, Guesdon JL, Avrameas S, Talwar GP (1985) J Immunol Methods 83:159
8. Gupta SK, Talwar GP (1986) Scand J Clin Lab Invest 46:751
9. Chaicumpa W, Thin-Inta W, Khusmith S, Tapchaisri P, Echeverria P, Kalambaheti T, Chongsa-Nguan M (1988) J Clin Microbiol 26:1824
10. Sadallah F, Brighouse G, DelGiudice G, Drager-Dayal R, Hocine M, Lambert PH (1990) J Infect Dis 161:59
11. Qadri A, Gupta SK, Talwar GP (1988) J Clin Microbiol 26:2292
12. Qadri A, Ghosh S, Upadhyay S, Talwar GP (1989) Hybridoma 8:353
13. Qadri A, Ghosh S, Talwar GP (1990) J Immunoassay 11:235
14. Qadri A, Ghosh S, Prakash K, Kumar R, Moudgil KD, Talwar GP (1990) J Immunoassay 11:251
15. Stevens DL (1992) Clin Infect Dis 14:2
16. Gupta R, Rattan A, Prakash K, Talwar GP, Gupta SK (1993) Indian J Med Res 97:25
17. Gupta R, Talwar GP, Gupta SK (1990) J Immunoassay 13:441
18. Gupta R, Kalia A, Rattan A, Kumar R, Gupta SK (1997) Indian J Med Res 105:200
19. Bloom BR, Murray CJL (1992) Science 257:1055
20. Talib VH, Pandey JN, Khurana SK (1993) Indian J Pathol Microbiol 36:339
21. Katti MK (2001) FEMS Immunol Med Microbiol 31:59
22. Sumi MG, Mathai A, Sarada C, Radhakrishnan VV (1999) J Clin Microbiol 37:3925
23. Gupta S, Bhatia R, Datta KK (1997) Tuber Lung Dis 78:21
24. Reddi PP, Talwar GP, Khandekar PS (1988) Int J Lepr 56:592
25. Singh KK, Nair MD, Radhakrishnan K, Tyagi JS (1999) J Clin Microbiol 37:467
26. Singh KK, Muralidhar M, Kumar A, Chattopadhyaya TK, Kapila K, Singh MK, Sharma SK, Jain NK, Tyagi JS (2000) J Clin Pathol 53:355
27. Chakravorty S, Tyagi JS (2001) FEMS Microbiol Lett 205:113
28. Arora SK, Gupta V, Gupta A, Bambery P, Kapoor GS, Sehgal S (1999) Tuber Lung Dis 79:229
29. Nagesh BS, Sehgal S, Jindal SK, Arora SK (2001) Chest 119:1737
30. Kadival GV, D'Souza CD, Kolk AH, Samuel AM (1995) Zentralbl Bakteriol 282:353
31. Ahmed N, Mohanty AK, Mukhopadhyay U, Batish VK, Grover S (1998) J Clin Microbiol 36:3094
32. Dar L, Sharma SK, Bhanu NV, Broor S, Chakraborty M, Pande JN, Seth P (1998) Indian J Chest Dis Allied Sci 40:5
33. Nayak NC, Panda SK, Zuckerman AJ, Bhan MK, Guha DK (1987) J Med Virol 21:137
34. Panda SK, Gupta A, Dutta R, Nayak NK (1986) Lancet 2(8512):919
35. Nayak NC, Panda SK, Datta R, Zuckerman AJ, Guha DK, Madanagopalan N, Buckshee K (1989) J Gastroenterol & Hepatol 4:345
36. Banerjee K, Sarin SK, Naik SR, Khandekar P (1992) Indian J Med Res 95:173

37. Dash S, Rao KV, Joshi B, Nayak NC, Panda SK (1991) Hepatology 13:134
38. Dash S, Rao KV, Panda SK (1992) J Med Virol 37:116
39. Mishra A, Durgapal H, Manivel V, Acharya SK, Rao KV, Panda SK (1992) Clin Exp Immunol 90:194
40. Manival V, Ramesh R, Panda SK, Rao KV (1992) J Immunol 148:4006
41. Tripathy SP, Kumar A, Manivel V, Panda SK, Rao KV (1992) J Immunol 148:4012
42. Manivel V, Panda SK, Rao KV (1992) J Immunol 149:2082
43. Choo Q-L, Kuo G, Weiner AJ, Overby LR, Bradley DW, Houghton M (1989) Science 244:359
44. Panigrahi AK, Nanda SK, Dixit RK, Acharya SK, Zuckerman AJ, Panda SK (1994) J Med Virol 44:176
45. Panigrahi AK, Roca J, Acharya SK, Jameel S, Panda SK (1996) J Med Virol 48:191
46. Panigrahi AK, Nayak B, Dixit R, Acharya SK, Panda SK (1998) Trop Gastroenterol 19:105
47. Jameel S, Durgapal H, Habibullah CM, Khuroo MS, Panda SK (1992) J Med Virol 37:263
48. Nanda SK, Panda SK, Durgapal H, Jameel S (1994) J Med Virol 42:237
49. Nanda SK, Yalcinkaya K, Panigrahi AK, Acharya SK, Jameel S, Panda SK (1994) J Med Virol 42:133
50. Panda SK, Nanda SK, Zafrullah M, Ansari IH, Ozdener MH, Jameel S (1995) J Clin Microbiol 33:2653
51. Jameel S, Zafrullah M, Ozdener MH, Panda SK (1996) J Virol 70:207
52. Zafrullah M, Ozdener MH, Panda SK, Jameel S (1997) J Virol 71:9045
53. Ansari IH, Nanda SK, Durgapal H, Agarwal S, Mohanty SK, Gupta D, Jameel S, Panda SK (2000) J Med Virol 60:275
54. Panda SK, Ansari IH, Durgapal H, Agarwal S, Jameel S (2000) J Virol 74:2430
55. Nanda SK, Ansari IH, Acharya SK, Jameel S, Panda SK (1995) Gastroentrol 108:225
56. Weniger BG, Takebe Y, Ou CY, Yamazaki S (1994) AIDS 8:S13
57. Pfutzner A, Dietrich U, von Eichel U, von Briesen H, Brede HD, Maniar JK, Rubsamen-Waigmann H (1992) AIDS 5:927
58. Dietrich U, Grez M, von Briesen H, Panhans B, Geibendorfer M, Kuhnel H, Maniar J, Mahambre G, Becker WB, Becker MLB, Rubsamen-Waigmann H (1993) AIDS 7:23
59. Jameel S, Zafrullah M, Ahmed M, Kapoor GS, Sehgal S (1995) AIDS 9:685
60. Gupta SK, Sengupta J, Bisht R, Bhatnagar A, Kaul R (1997) Gene 190:27
61. Gupta A, Gupta S, Chaudhary VK (2001) J Immunol Methods 256:121
62. Gupta SK, Jethanandani P, Afzalpurkar A, Kaul R, Santhanam R (1997) Hum Reprod Update 3:311
63. Moudgal NR, Ravindranath N, Murthy GS, Dighe RR, Aravindan GR, Martin F (1992) J Reprod Fertil 96:91
64. Moudgal NR, Jeyakumar M, Krishnamurthy HN, Sridhar S, Krishnamurthy H, Martin F (1997) Hum Reprod Update 3:335
65. Moudgal NR, Murthy GS, Prasanna Kumar KM, Martin F, Suresh R, Medhamurthy R, Patil S, Sehgal S, Saxena BN (1997) Hum Reprod 12: 457
66. Talwar GP, Sharma NC, Dubey SK, Salahuddin M, Das C, Ramakrishnan S, Kumar S, Hingorani V (1976) Proc Natl Acad Sci USA 73:218
67. Shahani SM, Kulkarni PP, Patel KL, Salahuddin M, Das C, Talwar GP (1982) Contraception 25:421
68. Gaur A, Arunan K, Singh O, Talwar GP (1990) Int Immunol 2:151
69. Singh O, Rao LV, Gaur A, Sharma NC, Alam A, Talwar GP (1989) Fertil Steril 52:739
70. Kharat I, Nair NS, Dhall K, Sawhney H, Krishna U, Shahani SM, Banerjee A, Roy S, Kumar S, Hingorani V, Singh O, Talwar GP (1990) Contraception 41:293
71. Talwar GP, Hingorani V, Kumar S, Roy S, Banerjee A, Shahani SM, Krishna U, Dhall K, Sawhney H, Sharma NC, Singh O, Gaur A, Rao LV, Arunan K (1990) Contraception 41:301
72. Talwar GP, Singh O, Pal R, Chatterjee N, Sahai P, Dhall K, Kaur J, Das SK, Suri S, Buckshee K, Saraya L, Saxena BN (1994) Proc Natl Acad Sci USA 91:8532
73. Singh M, Das SK, Suri S, Singh O, Talwar GP (1998) Am J Reprod Immunol 39:395

74. Herzenberg LA, Tokuhisa T, Herzenberg LA (1980) Nature 285:664
75. Herzenberg LA, Tokuhisa T (1982) J Exp Med 155:1730
76. Schutze MP, Deriaud E, Przewlocki G, Lecrec C (1989) J Immunol 142:2635
77. Renjifo X, Wolf S, Pastoret PP, Bazin H, Urbain J, Leo O, Moser M (1998) J Immunol 161:702
78. Gupta A, Pal R, Ahlawat S, Bhatia P, Singh O (2001) Vaccine 19:3384
79. Singh M, Singh O, Talwar GP (1995) Pharm Res 12:1796
80. Monaco HL (1997) EMBO J 16:1475
81. Sooryanarayana, Sarkar S, Adiga PR, Visweswariah SS (1998) Biochim Biophys Acta 1382:230
82. Muniyappa K, Adiga PR (1980) Biochem J 186:537
83. Visweswariah SS, Adiga PR (1987) Biochim Biophys Acta 915:141
84. Muniyappa K, Adiga PR (1980) FEBS Lett 110:209
85. Murthy CVR, Adiga PR (1982) Science 216:191
86. Seshagiri PB, Adiga PR (1987) J Reprod Immunol 12:93
87. Karande AA, Adiga PR (1991) Ind J Biochem Biophys 28:476
88. Geysen HM, Rodda SJ, Mason TJ, Tribbick G, Schoffs PG (1987) J Immunol Methods 102:259
89. Subramanian S, Adiga PR (1998) Biochim Biophys Acta 1429:74
90. Karande AA, Velu NK, Adiga PR (1991) Mol Immunol 28:471
91. Koshy BT, Karande AA, Adiga PR (1996) Vaccine 14:307
92. Subramanian S, Karande AA, Adiga PR (2000) Vaccine 19:1172
93. Subramanian S, Karande AA, Adiga PR (2001) Biochem Biophys Res Commun 287:236
94. Adiga PR, Subramanian S, Rao J, Kumar M (1997) Human Reprod Update 3:325
95. Subramanian S, Karande AA, Adiga PR (2000) Am J Reprod Immunol 44:184
96. Bhat KG, Malhotra P, Karande AA, Adiga PR (1995) Ind J Exp Biol 33:12
97. Ingerslev HJ (1981) Acta Obstet Gynecol Scand (Suppl) 100:1
98. Shaha C, Suri A, Talwar GP (1990) Int J Androl 13:17
99. Shaha C, Taneja M, Seshadri T (1993) Hybridoma 12:709
100. Aravinda S, Gopalakrishnan B, Dey CS, Totey SM, Pawshe CH, Salunke D, Kaur K, Shaha C (1995) J Biol Chem 270:15675
101. Fernandes JE, Hegde UC (1997) Indian J Biochem Biophys 34:274
102. Skinner SM, Mills T, Kirchik HJ, Dunbar BS (1984) Endocrinology 115:2418
103. Dunbar BS, Lo C, Powell J, Stevens VC (1989) Fert Steril 52:311
104. Sacco AG, Yurewicz EC, Subramanian MG (1989) Am J Reprod Immunol 21:1
105. Jones GR, Sacco AG, Subramanian MG, Kruger M, Zhang S, Yurewicz EC, Moghissi KM (1992) J Reprod Fertil 95:513
106. Paterson M, Thillai-Koothan P, Morris KD, O'Byrne K, Braude P, Williams A, Aitken RJ (1992) Biol Reprod 46:523
107. Bagavant H, Thillai-Koothan P, Sharma MG, Talwar GP, Gupta SK (1994) J Reprod Fertil 102:17
108. Kolluri SK, Kaul R, Banerjee K, Gupta SK (1995) Reprod Fertil Dev 7:1209
109. Gupta SK, Sharma M, Behera AK, Bisht R, Kaul R (1997) Biol Reprod 57:532
110. Jethanandani P, Santhanam R, Gupta SK (1998) Mol Reprod Dev 50:229
111. Kaul R, Afzalpurkar A, Gupta SK (1997) Mol Reprod Dev 47:140
112. Gahlay GK, Srivastava N, Govind CK, Gupta SK (2002) 53:67
113. Govind CK, Gahlay GK, Choudhury S, Gupta SK (2001) Biol Reprod 64:1147
114. Patra AK, Gahlay GK, Reddy BVV, Gupta SK, Panda AK (2000) Eur J Biochem 267:7075
115. Govind CK, Hasegawa A, Koyama K, Gupta SK (2000) Biol Reprod 62:67
116. Govind CK, Gupta SK (2000) Vaccine 18:2970
117. Lou Y, Ang J, Thai H, McEleven F, Tung KSK (1995) J Immunol 155:2715
118. Afzalpurkar A, Shibahara H, Hasegawa A, Koyama K, Gupta SK (1997) Human Reprod 12:2664
119. Kaul R, Sivapurpu N, Afzalpurkar A, Srikanth V, Govind CK, Gupta SK (2001) Reproductive Biomedicine Online 2:33

120. Fayrer-Hosken RA, Grobler D, Van Altena JJ, Bertschinger HJ, Kirkpatrick JF (2000) Nature 407 (6801):149
121. Kirkpatrick JF, Liu IKM, Turner JW Jr (1990) Wildlife Society Bulletin 18:326
122. Srivastava N, Santhanam R, Sheela P, Mukund S, Thakral SS, Malik BS, Gupta SK (2002) Reproduction 123:847
123. Kirkpatrick JF, Turner JW Jr, Liu IKM, Fayrer-Hosken R (1996) J Reprod Fertil Suppl 50:183
124. Singh MM (2001) Med Res Rev 21:302
125. Guha SK (1999) Asian J Andrology 1:131
126. Lohiya NK, Manivannan, B, Mishra PK, Pathak, N, Balasubramanian SPA (1998) Contraception 58:119
127. Lohiya NK, Manivannan B, Mishra PK (2000) Int J Andrology 23:36
128. Koul V, Srivastava A, Guha SK (1998) Contraception 58:227
129. Sethi N, Srivastava RK, Singh RK (1989) Contraception 39:217
130. Sethi N, Srivastava RK, Singh RK (1990) Contraception 41:333
131. Sethi N, Srivastava RK, Singh RK, Bhatia GS, Sinha N (1990) Contraception 42:337
132. Sethi N, Srivastava RK, Nath D, Singh RK (1992) Int J Fertil 37:183
133. Guha SK, Singh G, Anand S, Ansari S, Kumar S, Koul V (1993) Contraception 48:367
134. Guha SK, Singh G, Ansari S, Kumar S, Srivastava A, Koul V, Das HC, Malhotra RL, Das SK (1997) Contraception 56:245
135. Guha SK, Singh G, Srivastava A, Das HC, Bhardwaj JC, Mathur V, Koul V, Malhotra RL, Das SK (1998) Contraception 58:165

Received: April 2002

Adv Biochem Engin/Biotechnol (2003) 85: 215–281
DOI 10.1007/b11049

Medical Biotechnology in India [1]

Braj B. Lohray

Zydus Research Center, Cadila Healthcare Ltd., Sarkhej-Bavla Highway, Moraiya, Ahmedabad – 382 213, India. *E-mail: braj.lohray@zyduscadila.com*

The potential of biotechnology has just began to emerge in the 20th century. After the full knowledge of human genomes is available, biotechnology is going to play a major role in shaping the concept of future drug discovery, drug delivery, diagnostic methodology, clinical trials, and to a great extent the major lifestyle of the human society. This article is a comprehensive review of the major impact of biotechnology in diagnostics, antibiotics, r-proteins, vaccines, and antibodies production. It also highlights the future aspects of gene therapy in the management of healthcare. A comprehensive list of biotech products in healthcare management has been given. Also, the growth of biotechnology throughout the world at large and in the Indian industries in particular has been highlighted. Constraints, concerns and difficulties in biotechnology in India have been addressed mainly related to human resources, training institutions in India, funding in biotechnology, patent-related issues and regulatory hurdles. Like in information technology, India has great potential in bioinformatics as well. Some of the recent information on bioinformatics centers in India has been summarized. Indian biotechnology industries have the potential to use the modern discoveries in life sciences to reach an enviable position in the world of biotechnology

Keywords. Biotechnology in India, Diagnostics, Antibiotics, Vaccine, Antibodies, Gene therapy

[1] ZRC Communication no. 111.

Abbreviations

AIDS	Acquired immunodeficiency syndrome
ARC	AIDS-related complex
AUG	Adenosine uracil guanine
BCG	Bacillus Calment Guanine (tuberculosis vaccine)
BPI	Bactericidal permeability increasing protein
BSA	Bovine serum albumin
cDNA	Complementary deoxyribonucleic acid
CDR	Complementary determining regions

CFTR	Cystic fibrosis transmembrane conductance regulator
CMV	Cytomegalovirus
Crore	Ten million
CSF	Colony stimulating factor
DBT	Department of Biotechnology
DNA	Deoxyribonucleic acid
DNase	Deoxyribonuclease
E. coli	*Escherichia coli*
EBV	Epstein-Barr virus
EGF	Epidermal growth factor
ELISA	Enzyme-linked immunosorbent assay
EPO	Erythropoietin
ESTs	Expressed sequence tags
FCA	Freund's complete adjuvants
FDA	Food and Drug Administration
FGF	Fibroblast growth factor
FIA	Freund's incomplete adjuvants
G-CSF	Granulocyte colony-stimulating factor
GM-CSF	Granulocyte/macrophage colony-stimulating factor
GP-IIb/IIIa	Glycoprotein IIb/IIIa
HAMA	Human anti-mouse antibodies
HbsAg	Hepatitis B surface antigen
HBV	Hepatitis B virus
h-CSF	human colony-stimulating factor
hEGF	Human epidermal growth factor
hGH	Human growth hormone
Hib	Hepatitis influenzae type b
HIV	Human immunodeficiency virus
Ifn	Interferon
Ifn-$\alpha/\beta/\gamma$	Interferon-$\alpha/\beta/\gamma$
IgE antibody	Immunoglobulin E antibody
Igf	Insulin-like growth factor
Igf-1	Insulin-like growth factor-1
IgG	Immunoglobulin G
Il	Interleukin
LCR	Ligase chain reaction
LDL	Low density lipoprotein
Lps	Lipopolysaccharide
MAbs	Monoclonal antibodies
MCSF	Macrophage colony-stimulating factor
MIM	Mendelian inheritance in man
m-RNA	Messenger ribonucleic acid
NATP	National Agricultural Technology Project
NCBI	National Centre for Bioinformatics
NK	Natural killer
PCR	Polymerase chain reaction
PDGF	Platelet-derived growth factor

PMV	Parvo megalo virus
rDNA	Recombinant deoxyribonucleic acid
rEPO	Recombinant human erythropoietin
rFGF	Recombinant fibroblast growth factor
rFSH	Recombinant follicle-stimulating hormone
rG-CSF	Recombinant granulocyte colony-stimulating factor
rGM-CSF	Recombinant granulocyte/macrophage colony-stimulating factor
Rh	Rhesus
rhGH	Recombinant human growth hormone
rInsulins	Recombinant insulins
RNase	Ribonuclease
rPDGF	Recombinant platelet-derived growth factor
rtPA	Recombinant tissue plasminogen activator
rTSH	Recombinant thyroid-stimulating hormone
S. cerevisiae	*Saccharomyces cerevisiae*
SNP	Single nucleotide polymorphism
SOD	Superoxide dismutase
TGF α & β	Transforming growth factors
TNF α & β	Tumor necrosis factors
tPA	Tissue plasminogen activator
t-RNA	Transfer ribonucleic acid
Uk	Urokinase
uPA	Urokinase plasminogen activator
US FDA	United States Food and Drug Administration
USPTO	United States Patent and Trademark Office
vWD	von Willebrand's disease

1 Introduction

Biotechnology has been conventionally practiced in India for the last several decades, however, during the last ten years, the concept of biotechnology has been changing rapidly. In India, products based on genetic engineering, immunological techniques, cell culture methods, r-DNA technology, and hybridoma technology have slowly been gaining importance during the last five years. Furthermore, human and animal healthcare products are growing substantially. Diagnostic tools are now based on biotechnology and are going to grow substantially, although most of the diagnostic kits are imported. Diagnostics based on monoclonal antibodies, synthetic peptides, recombinant antigens or antibodies could be made locally with the skills available in the country.

Biotechnology, in recent decades, has revolutionized research in biomedicine and has created an entirely new facet in the pharmaceutical industry. Today, several therapeutically useful proteins are produced through r-DNA technology for the treatment and prevention of diseases such as cancer, autoimmune disorders, neurological disorders, infections, cardiovascular disorders and genetic diseases [1].

r-DNA technology has wide applications in medicines, agriculture, aquaculture, food, and several other industries. This area offers unparalleled growth potential with the advent of bioinformatics, genetic engineering, cell fusion technology, bioprocess technology, and structure-based molecular design.

More recently, complete analyses of the genetic code of various species including bacteria, viruses, animals, and humans have opened yet another landmark discovery in biotechnology. It will not be long before scientists will be able to correlate every action of a living organism with the genes they have, and be able to modify, manipulate and change the behavioral pattern of every living being by genetic manipulation. The power of biotechnology has started to reveal itself and the 21st century will be the true age of biotechnology, when scientists will be able to achieve what has been so far a mystery of unknown knowledge. This article will be restricted only to healthcare products produced via biotechnology.

Many proteins, which are naturally present in humans, are found to be important and proven drugs. Several proteins, antibodies, and enzymes are useful in diagnostic kits and as reagents. It is now possible to produce these proteins in large amounts by isolating the target gene, encoding the target protein and expressing them into microbial or other hosts. Proteins produced from recombinant DNA in heterologous hosts are known as recombinant proteins. This is just to distinguish them from the naturally occurring proteins. These proteins have been produced in a variety of host vectors such as bacteria, mammalian cells, yeast, or viruses. More recently, transgenic plants and animals are also being used for producing specific proteins. For example, several vaccines have been produced in plant products. Cows, goats, and sheep have been used to produce milk with specific proteins in useful quantities.

In 1982, Genentech introduced the first recombinant human insulin in the market [2]. Since 1985 onwards, nearly twenty different proteins are available for a variety of indications. Proteins such as erythropoietin and growth factors like granulocyte colony-stimulating factors (G-CSF) are among the top sellers in the pharmaceutical biotech products. Viral and bacterial proteins are used for the mass production of vaccines using r-DNA technology. r-Hepatitis B vaccine is one of the important examples. Several other enzymes, hormones, blood factors, cell proteins, and viral proteins can be produced by recombinant bacteria and other eukaryotic cells. Recombinant antibodies, vaccines, enzymes, and growth factors targeting a variety of medical conditions have reached the stage of clinical evaluation and many have even reached the market.

The present article will be restricted to the evolution of medical biotechnology, which has taken place in unraveling newer areas in medical research, however, not to the development of newer instrumentation technology. Thus, this review will be restricted to the use of biotechnology in developing newer drugs for human healthcare in the world at large, and with a particular emphasis on the Indian biotech industries that have evolved in healthcare management.

2 Diagnostics

In the healthcare sector, diagnostics has played a crucial role. Development of newer tools to detect, analyze and understand the etiology of various organisms, disease states and human physiology has helped scientists to unravel the secret of biological science. Since the discovery of the microscope by Leeuwenhock, more and more refined techniques have been developed to understand the microworld. With hemocytometry, the physical characteristics of the cells could be investigated. The development in immunology and immunochemistry with flow cytometry has become one of the most important techniques in cell diagnostics. With multidimensional flow cytometric analysis, it is possible to detect leukemic cells in bone marrow samples. The assessment of the antigen profile of leukemic cells can be used to monitor the effect of therapy on leukemic cell populations

Similarly, advances in diagnostic technology already has blunted significantly the public health threat from HIV, particularly in blood and blood-derived products. The ligase chain reaction (LCR) and the polymerase chain reaction (PCR) are both nucleic acid probe-based technologies, and have been found to be highly accurate in diagnostic applications [3]. Genetic information is contained within the nucleic acids such as DNA or RNA. Therefore, determination of the presence of disease-related genetic information in the form of nucleic acid is the most fundamental and correct method for determination of an active or a latent disease state. Disease-related genetic material is typically present at levels that are not detectable without amplification until other disease manifestations are apparent [4]. Therefore, these diagnostic technologies hold great promise. Similarly, there are other very sensitive techniques based on monoclonal antibodies for the detection of diseases and genetic changes.

2.1 Antibodies in Diagnostic Kits

Antibodies are special proteins which only bind to specific antigens [5]. They are made by white blood cells in response to any substance, including disease-causing organisms, that the body recognizes as a foreign substance.

Antibodies bind to these foreign substances and prevent them from harming the body. Because antibodies bind specifically to the substance that caused them to be produced, the past history of infectious diseases can be determined by examining the antibodies in the blood. Therefore, antibodies obtained from the blood of immunized animals are used widely for both diagnostic as well as vaccine purposes. These are polyclonal antibodies and have a wide diversity. In 1975, it was found that the antibody-producing cells from a mouse could be fused with mouse tumor cells to produce unlimited generations of new cells, called "hybridomas", which could make antibodies [6]. These antibodies are called monoclonal antibodies and are highly specific for their target substance. Today, monoclonal antibodies can be produced on an industrial scale and used in various diagnostic kits for diagnosing diseases. The most useful diagnostic kits based on monoclonal antibodies are pregnancy testing, cancer screening, detection of viral gastroenteritis, hepatitis B, cystic fibrosis, and sexually transmitted diseases including AIDS [7]. These kits are economical and very fast, giving results in a few minutes, and are extremely specific and sensitive. These diagnostic methods also need a very small amount of sample. Monoclonal antibodies are being used to deliver chemically bound radioactive material or X-ray dense labels to the site of tumor, so that the tumor can be located using modern computerized scanning and imaging processes.

The enzyme-linked immunosorbent assay (ELISA) test is a very sensitive serological assay that detects the presence of virus and other pathogenic organisms [8]. This technique again uses highly purified protein, mostly antibodies, which are very specific for the detection of antigens. ELISA-based assay kits are available for detection of low density lipoprotein (LDL) or bad cholesterol, total lipid protein, blood sugar, triglycerides, HDL etc., which can now be performed at home and are not expensive.

2.2 DNA Probes in Diagnosis

Each cell contains all the genetic information to make a human being. Millions of genes, which regulate various vital functions, are present in each cell. Defects, absences and infusion of external genes with the vital genes may alter the function of genes entirely, leading to disastrous effects. Gene defects cause potentially lethal diseases such as hemophilia, sickle cell anemia and other metabolic disorders such as phenylketonemia, diabetes, Huntington's disease, disease of nervous systems and cardiovascular diseases. Detecting genetic diseases, therefore, plays a very crucial role in healthcare management. Medical practitioners usually are not able to say whether a patient has inherited a genetic disease until the symptoms begin to develop. However, now using new genetic engineering techniques, scientists can locate and analyze a single gene among thousands of genes using "gene probes" [9]. In a "gene probe", the segment of DNA that matches with the DNA segment of individual genes, will bind with that gene and can be detected by attaching easily recognizable chemical labels to the DNA segment, such as radioactive, fluorescence or phosphorescence probes. By making probes that recognize the DNA sequence associated with genetic diseases, scientists can now detect genes for a number of genetic conditions in a minute tissue sample collected from adults or taken from embryos by amniocentesis. This information can be used for genetic counseling aimed at reducing the incidence of crippling genetic defects and allowing early treatment and management of disorders, if possible. DNA probe can also be used for identifying pathogenic organisms, especially where antibodies are unstable, unsuitable, or unavailable.

2.3 DNA Probe for Forensic Use

No two individuals including twins have the same genetic composition. This knowledge is used by forensic scientists to identify individuals from small traces of tissues, blood etc. left at the site of crime. Now, a number of bacterial enzymes is available that can cut DNA's chemical structure. As the chemical structure of DNA reflects the genetic composition of the individual, the range of DNA fragments produced when DNA is treated with these enzymes is characteristic of that individual. This technique is popularly known as DNA-fingerprinting [10].

By treating DNA from tissues or blood samples with various restriction enzymes, separating pieces of DNA by size, and then using labeled DNA probes to locate one or more specific genes, forensic scientists have developed an extremely sensitive method of genetic fingerprinting. This technique is also useful in determining parenthood in controversial cases.

3 Antibiotics

Microbial transformations of organic compounds have been known since the dawn of human civilization. Fermentation of fruit, grain or milk to obtain intoxicating and nourishing dietary factors has long been practiced. Wine was produced as early as around 3000 BC. However, modern microbiology and its application in industry started with the work of Louis Pasteur. He introduced the concept of fermentation to convert sugar into alcohol. He also found that these fermentation processes could be contaminated with rod-shaped organisms, which can oxidize alcohol to acid and affect the wine industry severely. He thus introduced the concepts of sterilization and "Pasteurization" [11]. Pasteur and Joubert (1877) also discovered that *B. anthracis* was killed in the presence of other bacteria that contaminated the culture. This was actually due to the chemicals (later on known as "antibiotics") produced by the other bacteria, which killed *B. anthracis*. Later, an antibiotics known as pyocyanin was discovered from *B. aeruginosa* (*Bacillus pyocyaneus*) which causes lysis of many bacteria. It also liberated other antibiotics known as pyocyanin and hemipyocyanin, which are quite active against fungi.

Penicillin liberated by *P. notatum* was the landmark discovery of Fleming in 1928 [12]. These findings led to the discovery of several antibiotics produced by several microorganisms which continue to be used for the management of bacterial infections. Antibiotics are classified according the their site of action.

These antibiotics inhibit the growth of bacteria by various mechanisms, e.g., penicillin, cycloserine, bacitracin, vancomycin effect the cell wall synthesis. Other antibiotics affect the cell membrane or protein synthesis. Yet another class of antibiotics inhibit the nucleic acid metabolism and act on the bacterial ribosomes. Thus, an antibiotic may be defined as a chemical substance produced by a living organism that demonstrates inhibitory or germicidal activity towards microorganisms in vivo or in vitro. Antibiotics can be produced by fungi, bacteria, algae and lichens. Some antibiotics are also produced by some seed plants. Then, the major challenge was to produce these antibiotics in large scale for commercial uses [14]. In the course of time, the fermentation technique was developed to produce and purify antibiotics on a large scale. Biotechnology played a crucial role in the production of life-saving antibiotics although

Table 1. Antibiotics for various therapeutic indications [116]

No.	Company	Trade-name	Generic name	Therapeutic indications	FDA approval
1	The Liposome company	Abelcet	Liposomal amphotericin B	Treatment of aspergillosis in refractory or intolerant patients	November 1995
2	Medimmune	RespiGam	Anti-RSV antibiotics	Treatment of respiratory syncytial viral infections in children	January 1996
				Prevention of lower respiratory disease in patients at risk of RSV	January 1998
3	Sequus	Amphotec	Liposomal amphotericin B	Second-line therapy for aspergillosis infections	November 1196
4	NeXstar	Ambisome	Liposomal amphotericin B	Empirical therapy for patients with presumed fungal infections	August 1997
5	Atrix Pharmaceuticals	Atridox	Doxycycline hyclate	Site-specific antibiotic for periodontal disease	September 1998

many of the specific details regarding the plant-scale production of the given antibiotics remain as trade secrets. Some of the antibiotics recently approved for the treatment of infection are listed in Table 1. These antibiotics require extensive biotechnological support for their preparation on a large scale.

However, the general methods used are a matter of common knowledge to those concerned with the fermentation industry. Presently, a whole range of antibiotics is prepared by fermentation. Some of the semi-synthetic antibiotics are also produced by minor chemical modification of antibiotics produced by fermentation. Even some peptides are being used as antibiotics against various organisms [15]. Several companies in India are producing these antibiotics in multiton levels (see Chap. 10).

4 Therapeutic Proteins

Most of the novel therapies developed over the past decade have been realized through the advances in biotechnology. These products are often referred as biotech or biopharmaceuticals. Numerous proteins occur naturally in the body. A deficiency of these proteins in the body may cause diseases. Thus, the exogenous supply of these proteins constitutes a class of biopharmaceuticals, which have minimal side effects if delivered in the right quantity and to the right target. For example, the use of insulin as a therapy for diabetes is not new. Gammaglobulin and protein-containing vaccines have been routinely used for decades. Major obstacles in protein therapy have been two-fold:

a) production of the desired protein in a large amount, and
b) delivery of the protein to the target site.

Advancements in biotechnology have largely solved the first problem. With the advent of recombinant DNA technology, virtually any protein can be produced in sufficient quantities for therapeutic use, although often with significant difficulties. Currently, most protein products are simply administered systemically, e.g., intravenously. If it were possible to actually deliver a protein to a particular site in a temporally defined manner, the utility of protein-based therapy would potentially be significantly enhanced. These recombinant proteins make viable medical products which can be reproducibly manufactured in purified form, many of them have now been shown to be capable of alleviating or preventing human diseases. Recombinant antibiotics, vaccines, enzymes, and growth factors targeting a variety of medical conditions have reached the clinic and many are now in the market. Recent sales of recombinant-biotech products worldwide are estimated to be in the range of US $ 18 billion (Fig. 1).

The diversity of product types reflects the wide range of therapeutic strategies that can be developed using recombinant proteins (Table 2).

Recombinant immunoglobulins could function passively to provide protective immunity against specific pathogens. Other proteins (antigens) function actively either by binding to cell surface molecules to regulate cellular function or by catalytically modifying specific substrates. Clearly, the biological functions that proteins perform can be both complex and specific, offering opportunities for affecting human health in ways which are not possible with simple small-molecule drugs. On the other hand, recombinant protein products present unique challenges in development as well as in their limitations in terms of drug delivery.

It may be noted that, unlike drugs derived from traditional organic chemistry, those biopharmaceuticals obtained by genetic engineering can, in the majority of cases, only be administered parenterally. This is a major disadvantage for single or subchronic administration of drugs, but becomes prohibitive for the long-term use of drugs.

The biopharmaceutical market for recombinant products is currently in a phase of active growth, since the release of first of these products into the market in the last decade. Thus r-DNA technology has a major impact on biotechnology research and the uses of protein as therapeutic products.

4.1 Recombinant Protein as Therapeutic Products

Reservations about the use of proteins as therapeutic products are slowly disappearing. Views were often expressed that the use of proteins as therapeutics would be reserved only for a handful of products. It was further speculated that these proteins were only a temporary generation of drugs that would be superseded by low molecular weight organic new chemical entities.

However, these views are rapidly changing due to the large availability of proteins by r-DNA technology. Several of these proteins which were isolated

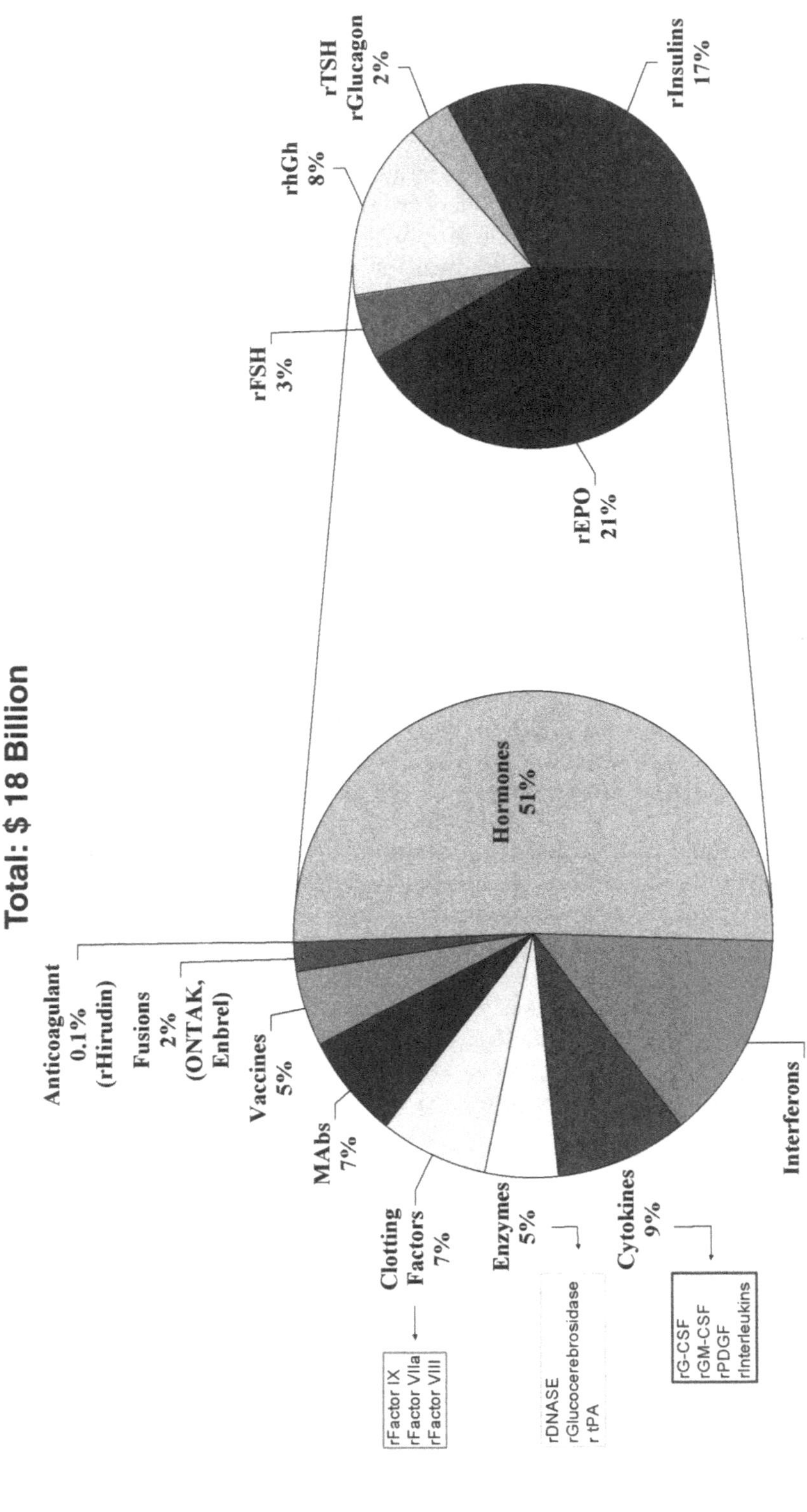

Fig. 1. Market share of recombinant protein biotech products

Table 2. Some of the r-DNA (or recombinant) proteins in therapeutic uses [117]

Diseases/targets	Therapeutic approaches	Products
Autoimmune disease		
Multiple sclerosis		IFN-β
Rheumatoid arthritis	Anti-cytokine	TNF-α antibody
Blood deficiencies		
Anemia	Boost blood-cell proliferation	Erythropoietin
Blood substitute	Replacement	Hemoglobin
Chemo-induced		G-CSF
Hemophilia		Factor VIII & IX
Cancer		
Bone marrow transplant	Immune cell proliferation	GM-CSF
Leukemia	Immune stimulation	IFN-α
T-cell lymphoma	Enhance tumor cell killing	IL-2 fusion toxin
Melanoma	Immune stimulation vaccination	Melanoma vaccine
Renal cancer	Immune stimulation	IL-2/IFN-γ
Cardiovascular diseases		
Myocardial infarction	Clot dissolution	tPA
Angina/restenosis	Block platelet aggregation	GP-IIb/IIIa antibody
Genetic diseases		
Cystic fibrosis	Mucous thinning	Dnase
Diabetes	Hormone replacement	Human insulin
Gaucher's disease	Enzyme replacement	Glucocerebrosidase
Growth deficiency	Hormone replacement	hGH
Infectious agents		
Hepatitis B virus	Immune stimulation vaccination	IFN-α
		Subunit vaccine
HIV	Immune stimulation vaccination	IFN-α/IL-2
Papilloma virus	Immune stimulations	IFN-α
Bordatella pertusis	Vaccination	Acellular vaccine
Inflammatory disorders		
Allergy	Mast-cell activation	
Graft-versus-host disease	Block IL-2 receptor	IgE antibody
Septic shock	Anti-endotoxin	Tac antibody
Nervous system disorders	Anti-cytokine	BPl
Amyelotrophic lateral sclerosis	Enhance neuron survival	IGF-1
Amyelotrophic lateral sclerosis	Block oxidative damage	PEG-SOD
Trauma		
Tissue damage		
Wound healing	Modulate cell function	TGF-β/PDGF

and used from natural sources are being replaced by recombinant proteins. For example, insulin, human growth hormones, tissue plasminogen are produced by r-DNA technology. Several interferons are produced by r-DNA technology, and many more proteins, which are produced by r-DNA technology, are under clinical trials and are expected to reach the market soon.

There are great advantages of using proteins as drugs. Many of these proteins are already present in the body and therefore are expected not to show side effects if used in the right dose. Secondly, unlike conventional drugs which are mostly synthetic chemicals, these biopharmaceuticals are highly specific. Therefore, these biopharmaceuticals can perform both complex and specific biological functions. Another advantage of proteins as drugs could be the lower toxicity of proteins compared to chemical substances. They are neither carcinogenic nor teratogenic and their degradation products are naturally occurring amino acids normally found in human body and are absolutely safe.

4.2 Advantages and Disadvantages of r-DNA Products as Therapeutics

There are a number of advantages of biopharmaceuticals [16]:

1) The r-DNA technology has led to the production of peptides and proteins such as erythropoietin, GMCSF, GCSF etc., that previously could not be isolated and purified for therapeutic uses.
2) Manufacture of sufficient quantities of protein drugs.
3) Safety concerns about the drug from natural resources.
4) Mutant protein.

4.3 Concerns about Biopharmaceuticals by r-DNA Technology [17]

1) The therapeutic protein may be contaminated with unwanted protein expressed from unwanted genes. Such gene products might arise because of (a) mutations, insertions, deletions or rearrangement in the coding region of the product occurring during fermentation; (b) mistakes in transcription initiated at several sites; (c) changes in transcription, initiation or termination of other genes in the vector or host cells.
2) Biologically active extraneous component such as DNA, protein etc. derived from host cell.
3) High cost of the protein.
4) The target protein produced by r-DNA technology is considered as new product due to new structural features (3D folding of protein) of the product. Therefore, the activity of the product may vary.
5) Modification of the protein in host cells may change the activity of the protein. Especially the N-terminal of the protein prepared by r-DNA technology is always methionine encoded by the translation initiation codon AUG, and therefore a method needs to be developed for preparing proteins with an NH_2 terminus as found in natural proteins with the derived N-terminus.

Despite all these concerns, scientists have developed methods to prepare proteins of acceptable purity and quality for therapeutic uses [18]. Several of these proteins synthesized by r-DNA technology have reached the market.

4.4 Hormones (Table 3)

4.4.1 *Insulin*

Insulin is an amino acid peptide hormone secreted by β-cells of the pancreas, in response to an elevation in blood glucose. Insulin also regulates the metabolism of lipids and proteins, the synthesis of nucleic acid and the expression of certain genes.

Insulin is used to treat patients producing insufficient amounts of this hormone from their islet β-cells. This is the sign of insulin-dependent diabetic mel-

Table 3. List of hormones approved for various indications [116, 118]

No.	Company	Trade-name	Generic name	Therapeutic indications	FDA approval
6	Eli Lilly/ Genetech	Humulin	Human insulin	Type I diabetes	October 1982
7	Genentech	Protropin	Human growth hormone	Hypopituitary dwarfism	October 1985
8	Genentech	Nutropin	Human growth hormone	Chronic renal insufficiency in children Growth failure associated with Turner's syndrome	March 1994 December 1996
9	Amgen	Epogen	Erythropoietin (EPO)	Anemia in kidney failure patients	June 1989
10	Amgen/ Ortho Biotech	Procrit	Erythropoietin (EPO)	Anemia in Retrovir-treated HIV-infected patients Chemotherapy induced anemia in patients with malignancies Reduction in the no. of blood transfusions in anemic surgery patients	December 1990 April 1993 December 1996
11	Bio-Technology General	Bio-Tropin	Human growth hormone	Growth hormone deficiency in children	May 1995
12	Serono Labouratories	Gonal-F	Follicle-stimulating hormone	Infertility	September 1997
13	Serono Labouratories	Geref	Sermorelin acetate	Growth hormone deficiency in children	October 1997

litus called type-1 diabetes. Nearly five million patients need daily insulin. Until 1982, all the insulin used for treating patients was prepared by extraction from the pancreata of cattle or pigs. In addition, insulin produced from pig pancreas differ from human insulin by the presence of different amino acids at the C-terminal end of the β-chain. Actually, this insulin is similar to human insulin in biological activities, however, in some patients it produces an immunogenic response and the development of antibodies, which makes insulin slowly ineffective in such patients. Although porcine insulin could be humanized by chemical transpeptidation, often by enzymatic cleavage of the terminal octapeptide (1980, Novo Nordisk A/S), the most lasting solution was achieved by an in vitro DNA recombination technique and expression of insulin in the microorganism *E. coli* and yeast (Eli Lilly) [19]. In addition, proinsulin, which is a long-acting insulin, can also be produced by r-DNA methodology.

4.4.2
Human Growth Hormone

Somatotrophin, a human growth hormone, is a 191 amino acid protein produced by the pituitary. It also promotes the synthesis of proteins and regulates the metabolism of carbohydrates and lipids.

Growth hormones are released from the pituitary in response to stimulation by hypothalamic peptide growth hormone releasing factor. This release is inhibited by somatostatin [20]. Growth hormones are also thought to promote the differentiation of fat cells and chondrocytes.

A deficiency of growth hormones in humans results in dwarfism and stunted growth. Nearly normal growth patterns can be restored in deficient children by administering growth hormone at the right time. However, the growth hormones isolated from cadaver pituitaries are not of acceptable quality and some patients developed Creutzfeldt-Jacob disease, a fatal neurodegenerative disease caused by retrovirus contamination [21]. Thus, the preparation of hGH by cloning and bacterial expression was highly desirable. Genentech (1979) first cloned hGH in *E. coli* and Sanofi made further modifications to prepare hGH on an industrial scale [22]. Clinical evidence suggests that hGH may be used in treating burns, peptic ulcer, osteoporosis, HIV and cancer. This hormone can be used to increase the growth rate of children of short stature. Similarly, growth hormones of bovine origin are used for increasing milk production. Several biotechnology companies have developed bovine somatotrophin preparations for use in dairy cattle.

4.4.3
Erythropoietin

Erythropoietin is a glycoprotein hormone produced by the kidneys. The hormone stimulates the production of red blood cells, the erythrocytes, from their precursor stem cells [23]. Human erythropoietin has a molecular weight in the region of 34,000 Da, 60% of which is carbohydrate. The hormone contains a high level of sialic acid and varying amounts of hexosamines and hexoses.

Patients with chronic renal failure develop a condition known as renal anemia because they are no longer capable of producing sufficient erythropoietin to stimulate the formation of reticulocytes. These patients require regular dialysis treatment as a substitute for their absent renal function. Regular blood transfusions are often also necessary in order to maintain at least a minimal oxygen transport function of their blood [24]. Supplementing the endogenous levels of erythropoietin with additional erythropoietin results in the production of a sufficient quantity of red blood cells so that blood transfusions are superfluous in all patients after several months at the latest.

Initially isolated in the urine of anemic patients, the purification of erythropoietin was long and difficult. With the use of recombinant technology, erythropoietin was successfully expressed as a heterologous protein in 1984 [25]. It is being marketed successfully by Amgen, USA. Some 3,00,000 patients with end-stage renal failure are currently being treated world-wide by conventional dialysis and 30–50% of these patients are receiving erythropoietin. Their quality of life has been considerably improved by using this recombinant protein [26].

4.4.4 *CSF (Human Colony-Stimulating Factors) [27]*

The term colony-stimulating factor reflects the ability of these substances to promote the growth in vitro of various leukocytes into clumps or colonies [28]. A number of different colony stimulating factors have been identified and characterized. These include granulocyte-macrophage colony-stimulating factor (GM-CSF < 30 kDa protein), granulocyte colony-stimulating factor (G-CSF = 20 kDa protein), macrophage colony-stimulating factor (M-CSF) and multipotential colony-stimulating factor (multi-CSF ≅ 30 kDa protein). CSFs exhibit a broad range of molecular weights depending upon their degree of glycosylation.

Colony-stimulating factors are produced by a variety of cell types, including lymphocytes and fibroblasts. The various CSFs exhibit little amino acids sequence homology and receptors for all four subtypes have been identified on responsive cells. All the receptors are transmembrane glycoproteins having an extracellular ligand binding domain, a transmembrane segment and an intracellular effector domain. Recombinant CSFs have also been produced in mammalian expression systems, which are capable of carrying out post-translational modifications, especially glycosylation [29].

Administration of colony-stimulating factors normally promotes a significant increase in the serum level of macrophages and granulocytes. Many clinical studies have successfully illustrated that administration of such cytokines can greatly augment the resistance of susceptible patients to infection. CSFs employed clinically thus far are normally administered to patients whose circulatory levels of leukocytes are diminished, due to diseases such as AIDS, or due to therapeutic regimens such as irradiation or chemotherapy associated with cancer treatments [30]. In many cases, accelerated granulocyte-macrophage production and enhanced leukocyte activity have been recorded.

4.5
Plasma Proteins

Specific plasma proteins that are of therapeutic use include a range of factors involved in the blood clotting process, fibrinolytic agents that degrades clots, serum albumin and immunoglobulin preparations. While these products have traditionally been obtained from blood donated by human volunteers, some are now produced by recombinant DNA technology.

4.5.1
Blood Clotting Factors

A variety of plasma proteins form an integral part of the blood clotting process. Either a genetic or an induced deficiency in any one blood factor results in severely impaired coagulation ability. The vast majority of hereditary diseases characterized by poor coagulation responses result from a deficiency of blood factors VIII and IX.

The process of blood coagulation is dependent upon a number of associated blood clotting factors. Some of the blood factors in the market are listed in Table 4. With the exception of factor IV (calcium ions), all the other blood factors are proteinaceous in nature.

Activated forms of the blood factors are designated by the addition of the letter "a" to the factor number, e.g., factor VIIa is activated factor VII.

More than 90% of all defects in clotting factors relate to a deficiency of factor VIII. Many of the remaining cases are due to a deficiency of factor IX. The clinical disorders associated with deficiencies of factors VIII or IX include hemophilia A, also known as von Willebrand's disease (vWD) or hemophilia B.

These blood factors have traditionally been purified from the blood of healthy human donors and subsequently administered to patients. Many such factors can be and are being produced by recombinant DNA methodology.

4.5.2
Factor VIII

This is a high molecular weight (2330 amino acids) serum glycoprotein, the deficiency of which is responsible for type A hemophilia. There are approximately 100,000 cases of this disease. The factor is normally produced by extraction from blood plasma at the same time when other serum proteins are extracted. However, this blood origin is responsible for the contamination of poly-transfused hemophiliacs by HIV virus present in the blood of certain donors. Demand for the extraction product market has therefore shifted towards the recombinant product.

The recombinant product must be glycosylated. The gene has been cloned and expressed in various mammalian cells (Chiron for Nordisk, Transgene, Green Cross), hamster kidney cells (Genentech for Cutter) and monkey kidney cells (Genetics Institute for Baxter) [31]. It should be noted that fragments of recombinant factor VIII have superiority both in biological activity and a longer half-life than those of the natural product (Biogen) [32]. The world-wide mar-

Table 4. Factors approved for various indications [116]

No.	Company	Trade-name	Generic name	Therapeutic indications	FDA approval
1	Baxter/ Genetics Institute	Recom-binate	Factor VIII	Hemophilia A	December 1992
2	Bayer (form-ely Miles)/ GNE	Kogenate	Factor VIII	Hemophilia A	February 1993
3	Genetic Institute	BeneFIX	Factor IX	Prevention and control of bleeding in hemophilia B patients	February 1997
4	Chiron	Regranex	PDGF (topical gel)	Treatment of diabetic foot ulcers	December 1997
5	Amgen	Neupogen	G-CSF	Chemotherapy-induced neutropenia	February 1991
				Bone marrow transplant	June 1994
				Chronic severe neutropenia	December 1994
				PBPC collection	December 1995
6	Immunex	Leukine	GM-CSF	Autologous bone marrow engraftment	March 1991
				Acute myelogenous leukemia	September 1995
				PBPC mobilization and post-transplantation support	November 1995
7	Alkermes/ Genentech	Nutropin Depot	Somatropin for injectable suspension	Pediatric growth hormone deficiency	December 1990
8	Genetics Institute	BeneFIX	Factor IX	Prevention and control of bleeding in hemophilia B patients	February 1997

ket for factors VIII and IX represents $ 300 million for a total demand of less than 1 kg.

4.5.3 Serum Albumin

Serum albumin constitutes the most abundant protein present in serum, representing approximately 60% of total plasma protein. It is also one of the smallest known plasma proteins with a molecular weight of approximately 69,000 Da. A major function of albumin is to provide most of the natural osmotic pressure of plasma. It also has a major transportation function and is especially important in transporting substances in aqueous media which are sparingly soluble. Human serum albumin preparations are administered to patients suffering from some forms of kidney or liver disease, and are also used as a plasma volume expander for patients suffering from shock as a result of a decrease in the volume of blood. Such decreases are often associated with surgery or occur subsequent to serious injury.

Albumin is normally purified from serum plasma or from placentas obtained from healthy donors. While commercially available human serum albumin is currently obtained from the above sources, the protein has also been produced as a heterologous product in a number of recombinant systems [33]. Systems employed have included bacteria such as *Bacillus subtilis* and *E. coli* [34] and yeasts such as *Pichia* etc. [35]. Recombinant plasma protein may have great therapeutic potential in replacing the human-derived sources.

4.6 Growth Factor(s)

Growth factors can be broadly defined as multifunctional, locally acting intercellular signaling polypeptides.

Growth factors are predominantly short-range, locally acting, intercellular signaling molecules. Growth factors share a number of common biological properties apart from their predominantly local mode of action. They often exert their biological actions at a very low (typically 10^{-9}–10^{-11} M) concentration. This is because their action is mediated by their association with specific, high affinity receptors expressed by the target cell type.

There are currently about 80 or more known genes in both vertebrates and invertebrates, encoding proteins that can be considered on the basis of their biological function to be growth factors. It is also very likely that a few more growth factor genes remain to be discovered. Growth factors, therefore, represent a large set of polypeptides. These proteins, as a whole, do not share any common structural features but many can be grouped on the basis of amino acid sequence and tertiary structure, into multi-gene families. Some of the growth factors that have generated considerable interest within the biopharmaceutical industry are listed in the Table 5.

4.6.1 *Epidermal Growth Factor Family (EGF)*

Epidermal growth factor (EGF) was one of the first growth factors to be discovered, as a result of its striking effects on the maturation of various epithelia [36].

Table 5. Some growth factors and their intended applications [117]

Growth factor	Intended applications
Epidermal growth factor	Wound healing, skin ulcers, corneal/cataract surgery [36]
Fibroblast growth factor	Chronic soft tissue ulcers, leg and foot ulcers, venous stasis [37]
Insulin-like growth factor 1	Type II diabetes [38, 39]
Platelet-derived growth factor	Skin ulcers, diabetics ulcers [43, 44]

EGF is a compact, highly folded molecule composed of 53 amino acids containing six cysteine sulfhydryl groups that form intramolecular disulfide bonds. EGF is perhaps the most extensively characterized of all growth factors. Human EGF was first isolated from urine. EGF has been proved to exert a powerful mitogenic influence over a wide variety of cell types, including epithelial and endothelial cells and fibroblasts. The skin represents the major target tissue of EGF, where it plays an important role in growth and development of the epidermal layer. In-addition to its presence in urine, human EGF has also been detected in serum, in the duodenum and in the salivary glands.

Human EGF has been produced as heterologous proteins in recombinant bacterial systems. The cDNA may be expressed in *E. coli* systems [40, 41].

EGF has received considerable attention as a potential therapeutic agent. It possesses potent gastric antisecretory activity and also stimulates the proliferation of epithelial cells from a variety of tissues. Consequently, it can be used for accelerating the healing process of skin and cornea. Its indication in the healing of ulcers of various origin (particularly diabetic) appears to have a very promising future.

4.6.2 *Fibroblast Growth Factor (FGF)*

Fibroblast growth factor (FGF) [37] was discovered from its ability to induce the proliferation of fibroblast cells in vitro. FGF was first isolated from bovine brain and pituitary glands. This growth factor was found to exhibit a marked mitogenic effect on fibroblast cell lines in culture. It has since been demonstrated that FGF induced the growth and division of numerous cells and vascular endothelial cells. Its ability to induce blood vessel growth in vivo heightens expectations of its potential therapeutic applications. Since this factor stimulates fibroblasts and capillary angiogenesis, it can be used for all situations of skin healing following burns, wounds and ulcers. Calbio, Synergen, Merck and Chiron are pursuing the production of rFGF by cloning in *E. coli* or in mammalian cells [42].

4.6.3 *Platelet-Derived Growth Factor (PDGF)*

Platelet-derived growth factor (PDGF) [43] was discovered as the major component of serum required for the multiplication of a variety of cell types in culture. PDGF was first isolated from platelets [44]. Subsequently, however, it has been established that this growth factor is synthesized and secreted by a number of other cell types. This growth factor has a molecular weight of 30,000 Da, and consists of two polypeptide chains termed A and B, which are covalently linked by an interchain disulfide bond. Three isoforms of the growth factor have been isolated – the homodimers AA and BB in addition to the heterodimeric form AB. The A chain consists of 124 amino acids, whereas the B chain contains 140 amino acids.

PDGF is a glycoprotein that stimulates the division of muscle cells and the collagenase activity of fibroblasts. The potential therapeutic indications concern healing. The gene has been expressed in yeast (Zymogenetics) [45] and in *E. coli* (Amgen and Creative Biomolecules) [46].

4.6.4
Transforming Growth Factor (TGF-α and -β)

Various transforming growth factors (TGFs) [47] which may be of potential therapeutic significance have also been identified. TGF-α exhibits similar biological activities to EGF. In addition to its presence in a variety of tumor types, TGF-α is synthesized by several normal cell types such as activated macrophages. A distinct group of transforming growth factors termed TGF-βs has also been characterized. These factors are generally dimeric proteins consisting of two identical polypeptide chains. The two TGF-βs that have been characterized in greatest detail are TGF-β1 and TGF-β2. These two factors exhibit a significant degree of amino acid sequence homology.

The TGFs exhibit a variety of biological effects, most of which relate to the modulation of growth of a number of cell types. TGF-β1 exhibits numerous immunoregulatory effects, including inhibition of B and T cell proliferation. TGF-βs has been found in the synovial fluid of arthritic joints. Indeed, recent experiments suggest that TGF-βs may arrest the progression of arthritis. These growth factors are also the subject of intensive academic and industrial research [47].

4.6.5
Insulin-Like Growth Factor (IGF-I)

IGF is a factor inducing general stimulation of bodily growth [38, 39] and which may, in turn, promote the regeneration of bone and cartilaginous tissues. IGF has been expressed in *E. coli* by Amgen [48], in yeast by Chiron [49] and in mouse fibroblasts by Merck [50, 51].

Many of the growth factors mentioned above are currently undergoing clinical evaluation in order to assess their potential as therapeutic agents and some of them are already in clinical use. Most clinical trials concentrate on their effectiveness in accelerating the wound healing process and treating ulcers [74]. Natural wound healing depends upon the release of growth factors that initiate the tissue repair process at the site of damage. Thrombin, factor IIa of the blood coagulation pathway, stimulates the release of granules from blood platelets. These granules contain a variety of growth factors including FGF, PDGF and TGF. Such growth factors exert mitogenic effects on certain cell types essential for tissue repair. They also exhibit chemotactic effects and, in this way, attract a number of cell types to the site of damage.

Clinical studies have confirmed that application of exogenous growth factors rapidly accelerates the wound healing process [52]. The encouraging results obtained seem to be reproducible in clinical trials involving humans. Furthermore, it is becoming apparent that transport of such growth factor types to the site of damage may thus yield the most effective response.

4.7 Cytokines

4.7.1 *Interferons (Ifn)*

Interferons are multimeric glycoproteins produced by cells in response to viral aggression [53]. The chemical structure of interferons may vary in amino acids with molecular weight ranging from 12 kDa to 20 kDa. The biological properties of interferons may also vary. In addition to their common antiviral properties, various interferons act on phagocytosis, cytotoxicity of T-lymphocytes, antibody and lymphokine production by lymphocytes, expression of cellular antigens, antineoplastic properties of macrophages, and regulation of cellular synthesis of macromolecules cell growth.

There are several types of interferons, however, the major ones are of three types: α-interferon of leukocytic and lymphoblastoid origin (166–172 amino acids, 16–27 kDa); [54] β-interferon of fibroblastic origin ($\cong$20 kDa), and γ-interferon (143 amino acids, 50 kDa as dimer) [55]. The biological activities of interferons are measured in terms of their inhibition of viral culture of the vesicular stomatiitis virus in various human cell lines.

Ifns are relatively small proteins of the size ranging between 146–166 amino acids. They are potent anti-viral materials and have been shown to be immunomodulators and inhibitors of cellular proliferation.

4.7.1.1 *Antiviral Properties*

Ifn is naturally produced in response to viral infections. The supplementation of endogenous Ifn, by means of exogenous interferon, is essentially justified by the possibility of obtaining much higher concentrations.

4.7.1.2 *Antineoplastic Properties*

The inhibitory effect of interferons on the growth of normal or neoplastic human cells [56] justifies its clinical trials for cancer treatment. The first successes were achieved with α- and β-Ifn in cases of osteosarcoma, Hodgkin's disease, multiple myeloma, certain forms of leukemia and malignant melanoma [57].

4.7.1.3 *Other Indications*

Clinical studies have been conducted with interferons in diseases affecting the immune system [56], rheumatoid arthritis, multiple sclerosis and in the treatment of AIDS with Kaposi's sarcoma, which has a simultaneous viral, cancerous and immunological etiology. Combinations with tumor necrosis factor and with interleukins (Il-2) have also been proposed.

Table 6. Various interferons in clinical application [116]

Product name	Company	Intended applications
Alferon N (IFN-α n3)	Interferon sciences	Genital warts
Intron A (IFN-α 2b)	Schering Plough	Hairy cell leukemia Genital warts, AIDS related Kaposi's sarcoma, non-A, non-B hepatitis; various cancer types, chronic hepatitis B, delta hepatitis, acute hepatitis B, chronic myelogenous leukemia
Roferon-A (IFN-α2a)	Hoffmann-La Roche	Hairy cell leukemia AIDS-related Kaposi's sarcoma
Actimmune (IFN-λ 1b)	Genentech	Small-cell lung cancer, atopic dermatitis, trauma-related infections, renal cell carcinoma, asthma and allergies, management of chronic granulomatous disease
Alferon LDO (IFN-α n3)	Interferon Sciences	AIDS and ARC
Betaseron (IFN-β)	Berlex Laboratories	Multiple sclerosis, cancer
Immuneron (IFN-λ)	Biogen	Rheumatoid arthritis; venereal warts
Interferon gamma (IFN-λ)	Amgen	Cancer; infectious diseases
R-Frone (IFN-β)	Serone Lab	Various malignancies

Interferon preparations currently produced and their intended clinical applications are summarized in Table 6.

In the late 1970s, bulk quantities of interferons were first produced by mammalian cell culture. A specific strain of a human lymphoblastoid cell line termed the Namalwa cell line was most often employed in this regard. Production has been undertaken in large culture vessels, many of which have a capacity in excess of 80,000 liters. The purified product has been shown to consist of at least eight distinct molecular species of Ifn-α. Interferons are also produced in large quantities by recombinant DNA methodologies. Several recombinant human Ifn-α, in addition to Ifn-β and Ifn-γ have been produced in *E. coli* [58], yeast [59] and in some eukaryotic cells such as cultured monkey cells and Chinese hamster ovary cells [60]. Gene cloning and bacterial expression have been indispensable for characterizing the molecules and for producing sufficient quantities of material to allow clinical use.

The essential advantage of recombinant Ifn concerns the high yields obtained leading to the reduction of the cost of Ifn and the availability of required quantities for therapeutic trials. Recently, various interferons have been approved for different therapeutic indications, some of which have been listed in Table 7.

Table 7. Various interferons approved and their therapeutic applications [116]

No.	Company	Trade-name	Generic name	Therapeutic indications	FDA approval
1	Roche/ Genentech	Roferon-A	Interferon alpha-2a	Hairy cell leukemia, AIDS-related Kaposi's sarcoma	June 1986 November 1988
2	Schering-Plough/ Biogen	Intron A	Interferon alpha-2b	Hairy cell leukemia Genital warts AIDS-related Kaposi's sarcoma Hepatitis C Hepatitis B Non-Hodgkin's lymphoma (in conjugation with chemo-therapy)	June 1986 June 1988 November 1988 February 1991 July 1992 November 1997
3	Interferon Sciences	Alferon N	Interferon beta-1b	Genital warts	October 1989
4	Genentech	Actimmune	Gamma-interferon	Chronic granulomatous disease	December 1990
5	Berlex/ Chiron	Betaseron	Interferon beta	Relapsing, remitting multiple sclerosis	July 1993
6	Biogen	Avonex	Interferon beta-1a	Relapsing, remitting multiple sclerosis	May 1996
7	Genetics Institute	Neumega	Il-11	Treatment of thrombo-cytopenia in chemotherapy	November 1997
8	Amgen	Infergen	Concensus alpha Interferon	Treatment of hepatitis C	October 1997

4.7.2 *Interleukin (Il)*

At least ten different interleukins have thus far been characterized and many of these preparations are currently the subject of clinical investigation [27]. Some interleukins such as interleukin-1 and interleukin-2 are already employed in the treatment of a number of medical conditions [61].

4.7.2.1 Interleukin-1

There are two types of Il-1, namely α and β which are essentially produced by macrophages activated by various mitogens, endotoxins, immune complexes and other lymphokines.

The cloning and expression of the Il-1 cDNA in *E. coli* has facilitated the large-scale production of biologically active Il-1 [62]. This, in turn, has rendered practical the widespread medical application of this cytokine. Large-scale clinical trials have been conducted to study the effectiveness of Il-1 in treating a variety of cancers in addition to promoting wound healing.

4.7.2.2
Interleukin-2

Il-2, which is secreted by helper T-lymphocytes, is probably the most important and is certainly the most extensively studied. This molecule plays a pivotal role in normal immunological functioning, and is likely to become one of the most widely used biopharmaceutical products. Il-2 is a single-chain glycosylated polypeptide of molecular weight 15,500 Da [63].

Amongst a range of activities, Il-2 has the ability to act as a growth factor for natural killer (NK) cells. The potential therapeutic applications include immunotherapy of various cancers, e.g., Kaposi's sarcoma, osteosarcoma, chronic leukemia, all types of solid cancers and the restoration of immunity in AIDS patients.

4.7.2.3
Other Interleukins

A number of other interleukins is known. Il-3 is a glycoprotein consisting of 133 amino acids. It is produced predominantly by activated T lymphocytes and thus may be termed a lymphokine. Biologically active Il-3 has been produced in a number of recombinant systems, including recombinant bacterial, yeast and mammalian cells. Il-3 is produced as a heterologous protein product in the bacterium *Bacillus licheniformis*. Il-3 exhibits significant therapeutic potential in the treatment of a number of medical conditions, including bone marrow failure or status after bone marrow transplants [27].

Interleukin 4 (Il-4) is synthesized by T helper lymphocytes. This glycoprotein, of molecular weight 20,000 Da, is also termed B cell growth factor. Il-4 exhibits a broad range of biological activities, including stimulating the proliferation of a number of cell types such as B and T lymphocytes, thymocytes and mast cells. Il-4 promotes enhanced secretion of IgG and IgE from stimulated B cells and also enhances antigen presentation by monocytes. A soluble form of this receptor has been produced by recombinant DNA methods [27].

Interleukin-6 (Il-6), Il-10 and Il-12 are other cytokines that may well prove to be therapeutically beneficial [27]. Some of the interleukins that have recently been approved are listed in Table 8.

4.7.3
Tumor Necrosis Factor

Tumor necrosis factor (TNF) [27] is an important member of the cytokine family of regulatory proteins. Two forms of TNF are now recognized, TNF-α and TNF-β. Although both proteins bind to the same receptors and elicit broadly similar biological responses, they are distinct molecules and share less than 30% homology. The original protein termed tumor necrosis factor, although

Table 8. Interleukins approved for various indications [116]

No.	Company	Trade-name	Generic name	Therapeutic indications	FDA approval
1	Chiron	Proleukin	Interleukin-2	Renal cell carcinoma Metastatic melanoma	May 1992 January 1998
2	Ligand Pharmaceuticals	Ontak	Interleukin-2 fusion protein	2nd line treatment of recurring/advanced cutaneous T-cell lymphoma	February 1999

still referred to as TNF, is properly termed TNF-α; it is also known as cachectin. TNF-β is also referred to as lymphotoxin. Tumor necrosis factor α is synthesized by a wide variety of cell types, most notably activated macrophages, monocytes, certain T lymphocytes, and NK cells, in addition to brain and liver cells. Although lipopolysaccharide (LPS) is the most potent inducer of TNF-α synthesis, various other agents such as some viruses, fungi and parasites also stimulate the synthesis and release of the cytokine. Furthermore, TNF-α may act in an autocrine manner, stimulating its own production.

Native human TNF-α is a homotrimer, consisting of three identical polypeptide subunits tightly associated about a three-fold axis of symmetry. The molecule has a molecular weight of 17,300 Da (consists of 157 amino acids), and contains one intrachain disulfide linkage

The TNF-α gene has been cloned and inserted in a variety of recombinant expression systems, both bacterial and eukaryotic. The resultant availability of large quantities of purified, biologically active TNF has facilitated clinical evaluation in a number of diseases, most notably cancer [64].

TNF seems to play a central role in several other physiological processes such as inflammation, immune response, and septic shock [46]. It has been demonstrated clinically that administration of exogenous TNF-α can induce symptoms identical to those seen in patients suffering from septic shock. Furthermore, it has been shown that pretreatment with anti-TNF-α antibodies exerts a protective effect on animals subsequently challenged with potentially lethal doses of lipopolysaccharide.

Prolonged production of inappropriately elevated levels of TNF-α has also been implicated in the development of cachexia, a syndrome often associated with chronic parasitic or other infections, and with cancer.

TNF-α modulates the immune response by a number of means [65]. It stimulates the production of a variety of other cytokines, in particular interleukins, Il-1 and Il-6, and various colony-stimulating factors. It also promotes synthesis of platelet-derived growth factor. TNF-α exhibits many biological activities similar to those of Il-1. It also stimulates the proliferation of a variety of T lymphocytes and plays a role in antibody production of B lymphocytes. TNF-α enhances the activity of various phagocytic cells such as macrophages. It has also been implicated in the development of certain autoimmune disorders, including that of rheumatoid arthritis [66]. In addition, it has been shown that administration of anti-TNF-α antibodies may decrease or prevent rejection of grafted or transplanted tissues.

TNF-β, previously known as lymphotoxin, was actually discovered prior to TNF-α, but at that time it was not recognized as a TNF. TNF-β is produced by T lymphocytes and is a glycoprotein consisting of 171 amino acids. Although TNF-α and TNF-β exhibit only limited amino acid sequence homology, they both bind to the same receptors and exhibit broadly similar biological activities. The TNF-β cDNA has been cloned and expressed in recombinant *E. coil* [67]. Its resultant availability in large quantities has allowed its biochemical characteristics and clinical potential to be more fully accessed.

4.8 Tissue Plasminogen Activators (tPA)

Plasminogen activators constitute a group of enzymes with a similar structure of the serine protease type. Their essential physiological role is in transforming a circulating proenzyme, plasminogen, into plasmin, which is itself a proteolytic enzyme capable of lysing fibrin clots [68]. The normal mechanism possesses a remarkable specificity of action in that fibrinolysis only occurs locally. However, the protease properties of plasmin are such that when it is generated not locally but systemically, they can induce marked breakdown of other serum proteins such as fibrinogen, prothrombin and factors V and VIII and this results in the development of hemorrhagic side effects [69].

Human serine-protease type plasminogen activators are classified into two groups: urokinase type plasminogen activators (uPA) and tissue plasminogen activators (tPA). The fundamental difference between uPA and tPA is that tPA retains its specificity for local fibrinolysis after cleavage, while the activity of uPA becomes systemic, although, the molecular structures of both are similar.

Four plasminogen activators can be used therapeutically, i.e., tissue plasminogen activators (tPA), streptokinase, prourokinase and urokinase, in the treatment of thrombolytic disorders. They have been administered in a variety of situations including treatment of myocardial infarction, embolisms, strokes, and deep vein thrombosis (which often develops subsequent to major surgery).

The oldest natural product in this category is streptokinase produced from the cultures of beta-hemolytic streptococci. When injected intravenously or in-

Table 9. Tissue plasminogen activators approved for various indications [116]

No.	Company	Trade-name	Generic name	Therapeutic indications	FDA approval
1	Genentech	Activase	Tissue-plasminogen activator	Acute myocardial infarction	November 1987
				Acute massive pulmonary embolism	June 1990
				Acute ischemic stroke	June 1996
2	Genentech	TNKnase	Tenectoplase	Treatment of heart attack	June 2000

traarterially, streptokinase has the disadvantage of acting systemically and of being immunogenic. It is used at doses of 10^6 to 3×10^6 units over 24 hours.

Regarding recombinant products, the cloning of *E. coli* [70, 71] produces non-glycosylated but nevertheless active products, while cloning in eukaryotic cells (myeloma, Chinese hamster ovary cells, etc.) [72, 73] produces an activator analogous to the natural product. Although the efficacy of this product is beyond doubt, it is quite expensive, especially when compared to streptokinase [74]. Some of the tPA activators that have been approved recently are listed in Table 9.

5 Vaccines

The history of vaccine dates back to the sixth century. The first written record of immunization, "*The correct treatment of Smallpox*", traced the art to a Buddhist nun practicing during the eleventh century. The modern understanding of vaccine in fact started with Dr. Edward Jenner, an English physician (May 1796) [75], who injected a few drops of a fluid from the skin sore of a woman infected with cowpox, into the arm of a healthy young boy James Phipps, who never had cowpox or smallpox. Six weeks later, Jenner injected the boy with a fluid from smallpox pore, but the boy remained free from the deadly smallpox. This was the foundation of immunization. One can use a relatively harmless foreign substance to evoke an immune response that would protect one from disease-causing microbes. Nearly a century later, Louis Pasteur in France found that one could be protected against infections by being injected with a weak pathogen of similar origin, which causes a relatively harmless infection. Pasteur also introduced a direct application of microbiology, when he introduced the concept of vaccine in 1877. He investigated the cause of sheep death in France, which was due to anthrax. He prepared an anthrax vaccine from attenuated *Anthrax bacillus* (1881), and it was found to be quite effective in protecting sheep. In 1885, Pasteur treated Joseph Meister, bitten by a rabid dog, with attenuated rabies virus. Later, several vaccines were developed for typhoid, cholera, plague, polio, measles, mumps, rubella, hepatitis, pneumonia and influenza [75a, b]. Some of these diseases were the greatest killers of human society in the past. Some of the veterinary vaccines also play a crucial role in human healthcare.

Some diseases such as rabies, anthrax, encephalitis etc., are transmissible from animal species to humans and in such cases animals need to be vaccinated to prevent the transmission of these diseases.

Although no vaccine can be 100% safe and protective, it is one of the most economical medical interventions available to mankind. Secondly, vaccine is a preventive measure for several diseases. Some of the criteria for a vaccine are as follows:

a) The vaccine must be safe. Since immunization is carried out generally in healthy individuals and is a preventive measure, the safety of vaccine is of the greatest concern. However, minor side effects such as redness or soreness at the vaccination site are tolerable.

b) The vaccine must be immunogenic. Vaccine usually contains antigens which can elicit immune responses by the host, create antibodies that can recognize the disease-causing microbes, and launch a counterattack before the illness can occur.
c) The vaccine must be stable during its shelf-life,

The concept of vaccine is based on the defense mechanism that the body uses against foreign invaders – a branch of science well developed as immunology. It is beyond the scope of this article to elaborate more on this, however, it suffices to state that there are different kinds of B and T cells which participate in the immunogenic response to create antibodies. The latter can recognize the disease-causing organisms and destroy them.

5.1 Different Types of Vaccines

5.1.1 *Inactivated Vaccines* [76]

Inactivated vaccines are produced by killing the disease-causing microorganisms with heat, light or chemicals. They are quite safe but less effective and may need repeated dosing. Examples of this class are vaccines for cholera, plague and hepatitis A.

5.1.2 *Live-Attenuated Vaccines* [77]

These vaccines are produced by growing the disease-causing organisms under special laboratory conditions that cause it to lose virulence. Although live vaccines require special handling and storage in order to maintain their potency, they produce both antibody-mediated and cell-mediated immunity and generally are very potent and need only one booster dose or at the most an additional dose. Examples are polio vaccine, flu, yellow fever, measles, rubella, and mumps.

5.1.3 *Hepatitis B Vaccines* [78]

Recent advances in vaccine development center around recombinant DNA technologies. Such technologies allow the production of specific protein antigens identical to the antigen sourced from the wild-type pathogen, which are then purified and used as a vaccine. As compared to the conventional vaccine, the recombinant vaccine is extremely safe and consists of a single antigenic constituent of the pathogen against which immunity is desired. It is thus impossible accidentally or otherwise to induce the disease state with such vaccine preparations. Administration of such a defined and structurally less complex vaccine is also less likely to induce unexpected adverse clinical reactions, besides ensuring a continuous and convenient supply of material from the safe source.

The first vaccine product produced by recombinant DNA technology approved for human use was that for hepatitis B. Individuals infected with the hepatitis B virus (HBV) have in their circulation particles consisting of multimers of the viral coat protein, referred to as surface antigen (HBsAg). These highly immunogenic particles have been isolated from carriers of the hepatitis B virus and have been utilized as a vaccine because the virus cannot be grown in culture to produce the protein. Even though the vaccine is effective there are serious limitations with a vaccine obtained from the blood of HBV carriers. There is a risk of the final preparations being contaminated with active hepatitis virions or other pathogenic viruses such as HIV. Furthermore, production of the vaccine depends upon a constant supply of plasma from infected individuals. Therefore, extensive efforts have been focused on producing HBsAg by recombinant means.

HBsAg is a 25 kDa protein and its gene has been cloned and expressed in a number of recombinant systems such as *E. coli*, *S. cerevisiae*, monkey kidney cells, mouse cells, human hepatoma cell line, monkey cell line etc. [79]. The protein obtained from transformed yeast was the first recombinant vaccine made available for general medical use. It may be purified from yeast extract by immunoaffinity chromatography, utilizing anti-HBsAg antibodies. It was approved for sale by the regulatory authorities in 1986, and is cheaper than the conventional vaccine preparations. Other HBsAg have also been successfully produced in recombinant systems. A number of companies are already marketing this vaccine.

5.1.4 Toxoids

These are inactivated toxins produced by microbes. Some of the diseases are caused by the harmful toxins produced by the microbes, rather than the microbes themselves. For example, the bacterium that causes tetanus is due to the tetanus toxin. The latter is inactivated by treatment with formalin, and is used as a toxoid. Toxoids are used to immunize people against tetanus and diphtheria [80].

5.1.5 *Conjugate Vaccines*

In a conjugate vaccine, proteins or toxins from a second type of organism, one that an immature immune system can recognize, are linked to the outer coats of the disease-causing bacteria. This enables a young immune system to respond and defend against the disease agent.

A conjugate vaccine of this kind is used for the protection against bacterial meningitis caused by H. influenzae type b (Hib) [81].

5.1.6 *Peptide Vaccines* [81]

An alternative approach to the production of subunit vaccines entails their direct chemical synthesis. Peptides identical in sequence to short stretches of pathogen-derived polypeptide antigens can be easily and economically synthesized. The feasibility of this approach was first verified in the 1960s when a hexapeptide purified from the enzymatic digest of tobacco mosaic virus was found to confer limited immunological protection against subsequent administration of the intact virus [the hexapeptide hapten was initially coupled to bovine serum albumin (BSA), used as a carrier to ensure an immunological response].

Similar synthetic vaccines have also been constructed which confer immunological protection against bacterial toxins, including diphtheria and cholera toxins. While coupling to a carrier is generally required to elicit an immunological response, some carriers are inappropriate due to their ability to elicit a hypersensitive reaction, particularly when repeat injections are undertaken. Such difficulties can be avoided by the judicious choice of carrier. Often a carrier, normally used for vaccination itself, is used. For example, tetanus toxoid has been used as a carrier for peptides, derived from *Influenza haemaglutanin* and *Plasmodium falciparum.*

5.1.7
Vaccine Vector

An alternative approach to the development of novel vaccine products entails the use of live vaccine vectors [82a]. The strategy followed involves incorporation of a gene/cDNA coding for a pathogen-derived antigen into a non-pathogenic species [82b]. If the resultant recombinant vector expresses the gene product on its surface, it may be used to immunize against the pathogen of interest.

Most vaccine vectors developed to date are viral based with poxviruses as well as picornaviruses and adenoviruses being used most often. In general, such recombinant viral vectors elicit both strong humoral and cell-mediated immunity. The immunological response (particularly the cell-mediated response) to subunit vaccines is often less pronounced.

Poxviruses and, more specifically, the vaccinia virus remain the most thoroughly characterized vector systems developed [83]. These are large, enveloped double-stranded DNA viruses. They are the only DNA-containing viruses that replicate in the cytoplasm of infected cells. The most studied members of this family are variola and vaccinia. The former represents the causative agent of smallpox, while the latter, being antigenically related to variola but non-pathogenic, was used to immunize against smallpox. Vaccinia-based vaccination programs led to the global eradication of smallpox, finally achieved by the early 1980s.

A number of factors render vaccinia virus a particularly attractive vector system. These include:

- Capacity to successfully assimilate large quantities of DNA in its genome.
- Prior history of widespread and successful use as a vaccination agent.
- Ability to elicit long-lasting immunity.
- Ease of production and low production costs.
- Stability of freeze-dried finished vaccine products.

The ability of vaccinia (and other poxviruses) to accommodate large sequences of heterologous DNA into its genome without adversely affecting its ability to replicate remains one of its most attractive features. Integration of foreign genes must occur in regions of the viral genome not essential for viral replication.

Early animal experiments have underlined the potential of vaccinia-based vector vaccines. Vaccinia virus systems housing genes from HIV have clearly been found to elicit both humoral and cell-mediated immune responses in monkeys [84]. Similar responses in other animals have been reported when surface polypeptides from a variety of additional pathogens have been expressed in recombinant vaccinia systems (Table 10). Human clinical trials are now in progress.

Adenoviruses also display potential as vaccine vectors [85]. These double-stranded DNA viruses display a genome consisting of a region of 36 kb, encoding approximately 50 viral genes. Several antigenically distinct human adenovirus serotypes have been characterized.

They have been proven to be very safe and effective in adenovirus vaccines. Unlike vaccinia, few sites exist in the adenoviral genome into which foreign DNA can be integrated without compromising viral function.

Recombinant adenoviruses containing the HBsAg gene, the HIV P160 gene, the respiratory syncytial virus F gene, as well as the herpes simplex virus glycoprotein B gene, have all been generated using this approach. Many have been tested in animal models and have been found to elicit humoral and cell-mediated immunity against the pathogen of interest [86].

Table 10. Recombinant vaccine vector against various pathogen protected species [117]

Pathogen	Protected species
Bovine leukemia virus	Sheep
Bovine papilloma virus	Rats
Epstein-Barr virus	Cotton-top tamarins
Equine leukemia virus	Hamsters
Friend leukemia virus	Mice
Hepatitis B virus	Chimpanzees
Herpes simplex virus	Mice
Human papilloma virus	Mice, rats
Human parainfluenza virus	Monkeys
Leishmania	Mice
Measles	Mice, rats, dogs
Polyomavirus	Rats
Pseudorabies virus	Mice, pigs
Rabies	Mice, foxes, raccoons, dogs
Respiratory syncytial virus	Rats, mice, monkeys
Yellow fever	Mice

Some pathogens against which protective immunity was elicited by recombinant vaccinia vector systems. The virus invariably expressed a gene coding for a pathogen-derived surface polypeptide. The animal species in which the experiments were carried out are also listed.

Picornaviruses are also being evaluated as potential vaccine vectors [87]. Unlike the large pox and adenoviruses discussed above, these are small viruses, incapable of carrying a gene coding for a complete foreign protein. However, such viral particles could easily house nucleotide sequences coding for short peptides representative of specific antigenic sites/epitopes present in pathogen-derived polypeptides.

5.1.8
Development of an AIDS Vaccine

AIDS (acquired immune deficiency syndrome) commonly known as HIV virus has a viral surface protein, gp120, which is capable of binding to a specific site on the CD4 molecule, found on the surface of susceptible cells [88]. Some CD4-negative cells may (rarely) also become infected, indicating the existence of an entry mechanism independent of CD4.

Infection of $CD4^+$ cells commences via interaction between gp120 and the CD4 glycoprotein, which effectively acts as the viral receptor. Entry of the virus into the cell, which appears to require some additional cellular components, occurs via endocytosis and/or fusion of the viral and cellular membranes. The gp41 transmembrane protein plays an essential role in this process.

Once released into the cell, the viral RNA is transcribed (by the associated viral reverse transcriptase) into double-stranded DNA. The retroviral DNA can then integrate into the host cell genome (or, in some instances, remain unintegrated). In resting cells, transcription of viral genes does not usually occur to any significant extent. However, commencement of active cellular growth/differentiation usually also triggers expression of proviral genes and, hence, synthesis of new viral particles. Aggressive expression of viral genes usually leads to cell death. Some cells, however (particularly macrophages), often permit chronic low-level viral synthesis and release without cell death.

Entry of the virus into the human subject is generally accompanied by initial viral replication, lasting a few weeks. High-level viremia (presence of viral particles in the blood) is noted and p24 antigen can be detected in the blood. Clinical symptoms associated with the initial infection include an influenza-like illness, joint pains, and general enlargement of the lymph nodes. This primary viremia is brought under control within three to four weeks. This appears to be mediated largely by HIV-specific cytotoxic T lymphocytes, indicating the likely importance of cell-mediated immunity in bringing the initial infection under control. While HIV-specific antibodies are also produced at this stage, effective neutralizing antibodies are detected mainly after this initial stage of infection [89].

After this initial phase of infection subsides, the free viral load in the blood declines, often to almost undetectable levels. This latent phase may last for anything up to 10 years or more. During this phase, however, there seems to be continuous synthesis and destruction of viral particles. This is accompanied by a high turnover rate of ($CD4^+$) T-helper lymphocytes. The levels of these T lymphocytes decline with time, as do antibody levels specific for viral proteins. The circulating viral load often increases as a result and the depletion of T-helper cells compromises the general immune function. As the immune system fails, classical symptoms of the AIDS-related complex (ARC) and, finally, full-blown AIDS begin to develop.

5.1.8.1
Difficulties Associated with AIDS Vaccine Development

A number of attributes of HIV and its mode of infection render the development of an effective vaccine difficult. Some of the factors which need consideration are:

- HIV has extensive genetic variation, even within a single individual. Such genetic variation is particularly prominent in the viral *env* gene whose product, gp160, is subsequently proteolytically processed yielding gp 120 and gp41.
- HIV infects and destroys T-helper lymphocytes, an essential component of the immune system itself.
- Although infected individuals display a wide range of antiviral immunological responses, they fail to destroy the virus. Thus, a greater understanding of most effective elements of immunity in combating HIV infection is required.
- After the critical virulence subsides, large numbers of cells harbor unexpressed proviral DNA. The immune system has no way of identifying such cells. Therefore, an effective vaccine must induce the immune system to bring the viral infection under control before cellular infection occurs or destroy cells once they begin to produce viral particles and stop the viral particles from being released.
- The infection may often be spread, not via transmission of free viral particles, but via direct transmission of infected cells harboring the proviral DNA. There are several candidate vaccines in clinical trials that are being developed. Some of the approaches include inactivated viral particles; or r-gp-120 subunit vaccine or r-gp-160 subunit vaccine or r-p24-subunit vaccine or live vaccine based on viral vectors. Although much progress has been made, the complexity of the disease has made the development of a truly effective vaccine difficult so far.

5.1.9
Cancer Vaccines

The identification of tumor-associated antigens could pave the way for the development of a range of cancer vaccines. Several tumor-associated antigens have already been characterized. Theoretically, administration of tumor-associated antigens may effectively immunize an individual against any cancer type characterized by expression of the tumor-associated antigen in question [82]. Co-administration of a strong adjuvant (see next section) would be advantageous, as it would stimulate an enhanced immune response [90]. This is important, as many tumor-associated antigens appear to be weak immunogens. Administration of subunit-based tumor-associated antigen vaccines would primarily stimulate a humoral immune response. The use of viral vectors may ultimately prove more effective, as a T cell response appears to be central to the immunological destruction of cancer cells.

This approach has been adopted in experimental studies involving malignant melanoma. These transformed cells express significantly elevated levels of a surface glycoprotein, p97. p97 is also expressed, but at far lower levels, on the surface of many normal cell types. Initial animal studies have indicated that administration of a recombinant vaccinia vector expressing p97 has a protective effect against challenge with melanoma cells. However, protracted safety studies would be required in this, or similar, instances to prove that such vaccines would not, for example, include an autoimmune response if the antigen were not wholly tumor specific. The development of truly effective cancer vaccines probably requires a more comprehensive understanding of the transformed phenotype and how these cells normally evade immune surveillance in the first place. Limited clinical studies of these vaccines have already begun.

5.1.10
DNA-Based Vaccines [91]

Among the most recent and innovative developments in vaccine technology is the construction of a putative nucleic acid-based vaccine (a form of gene therapy). This avenue of vaccine research was first opened in 1990, when it was shown that naked plasmid DNA was expressed in mice muscle cells subsequent to its intramuscular (i.m) injection. The concerned plasmid DNA housed the β-galactosidase gene as a reporter. Subsequent expression of β-galactosidase activity could persist for anything from a few months to the remainder of the animal's life. The transfection rate recorded was low (1–2% of muscle fibers assimilated the DNA), and the DNA was not integrated into the host cell's chromosomes.

Up to this point, it was assumed by most that naked DNA injected into animals would not be spontaneously be taken up and expressed in host cells.

Scientists have also since demonstrated that DNA (coated on microscopic gold beads) propelled into the epidermis of test animals with a "gene gun" is expressed in the animal's skin cells. Furthermore, the introduction in this fashion of DNA coding for human influenza viral antigens resulted in effective immunization of the animal against influenza. Similar results, using other pathogen models, have also now been generated. It is assumed that the expressed antigen is secreted by the cell and in this way, is exposed to immune surveillance [92].

While the efficacy of the nucleic acid-based vaccine approach has been clearly demonstrated in animals at least, a number of safety questions need to be addressed before widespread human trials can be contemplated. Chief among these is the issue of whether any naked DNA could become randomly integrated into the host's chromosomal DNA. Such an event, although unlikely, could disrupt an essential gene or perhaps induce cancer.

Other issues that need to be investigated include the possible impact on the immune response of prolonged expression of the foreign DNA. It must be demonstrated that this does not induce a hyperimmune/autoimmune response. On the other hand, it must be shown that prolonged but perhaps low-level expression would not be likely to induce immunological tolerance to the antigen.

However, while such issues need to be addressed, DNA-based vaccines appear to have a better than ever chance of becoming a clinical reality.

5.1.11
Adjuvant Vaccine [93]

Administration of many vaccines on their own stimulates a poor host immunological response. This is particularly true for the more recently developed subunit vaccines. An adjuvant is defined as any material that enhances the cellular and/or humoral immune response to an antigen. Adjuvants thus generally elicit an earlier, more potent and longer-lasting immunological reaction against a co-administered antigen. In addition, the use of adjuvants can often facilitate the administration of reduced quantities of antigen to achieve an adequate immunological response. This implies economic savings, as vaccines (particularly subunit and vector vaccines) are far more expensive to produce than the adjuvant.

A number of different adjuvant preparations has been developed (Table 11). Most preparations also display some associated toxicity and, as a general rule, the greater the product's adjuvanticity, the more toxic it is likely to be. A few different adjuvants may be used in veterinary medicine; however (for safety reasons), aluminum-based products are the only adjuvants routinely used in human medicine. Application of many of the aggressive adjuvant materials is reserved for selected experimentation purposes in animals.

The concept of enhancing the immune response against an antigen by co-administration of an immunostimulatory substance dates back to the beginning of the 20th century. Oil-based emulsions were used from 1916. An ideal adjuvant should display several specific characteristics. It should be safe (no unacceptable local/systemic responses), elicit protective immunity, even against weak

Table 11. Various adjuvants used in vaccines [117]

Mineral compounds	Aluminum phosphate ($AlPO_4$) Aluminum hydroxide ($Al(OH)_3$) Alum [$AlK(SO_4)_2 \cdot 12H_2O$] Calcium phosphate ($CaPO_4$)
Bacterial products	Mycobacterial species Mycobacterial components (e.g., trihalose dimycolate, muramyl dipeptide) *Corynebacterium* species Bordetella pertussis Lipolysaccharide
Oil-based emulsions	Freund's complete/incomplete adjuvants (FCA/ FIA)
Saponins	Starch oil
Liposomes	Quil A
Immunostimulatory complexes (ISCOMs)	
Some cytokines	Interleukins 1 and 2, interferons

Table 12. Vaccines approved for various indications [116]

No.	Company	Trade-name	Generic name	Therapeutic indications	FDA approval
1	Merck/ Chiron	Recom-bivax HB	Hepatitis B vaccine	Hepatitis B vaccination	July 1986
2	Smithkline Beecham	Engerix-B	Hepatitis B vaccine	Hepatitis B vaccination	September 1989
3	Chiron	RabAvert	Rabies vaccine	Use in pre- and post-exposure prophylaxis against rabies in humans	November 1997
4	North American Vaccine	Certiva	Acellular pertussis vaccine	Protection against diphtheria, tetanus and pertussis	July 1998
5	BioChem Pharma	Pacis BCG	Live, atten-uated strain of BCG	Bladder cancer	March 2000

immunogens, non-pyrogenic, chemically defined, be effective in infants/young children, lead to stable formulation with antigen, be biodegradable and be non-immunogenic itself.

Various vaccines that have been recently approved for various therapeutic applications are listed in Table 12.

6 Antibodies

In recent years, antibody therapy has made a great positive impact upon human healthcare management. Polyclonal antibody preparations have been used to induce passive immunity against a range of foreign agents. Monoclonal antibodies find a range of therapeutic applications [94].

6.1 Polyclonal Antibodies

Polyclonal antibody preparations have been used for several decades to induce passive immunization against infectious diseases and other harmful toxins [94]. The antibody preparations are usually administered by direct injection. While this generates immunological protection it lasts only for a short time. Antibody preparations from animal origin are generally termed "antisera" while those from humans are called "immunoglobulin preparations". Predominantly the latter are IgG. While "antisera" have been quite valuable in the treatment of a variety of medical conditions, they can also introduce unwanted side effects such as, e.g., hypersensitivity reactions and some of these may be life-threatening. Therefore, antibody preparations derived from human donors (immunoglobulin) are preferred. These antibodies are raised against specific targets such as microbial or viral pathogens, microbial toxins, or venoms [95].

6.2 Anti-D-Immunoglobulin

Antibodies raised against alloantigen (the antigen that differs between individuals of the same species) are known as anti-D-immunoglobulin, and the antigen as D-antigen (e.g., Rh antigen). These antibodies are usually used in cases of pregnancy where a Rh (–)ve female becomes pregnant with a Rh (+)ve male, and the fetal erythrocytes are also Rh (+)ve. The antibodies bind to fetal Rh (+)ve erythrocytes, marking them for destruction before the maternal immunological reaction is triggered [96].

Similarly, antibodies can be produced from the plasma, serum or placentas of normal healthy donors as normal immunoglobulin. They generally contain antibodies against diphtheria, measles, poliomyelitis, hepatitis A, rubella, and varicella. These antibodies are used to provide passive immunization against these diseases.

6.3 Monoclonal Antibodies

During the last decade, antibody-based medical application has been focused on monoclonal antibodies. In 1970, Kohler and Milstein fused immortal myeloma cells with antibody-producing B-cell lymphocytes to produce stable antibody-producing cells. This hybridoma technology facilitates the relatively straightforward production of monospecific antibodies against virtually any desired antigen.

Although numerous applications of monoclonal antibodies are under clinical investigation, some of the clinical applications of antibodies are induction of passive responses, diagnostic imaging, e.g., of cancer, infectious diseases, cardiovascular disease and deep vein thrombosis and therapy for cancer, transplant and cardiovascular diseases [97].

6.4 Murine Monoclonals

Antibody immunogenicity is one of the limitations associated with the administration of murine monoclonals in human subjects. In most of the cases, a single injection of a murine monoclonal will elicit a response in 50–80% of the patients. Human anti-mouse antibodies (HAMA) will be detected within 14 days of administration. Sometimes repeated dosing is required. The HAMA response will effectively and immediately destroy subsequent doses of the monoclonal administered. Therefore, the therapeutic efficacy of murine monoclonals is limited to the first, or at the most the second dose administered.

This problem can be overcome by using monoclonal antibodies of human origin. Human antibody-producing lymphocytes can be immortalized by transformation with Epstein-Barr virus (EBV), by fusion with a murine monoclonal antibody, or by fusion with human lymphoblastoid cell lines [98].

6.5 Chimeric and Humanized Antibodies

Recombinant DNA-technology has provided an excellent tool for making various hybrid antibodies [98]. In this strategy, "chimeric" antibodies consisting of mouse variable regions and human constant regions are produced. The chimeric antibody should display the specificity of original murine antibody, but will largely be human in sequence. Such "chimeric" antibodies will be significantly less immunogenic, possess prolonged serum half-lives and allow the activation of various Fc-mediated functions. Clinical trials with chimerics have shown them to be safe and non-toxic, but the immune response after single dose administration was very poor, and repeated doses of chimerics are needed to raise the immune response. An alternative method is to humanize the murine antibodies. This is done by transferring the nucleotide sequence coding for the six complementarily-determining regions (CDR) of the murine antibody to the desired specificity into the human antibody gene. The resulting hybrid will be entirely human in nature, except for the CDR region. This process may facilitate the folding of the CDRs into a true natural conformation. This, in turn, normally restores the antibody-antigen binding affinity. Most of such hybrid antibodies are human in sequence. Clinical trials indicate that such proteins do indeed behave similarly to native human antibodies.

6.6 Applications of Monoclonal Antibodies

Apart from medical uses of monoclonal antibodies for the treatment of cancer, recently several monoclonal antibodies have been approved by the US FDA for their use in the detection and treatment of cardiovascular diseases, infectious agents, arthritis, allergy, acute transplant rejection, prevention of blood clotting in high risk angioplasty, prevention of respiratory disease, Crohn's disease etc. [97]. A list of various antibodies and immunoglobulins approved for medical uses is given in Table 13.

7 Gene Therapy – Future in Healthcare Management

The therapeutic uses of nucleic acid-based biopharmaceuticals have a great potential in the present century. With the advent of the Human Genome Project, the potential of gene therapy and antisense technology has become very bright [99], although the full potential of this therapy will only be utilized after the satisfactory resolution of several technical difficulties currently impeding their routine medical application. One of the difficulties in gene therapy is the introduction of the gene into the genetic complement of the cell, such that subsequent expression of the gene achieves a therapeutic goal. This technique could

Table 13. Antibodies and immunoglobulins approved, and their indications [116]

No.	Company	Trade-name	Generic name	Therapeutic indications	FDA approval
1	J & J/Ortho-Biotech	OKT-3	Anti-CD3 Mab	Acute transplant rejection Heart/liver transplant rejection	June 1986 June 1993
2	Medimmune	CytoGam	CMV Immune globulin	CMV infection in transplant patients	N/A
3	Cytogen	Oncoscint	MAb	Imaging agent for colorectal and ovarian cancer	December 1992
4	Centocor	ReoPro	Anti-platelet Mab	To prevent blood clots in high risk angioplasty patients Blood clots in all percutaneous coronary intervention procedures	December 1994 November 1997
5	Immunomedics	CEA-Scan	Radio-labeled antibody imaging agent	Detection of recurrent colorectal cancer	July 1996
6	Cytogen	ProstaScint	Antibody imaging agent	Imaging agent for prostate cancer	October 1996
7	Idec pharmaceuticals	Rituxan	Anti-CD20 Mab	Non-Hodgkin's B-cell lymphoma	November 1997
8	Protein Design Labs	Zenapax	Anti-Tac Mab	Treatment of organ rejection in renal transplantation	December 1997
9	Medimmune	Synagis	Palivizumab	Prevention of respiratory disease caused by RSV in children	June 1998
10	Centocor	Remicade	Infliximab	Crohn's disease Rheumatoid arthritis	August 1998 November 1999
11	Nabi Pharmaceuticals	Nabi-HB	Hepatitis B immune globulin	Treatment of hepatitis B	February 1999
12	Genzyme	Thyrogen	Diagnostic	Test serum thyroglobulin levels in thyroid cancer patients	December 1998
13	SangStat Pharmaceuticals	Thymoglobulin	Anti-thymocyte globulin	Treat acute organ rejection in kidney transplant recipients	January 1999
14	NABI/Cangene	Nabi-HB	Hepatitis B Immune globulin	Treatment following exposure to hepatitis B virus	March 1999
15	Celltech Group	Mylotarg	Gemtuzumab ozogamicin	Treatment of patients >60 years in first relapse with CD33+AML	May 2000

be utilized as a curative therapy for inborn errors of metabolism and other conditions induced by the presence of a defective copy of a specific gene.

With the increasing understanding of the molecular basis of diseases, it should be possible to combat diseases such as cancer [100], AIDS or other neurological disorders by gene therapy. The basic approach in gene therapy has the following steps. Firstly, the desired gene must be packed into a vector system capable of delivering it safely inside the intended recipient cells [101]. A variety of vectors can be used to effect such a gene transfer. Once the gene is delivered inside the cell, the desired gene must be now integrated inside the nucleus in order to treat the desired gene. Three different strategies so far have been employed:

a) The target cells which have to be treated are removed from the body, cultured in vitro and incubated with a vector containing the nucleic acid to be delivered. The genetically altered cells are then reintroduced into the patient's body. Such an approach has successfully been undertaken to deliver genes in blood cells, stem cells, epithelial cells, muscle cells and hepatocytes.
b) A second approach involves direct injection of the nucleic acid-containing vector to the target cell in situ in the body. For example, direct injection of the vectors into tumor cells or aerosol administration of the vectors to the respiratory tract epithelial cells (containing the cystic fibrosis gene).
c) A third approach is the injection of the vector into the immediate vicinity of the target cells. This approach is suitable especially for the target cells that are not localized to one specific area of the body such as blood cells. Alternatively, a vector capable of recognizing and binding only to a specific, pre-defined cell type can be used for gene delivery. Such a vector will only deliver its nucleic acid payload to the specific target cells.

All the above approaches are being pursued specific nucleic acid delivery to various cells. For example, the inclusion of an antibody on the sample of a vector will allow specific binding of the surface antigen uniquely associated with the target cell and therefore a very selective delivery of the nucleic acid.

The choice of vector, target cell and protocol used will depend upon a number of considerations. The major consideration is obviously what the ultimate goal of the gene therapy is in any given case. For example, in some instances it may be to correct an inherited genetic defect, whereas in other instances it may be to confer a novel function upon the recipient cell. An example of the former would be the introduction of the cystic fibrosis transmembrane conductance regulator (CFTR) gene (the cystic fibrosis gene) into the airway epithelial cells of CF sufferers. An example of the latter would be the introduction of a novel gene into white blood cells whose protein product is capable of interfering in some way with HIV replication. Such an approach might prove to be an effective therapeutic strategy for the treatment of AIDS.

An additional consideration of the gene product, which may influence the protocol used, is the desired duration of subsequent expression of the gene product. In most cases of genetic disease, long-term expression of the inserted gene would be required. In other instances (e.g., some forms of cancer therapy, or the use of gene therapy to deliver a DNA-based vaccine) short-term expression of the gene introduced would be sufficient [102].

For most applications of gene therapy, straightforward expression of the gene product itself will suffice. However, in some instances, regulation of expression of the transferred gene would be required (e.g., if gene therapy combating insulin-dependent diabetes mellitus were to be considered). Thus far, it has not proven possible to exert such expressional control over transferred genes [103].

The choice of the target cells is another point worthy of discussion. In some instances, this choice is pre-determined, e.g., treatment of the genetic condition, familial hypercholesterolemia, would require insertion of the gene coding for the low-density lipoprotein receptor specifically into hepatocytes [104].

In other cases, however, some scope may be available to choose a target cell population. Even in the case of redressing some genetic diseases, it may not be necessary to genetically correct the exact population of cells affected. For example, a hallmark of several of the best-characterized genetic diseases is the exceedingly low production of a circulatory gene product. Examples include clotting factors VIII and IX, a lack of which leads to hemophilia. It may be possible to correct such defects by introducing the appropriate gene into any recipient cell capable of exporting the gene product into the blood. In such cases, the choice of a target cell could be made upon practical considerations, such as their ease of isolation and culture, their capacity to express (and excrete) the protein product, and their half-lives in vivo.

Several cell types, including keratinocytes, myoblasts and fibroblasts have been studied in this regard. It has been shown, for example, that myoblasts, into which the factor IX gene and the growth hormone gene have been introduced, could express their protein products and secrete them into the circulation.

Vectors capable of introducing genes into recipient cells are mostly virus-based such as retrovirus, adenovirus, adeno-associated virus, herpes virus, polio virus and vaccinia virus [105]. These vector systems are used to deliver genes into mammalian cells. To date, the majority of clinical trials undertaken have utilized retroviral vector systems. Non-viral systems have generally been employed least often, although some, e.g., nucleic acid-containing liposomes, may be used more extensively in the future. Some of the methods tested, e.g., calcium phosphate precipitation, electroporation and particle acceleration, are unlikely to be employed to any great extent in gene therapy protocols.

To date over 4000 genetic diseases have been characterized. They may be either due to the lack of production of a single gene product or due to production of a mutant gene product incapable of carrying out the natural function. Gene therapy is a simple and straightforward therapeutic option which would correct such genetic diseases [106]. This is possible by simple insertion of a healthy copy of the gene into appropriate cells of the patient. Although, the concept of gene therapy is quite simple, the success has been quite elusive. A number of factors may be associated with the gene therapy [107]:

- The number of genetic diseases for which the actual gene responsible has been identified and studied is relatively modest. However, the complete sequencing of all the genes in the human organism may slowly change this scenario as more and more functions of the gene will be identified.

- Availability of a suitable vector for delivery of the genes to the desired target site [107].
- Some genetic disorder are quite complex involving several organs and cell types. It has proved quite difficult to introduce the required gene into all the affected cell types.
- Regulation of expression levels of the gene transferred has proved problematic.
- Costs involved in gene therapy may be very high at present.

7.1 Future of Gene Therapy [108]

The importance of gene therapy has gained considerable attention, although there are several queries to be answered. Technical difficulties, the question of public perception, ethics, and costs are of major concern. In the longer run, the cost of gene therapy may be comparable to the cost of present day biopharmaceuticals. Many at present incurable disease conditions may become amenable to treatment. The ability to readily modify our genetic complement holds great therapeutic promise. However, there also exists a danger of misuse, such as the possibility that gene therapy could eventually be used to improve or change human characteristics. Genetic manipulation of human germ cells is a major ethical question. Any genetic alteration achieved will thus be transmitted to further generations.

In years to come, scientists would be able to correlate the function of each gene and the disease conditions, with a mutated or defective gene. Extensive studies are underway to pindown the function of genes, the single nucleotide polymorphism (SNP), and the possible change in the function due to SNP. This is expected to dramatically change the drug development protocol, clinical trial protocol etc., due to pharmacogenetic differences in the patients and their pharmacogenomic response towards the drug. It is expected that selected drugs will be made for the patients having similar genetic morphology, and these drugs are expected to show high response and low toxicity in these patient populations. Thus, the clinical trials may be based on the genetic morphology of the patient and the doctor may prescribe a particular drug to a particular patient only. This is expected to reduce the cost of clinical trials. However, such drugs that are tested in selected patient populations cannot be block bluster drugs as observed in the past. In fact, this may open up a vast opportunity to develop several analogues of the same drug, which may suit different sets of patient populations.

7.2 Antisense Technology

The antisense technology is based upon the generation of short, single-stranded stretches of nucleic acids, which can be DNA- or RNA-based, displaying a specific nucleotide sequence. These are generally termed antisense oligonucleotides [109]. These oligos are capable of binding to DNA (at specific gene sites) or more commonly to m-RNA derived from specific genes and are complementary to the nucleotide base pair present in DNA or m-RNA. Binding prevents expression of gene products by preventing either the transcription or translation process. There are three major class of antisense agents - antisense oligonucleotides, antisense sequences and ribozymes.

7.3 Antisense Oligonucleotides

The nucleotide sequence of an m-RNA molecule dictates the amino acid sequence of a protein. This m-RNA is complementary to an antisense DNA strand, which serves as a template for m-RNA synthesis. Even if part of nucleotide sequence of any m-RNA is known, it is possible to chemically synthesize an oligo (ribonucleotide or deoxyribonucleotide), whose base sequence is complementary to at least a section of the m-RNA sequence. As long as such an antisense oligo can enter the cell, the complementary nature of the sequence will promote hybridization between the m-RNA and the antisense oligonucleotide. This will block translation of the m-RNA and hence prevent synthesis of the mature gene product.

The prevention of m-RNA translation by duplex formation with antisense oligos appears to act as steric blockers or allows targeting by intracellular RNases such as RNase H. This enzyme is capable of binding to an RNA-DNA duplex and degrading the RNA portion of the duplex [110]. These antisense therapies are being assessed in pre-clinical and clinical studies for the treatment of cancer, viral diseases (e.g., HIV, hepatitis B, herpes, PMV etc.) [111]. Other disease treatments may include restenosis, rheumatoid arthritis, and other allergic disorders. Generally, oligonucleotides of 17 or more nucleotides length are used since they are extremely specific and are extremely unlikely to occur more than once in human cell nucleic acid complement.

The advantages of such approaches are that they are absolutely specific, show minimum toxicity because of the natural biomolecule, only low levels of oligos are needed inside the cell for therapeutic effects and oligos are easy to synthesize. Some of the shortcomings are sensitivity of oligos towards nuclease, low serum half-life, poor rate of cellular uptake and that they are orally inactive.

7.4 Antigene Sequences and Ribozymes [112]

Antigene sequences and ribozymes form two additional classes of antisense agents. Certain RNA sequences can function as catalysts. These so called ribozymes function to catalyze cleavage at a specific site in a specific m-RNA substrate. Many ribozymes will cleave the target m-RNA, where there exists a particular triplet nucleotide sequence. Ribozymes can be directed to a specific m-RNA by introducing short flanking oligonucleotides, which are complementary to the target m-RNA. The resultant cleavage of the target will prevent translation. A single ribozyme, as catalytic agent, could destroy thousands of copies of target mRNA. Such a drug will obviously be very potent.

Antigene (oligonucleotide) sequences function to inhibit transcription of a specific gene. These oligos hybridize with appropriate stretches of double stranded DNA, forming a triple helix. This inhibits initiation of transcription of the gene in this region.

So far none of the oligos with potential as a drug has reached clinical studies or the market. However, it is believed that the potential of gene therapy and antisense technology represents the most promising medical advance expected to be achieved in the present decade.

8 Bioproducts Market in India [113]

The biotech products market is expected to grow into a $ 3 billion industry in 2002, according to a Rabo India finance report. According to the latest estimate – the domestic biotech market in India which was at $ 500 million in 1997–1998 is expected to be $ 2.5 billion (>Rs 11000 crores) by 2001,with the product market for healthcare biotech products alone reaching $ 1.5 billion. According to a reference from Chemical Weekly of 26 March 1996 it is US $ 700 million (Rs 3500 crores), whereas Biotechnology Monographs of BCIL (of 1993) mentions it as US $ 1110 million (Rs 5547 crores). Nearly US $ 1 billion (5000 crores) in 1997 of which human healthcare biotech accounts for 60% of the sales and medical devices, contract R & D and reagents and supplies constitute the remainder [114] (Fig. 2).

Globally, during the last three decades biotechnology has rapidly been gaining ground [115]. The United States of America has been playing the most crucial role in developing biotechnology-based foods, drugs, pharmaceuticals, chemicals, effluent treatment systems, biodegradable wastes, composites, diagnostics etc. Some of the biotechnology products in the management of cancer, AIDS, cardiovascular diseases and various other indications, which are presently in the market are listed in Table 14.

According to the report of Frost and Salivan (U.K.), the European biotech market reached $ 106.5 billion by 2002. In 1995 industry reported revenues of $ 29.2 billion, up 17%. Human healthcare is the largest market segment, with a share of 65% in 1995 and $ 77.3 billion in 2002. The food-processing segment had 10% of the market with a market share of $ 18 billion in 2002. As new seeds and crop protection agents enter the agricultural sector it should grow at the rate of 23.6% per year. The smallest sector, waste management, should enjoy the largest yearly growth of all, at 38.2% and revenue should reach $ 7.2 billion by 2002 because many new anaerobic plants are being built [113].

The European biotech sector is second to that of USA. According to reasons stated in the survey, Europe introduced regulations that slowed down product testing, development and time to market, and increased costs. Also companies found that market conditions such as public acceptance and fiscal conditions like tax incentives for R & D were less favorable in Europe. The European market was worth around 27 billion pounds in 1995. 72% of the companies favored making R & D investments in USA with 28% choosing Europe. About 60% prefer to invest in manufacturing in USA compared with 23% in Europe and 18% in Asia. Also

Table 14. Biotech products approved for various indications [116]

No.	Company	Trade-name	Generic name	Therapeutic indications	FDA approval
1	Immunex	Thioplex	Sterile thiotepa	Cancer (tumors)	December 1994
2	QLT Photo-therapeutics	Photofrin	Porfimer sodium	Photosensitive drug for photodynamic therapy	September 1994
3	Genentech	Pulmozyme	DNase	Cystic fibrosis	December 1993
4	US Bio-science	Ethyol	Amifostine (WR-2721)	Reduction of kidney damage w/ovarian cancer treatment	June 1995
5	Sequus (Alza)	Doxil	Liposomal doxorubicin	AIDS-related Kaposi's sarcoma Refractory ovarian cancer	November 1995 January 1999
6	Immunex	Etoposide injection	Generic Cancer Chemo-therapeutic	Refractory testicular tumors and small cell lung cancer	March 1996
7	Ligand pharma-ceuticals	Targretin capsules	Bexarotene	2nd line treatment of cuta-neous T-cell lymphoma	December 1999
8	Alza	Ciadur	Leuprolide acetate implant	Advanced prostate cancer	March 2000
9	Neurex	Corlopam	Dopamine DA1agonist	Malignant hypertension	September 1997
10	Ligand Pharma-ceuticals	Panretin Gel	Topical 9-*cis*-retinoic acid	AIDS-related Kaposi's sarcoma	February 1999
AIDS					
11	US Bioscience	Neutrexin	Trimetrexate	PCP infection in AIDS Patients	N/A
12	Chiron	Vitrasert	Intravitreal implant with ganciclovir	Cytomegalovirus (CMV) retinitis in AIDS patients	March 1996
13	BioChem Pharma	Epivir Epivir-HBV	3TC 3TC	Combination therapy w/ AZT in AIDS Hepatitis B	November 1995 December 1998
14	NeXstar	DaunoXome	Liposomal daunorubicin	First line therapy against HIV-related Kaposi's sarcoma	April 1996
15	Gilead	Vistide	Injectable cidofovir	Cytomegalovirus (CMV) retinitis in AIDS patients	June 1996
16	Guilford Pharma-ceuticals	Gliadel	Carmustine wafer	Second line therapy for glio-blastoma multiforme (brain cancer)	September 1996

Table 14 (continued)

No.	Company	Trade-name	Generic name	Therapeutic indications	FDA approval
17	Immunex	Novatrone	Mitoxantrone	Combination treatment of acute non-lymphatic leukemia in adults	N/A
				Pain associated with hormone-refractory prostate cancer	November 1996
				Multiple sclerosis	March 2000
18	Agouron Pharmaceuticals	Viracept	HIV protease inhibitor	Treatment for HIV where antiviral therapy is indicated	March 1997
19	Vertex/Glaxo Wellcome	Agenerase	Amprenavir	HIV infection in combination with other antiretrovirals	April 1999
Pain					
20	Cytogen	Quadramet	Rasiopharmaceutical	Severe pain associated with cancer that has spread to the bone	March 1997
21	Biomatrix	Synvisc	Elastoviscous hylan biopolymer	Osteoarthritis of the knee	August 1997
22	Immunex	Embrel	Soluble tumor necrosis factor	Treatment of refractory, moderate-to-severe adult RA	November 1998
23	Cypress Bioscience	Prosorba column	Protein A filter	Treatment of RA in patients who have failed 2nd line therapy	March 1999
Cardiovascular					
24	Cor Therapeutics	Integrilin	Peptide GP IIb/IIIa antagonist	To prevent blood clots in patients with acute coronary syndrome	May 1998
25	GelTex	Welchol	Colesevelam hydrochloride	Reduction of elevated LDL	May 2000
Allergies					
26	Sepracor/ HMR	Allerga	Non-sedating antihistamine	Allergic rhinitis	July 1996
27	Sepracor	Xopenix	Active (*R*)-isomer of racemic albuterol	Treatment and prevention of bronchospasm	March 1999
Others					
28	Genzyme	Ceredase	Glucocerebrocidase	Gaucher's disease (type I)	April 1991
29	Genzyme	Cerezyme	Glucocerebrocidase	Gaucher's disease(type I)	May 1994

Table 14 (continued)

No.	Company	Trade-name	Generic name	Therapeutic indications	FDA approval
30	Integra life Sciences	Artificial Skin	Bovine tendon-derived matrix	Full thickness burns where conventional autograft is not desirable	March 1996
31	Genzyme	Seprafilm	Bioresorbable membrane	Prevention of adhesions in abdominal and pelvic surgeries	August 1996
32	Teva Pharmaceuticals	Copaxone	Amino acid polymer	Relapsing remitting multiple sclerosis	December 1996
33	Advanced Tissue Sciences	Dermagraft-TC	Engineered human tissue	Severe burns	March 1997
				Temporary wound covering for partial-thickness burns	October 1997
34	Organogenesis	Graftpatch	Porcine-source, Connective tissue	For use in surgical procedures for reinforcement of soft tissue	August 1997
35	Athena Neurosciences	Carbatrol	Extended release Carbamazepine	Anticonvulsant for use in patients with epilepsy	October 1997
36	Patho-Genesis	Tobi	Aerosol formulation of tobramycin	Treatment of chronic lung infections in cystic fibrosis patients	December 1997
37	Gliatech	ADCON-L	Carbohydrate gel	Reduction of post-surgical adhesions in lumbar surgery	May 1998
38	Organogenesis	Apligraf	Engineered human tissue	Treatment of venous leg ulcers	May 1998
39	Celgene	Thalomid	Thalidome	Erythema nodosum leprosum	July 1998
40	Bio-Technology General	BioLon	1% sodium hyaluronate	Protect corneal endothelium during ophthalmic surgery	July 1998
41	Isis Pharmaceuticals	Vitravene	Injectable fomivirsen sodium	CMV Retinitis	August 1998
42	SangStat Pharmaceuticals	SangCya	Oral generic cyclosporine solution	Prevention of solid organ transplant rejection	October 1998
43	CollaGenex Pharmaceuticals	Periostat	Sub-antibiotic dose of soxycycline	Adjunctive therapy in adult periodontitis	October 1998
44	GelTex/ Genzyme	RenaGel	Sevelamer hydrochloride	Reduction of serum phosphorus in patients with ESRD	November 1998
45	Cephalon	Provigil	Modafinil	Narcolepsy	December 1998

Table 14 (continued)

No.	Company	Trade-name	Generic name	Therapeutic indications	FDA approval
46	BioTime	Hexend	Blood plasma volume expander	Treatment of hypovolemia	March 1999
47	Skye-Pharma*/ Chiron	DepoCyt	Sustained release cytatabine	Neoplastic meningitis in lymphoma patients	April 1999
48	Glaxo Well-come/Biota Holdings	Relenza	Zanamivir	Treatment of influenza A&B	July 1999
49	Gilead Sciences	Tamiflu	Oseltamivir phosphate	Treatment of influenza A&B	October 1999
50	Elan	Zonegran capsules	Zonisamide	Adjunctive therapy for focal epileptic seizures	March 2000
51	QLT Photo-therapeutics	Visudyne	Verteporfin for injection	Wet form of age-related macular degeneration	April 2000
52	Texas Bio-technology	Novastan	Argotroban	Treatment of thrombosis in patients with HIT	June 2000

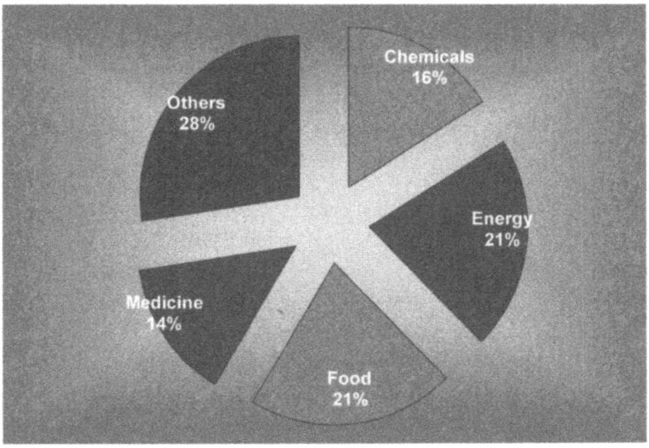

Fig. 2. Recent estimate of biotechnology market in 2000

84% would opt to launch biotech-based products in USA rather than 13% in Europe and 3% in Asia. North America was the favored location for developing biotech drugs in 1995. About 63% of 770 drugs were in development in USA companies compared to 25% in Europe, 7% in Japan and 5% in rest of the world. In agro-biotechnology, 67% of almost 9000 authorized field trials of transgenic crops took place in North America while only 22% were reported in Europe.

During the last decade, the Indian government has granted marketing licenses for about 25 recombinant protein therapeutics. Recombinant insulin, human growth hormone, interferon and hepatitis B vaccine are the products with a larger market share. Medical proteins such as relaxin, rennin, the interleukins and tumor necrosis factor also offer market opportunities.

In 1998, the Department of Biotechnology (DBT) had a budget of over $ 30 million. It has launched a $ 20 million Indian Genome Initiative (IGI) for a five year period to study the genetic variation of the diverse Indian populations [132].

The vaccine market in India is currently approximately $ 260 million growing at the rate of more than 20% per year. India's diagnostic market totals approximately $ 50 million. Monoclonal and polyclonal antibodies for disease immunodiagnosis, tissue typing, clinical assays and research constitute a huge portion of the market. The animal health biotech market is yet another expanding field. It is expected to reach $ 200 million by 2001, with increasing demand for veterinary vaccines, diagnostics, therapeutics, and protein feed.

The global market for genetically modified (GM) crops may soar to $ 25 billion by 2010 from an estimated $ 3 billion in 1999, according to a non-profit organization tracking developments on biotechnology in agriculture.

The biotechnology industry will grow to $ 35 billion by the year 2004 from the current (1999) sales of around $ 7 billion.

Since development of indigenous know how by developing state of the art R & D facilities will be of immense importance in the biotech sector, the 1999 financial budget has given proper weightage to it, and all biotechnology research over the next five years has been given 150 percent exemption over R & D spending.

The total market size for seeds in India is estimated at approximately US $ 1.0 billion (Rs 5000 crores) - comprising seed retained by the farmer US $ 700 million (Rs 3500 crores), public bred seeds US $ 260 million (Rs 1300 crores) and research hybrids US $ 40 million (Rs 200 crores).

Flowers of late have become a money-spinning activity with the Indian Floriculture Industry earning forex to the tune of $ 5 million (approx. Rs 25 crores), while the world floriculture market is worth $ 25 billion.

Demand for biotech products in human and animal healthcare in India was approximately US $ 70 million (Rs 352 crores) in the year 2000 and is expected to be US $ 115 million (Rs 574 crores) by 2005.

There are around 800 companies operating in all sectors of biotechnology. Only about 25 of them are working in modern biotech.

India's presently employed biotech workforce of 10,000 is expected to double in the next year. Half of the increase will be in research, 35% in the technical and services sector; and 15% in management.

Small and medium enterprises in plant-tissue culture and aquaculture can enter the biotech industry, especially in contract research.

The global bioinformatics industry clocked an estimated turnover of US $ 2 billion last year and this figure is expected to grow to US $ 60 billion by 2005. If industry and government work together, it is possible for India to achieve a 5% global market share by 2005. By 2010, the market is estimated to reach $ 4.5 billion in India.

The immunodiagnostic market is expected to increase four to five times by 2005. Diagnostics for malaria is expected to increase by more than 50 percent.

The World Bank has funded up to US$ 240 millions to the ICAR towards the National Agricultural Technology Project (NATP), a 5-year project to focus on plant and agriculture biotechnology research and private sector development.

A study by Rabo India Finance indicated that the investments in bioinformatics have increased from US $ 7.3 million (Rs 3.85 crore) in 1997–1998 to US $ 12 million (Rs 6 crore) in 1999–2000.

Biotechnology R & D in India has been largely dominated by Government-funded institutions that have absorbed nearly US $ 380 million (Rs 1900 crore) during the last five years. The total expenditure in 1999–2000 has been to the tune of US $ 920 million (Rs 459 crore).

In India, the vaccines market is estimated to be around US $ 260 million (Rs 1300 crores), growing at a rate of more than 35%.

8.1 Enzymes

90% of the industrial enzymes are produced in Europe, Japan, USA. China and India are expected to become significant producers. The Indian market of enzymes is US $ 40 million (Rs 800 million) and is expected to become US $ 75 million (Rs 1500 million) in 2002–2003. As regards the supply of enzymes, India imports 70% of its total enzyme consumption (detergent, textile, starch, pharmaceuticals etc.). Novo Nordisk & Genencor have a significant presence in India. Their market share is 30%.

8.2 Vaccines

The vaccine market in India is nearly US $ 260 million (Rs 13.00 billion) (Table 15).

8.3 Immunoglobulins

Some of the immunoglobulin sold in India are listed in Table 16.

8.4 Plasma

See Table 17.

Table 15. Companies producing and marketing vaccines in India (2001) [120]

Vaccine	Trade Name (company)	Unit in ('000)	Value in Rs (million)[a]	% Growth in 2001
MMR	Tremovax (AP)	0.2	0.02	62.7
	Tresvac (SI)	–	–	–
Measles	M-vac (SI)	131.8	3.8359	34.6
Chicken Pox	Varilrix (SKB)	13.7	2.1695	21.4
Polio	Polio Vaccine (SKB)	12.4	0.1532	16%
Rabies	IgG (Aventis)	2.6	0.6415	–
	IgG(Cadila Pharma)	2.0	0.4400	–
Hepatitis A	Havrix (SKB)	17.0	1.7871	1.7
Hepatitis B	Hepacine (ALK)	4.0	0.1378	59.1
	HB Vaccine (CPL)	32.4	1.3552	59.7
	Shanvac B (SB)	49.7	1.2650	48.5
	Engrix B (SKB)	308.7	12.5291	29.3
	Bevac (BE)	70.9	1.5615	432
	Genivac (SI)	4.7	0.0859	100
	Revac-B (INT)	49.7	2.3740	63.4
	Enivac-HB (PB)	64.9	1.8159	49
	Hepashield (PFZ)	38.5	1.0667	177.9
	Biovac (WD)	1.1	0.2100	100
DPT+Hepatitis B	Tritancix HB (SKB)	2.3	0.2967	100
D.T.	Double Antigen (SI)	40.5	0.0273	16.3
DPT	Tripvac (BE)	243.4	0.2044	13.5
	3 Vac (BE)	31.4	0.0259	14.4
	Triple Antigen (SI)	460.9	0.6204	16.3
DPT+Hib	Tetrahibest (AP)			
	Tetramunc (WL)			
Typhoid	VacTyph (CHL)	1.9	0.0618	78.3
	Typhim (CPL)	27.5	0.9392	22.9
	Tophorel (HMR)	30.2	0.9718	24.4
	Typherix Mono (SKB)	6.3	0.2008	100
	Typherix (SKB)	6.8	0.1877	100
Rabis	Rabipu (HMR)	1907.7	62.6134	13
	Verorab (CPL)	592.1	19.6283	6.8
Influenza	Vaxigrip (AP)			
Tetanus	Tet. Texoid (SI)	3355.7	2.0718	6.5
	Tet. Toxoid (BI)	144.1	0.0401	60.2
	Tet. Taxoid (BE)	17235.5	8.5928	8.7

[a] 1 US $ ≈ Rs 50.

Table 16. Various companies selling immunoglobin in India and their market value [121]

Human Anti. D. (Company)	Units		Value Rs (million)[a]
Rhoclone (B. serum)	150 mg	8500	1.2750
	300 mg	40,000	9.9000
Vinobulin (B. S.)	100 mg	887	1.5966
	300 mg	2800	0.6132
Rhogam (J & J)	300 mg	21,000	5.2500
Matergam P (Zydus Biogen)	300 mg	9400	2.4910
Tetanus IgG (Aventis)		61,400	2.1481
B. serum		622,000	7.4640

[a] 1 US $ ≈ Rs 50.

Table 17. Plasma market in India (2000) [121]

Product	Company (Market Share)	Rs. million[a]
Albumin	Serum Int (32%)	0.129
	Baxter (20%)	0.087
	Panacea Bio (7%)	0.028
	KEM Hop (6%)	0.024
	Korea Gree (4%)	0.016
	Aventis (20%)	0.080
IGIV	Korea Gree (49%)	13.965
	B. Serum (14%)	3.990
	Novartis (11%)	3.135
	Aventis (6%)	1.710
	KEM (6%)	1.710
	Panecea Biotech (5%)	1.425
IGIM	All companies	311.600
Factor VIII		34.150
Factor IX		17.850
Fibrinogen		6.400

[a] 1 US $ ≈ Rs 50.

8.5 Antibiotics

See Table 18.

8.6 Diagnostics

The present diagnostic market in India is approximately Rs 6500 crores which is expected to grow by 20% every year for the next few years (Table 19).

Table 18. Antibiotic producing companies in India [120]

Antibiotics	Major producer of antibiotics through fermentation in India
Penicillin G	Alembic, Feremntapharma, Glaxo, IDPL, Max India, Southern Petrochem, Torrent.
Cephalosporin C	Alembic, Kopran
Erythromycin	Alembic
Azithromycin	Alembic
Claithromycin	Alembic
Roxithromycin	Alembic
Cephalexin	Glindia, SOL Pharma, Glaxo.
Cloxacillin	SOL Pharma
Amoxycillin	Kopran
Rifampicin – S	Lupin, Themis, Unichem.
Ampicillin	SOL Pharma
Cloxacillin	SOL Pharma
Gentamycin	Themis Chemicals
Sisomycin	Themis

Table 19. Major Indian firms supplying or manufacturing diagnostic kits in India [120]

Diagnostic Kits, Reagents and Equipment	Major Indian Biotech companies in Diagnostic Business
Immunodiagnostic Kits ELISA Diagnostics	Anglo French Drug, Astra-IDL, Biotechnics Pvt, Biotron Healthcare, Cadila Healthcare, Chemech Laboratories, Dr. Reddy's Laboratories Ltd., Hoffkines Biopharmaceuticals, Herichson diagnostics, Infar (India) Ltd., Monoenzyme Indian Ltd., Omega Biotech, Ranbaxy, Span Diagnostics, Rashmi Diagnostics, Lupin laboratories, J. Mitra & Company, Rallis India Ltd.
Monoclonal Antibodies Based Diagnostics	Hindustan Ciba-Geigy, Merind Ltd., Mediclone Biotech, Ranbaxy Laboratories, Raj Biotech, Professional Biotech
Reagents and Equipment for Diagnostics	Biotron Healthcare, Chemech Laboratories, herichson Diagnostics and Biotech, Infar (India), KAPL, Bangalore, Banglore gene, Mediclone Biotech (PVT) Ltd., Merind Ltd., Omega Biotech, Yashraj Biotech Ltd., Hitech Bio-services Inc.

8.7 r-DNA Based Products

Several companies in India are engaged in making r-DNA-based proteins for human therapeutic uses. For example, Cadila Healthcare Ltd., Ahmedabad, has developed a novel process for interferon α-2b and human plasma proteins. Patents have been filed for the same. Similarly, Wockhardt has developed a process for the production of erythropoietin (EPO) and Dr. Reddy's has claimed that they have a process for GCSF, which they have launched in the Indian market. Similarly, several companies are making vaccines using biotechnology as a tool.

9 Bioinformatics and Information Service Support

With the recent advances in modern biology, especially due to the completion of the Human Genome Project, there is a great thrust on genomics and proteomics of life form. Bioinformatics now constitutes a separate branch of information technology and has progressed very rapidly in the last few years. The information related to gene sequences or protein sequences or their various structures is now available on world wide web (www) for study by biotechnologists. These data are crucial for comparison, confirmation, storage and analysis of study by scientists. There are a number of databases on specific genes and proteins pertaining to humans, animals, plants, bacteria and viruses. These databases are regularly updated by scientists who submit information to various agencies such as National Centre for BioInformatics (NCBI); EMBLnet, ICCBnet, MIM and USPTO.

Bioinformatics is a computer-assisted interface discipline dealing with the acquisition, storage, management, access and processing of molecular biology data. This discipline helps us to collect, compile, analyze, process and represent the information in order to understand processes of life in healthy and diseased states [122]. It is an interdisciplinary scientific tool without barriers among various disciplines of science like biology, mathematics, computer science and information technology.

9.1 Relevance of Bioinformatics

Bioinformatics is being practiced world-wide because of its great relevance in modern biology. Bioinformatics is a science typically associated with databases in genomics and proteomics and structure and function information of genes and proteins, of all forms of life on earth. The field of comparative genomics in biotechnology has provided a path of evolution to researchers through accumulated data on genomics and proteomics. Bioinformatics has made it possible to trace the migration pattern of ancient humans from the traces of chromosomal sequences left in the genomic patterns of modern-day society descendants. Bioinformatics also manages all the information and data on several aspects of transgenic plants. It has been reported that a number of US companies active in transgenic plant research would be spending more than US $ 1.5 billion, which they would be able to accumulate from sales of modified seeds in the available market of US $ 500 billion [123].

Biotechnology has been instrumental in developing databases on genomics and proteomics pertaining to such developments in agriculture as well as other forms of life in general. Scientists are now working to segregate specific databases of different life forms from the available information in different databases. These databases are the bedrock of current and future biotechnology research. It has paved the way for the progress of biotechnology with interdisciplinary scientific research becoming essential and the barriers between traditional disciplines disappearing. The biotechnological databases of discoveries in model organisms highlight the unity of life resulting from evolution by common descent and the data are readily applicable to human biology and other areas of research through bioinformatics. As genomics, proteomics and bioinformatics shift their focus of analysis from individual life component to complete biological systems, an informational science of the whole organism has come into being due to the merger of disciplines like molecular, cellular, developmental, computational and physiological sciences. The biotechnological databases are being utilized for designing new experiments for future research in genomics, proteomics, metabolic regulation, transgenics, horizontal gene transfer in plants and animals, computational approaches in structural biology, protein characterization, structural genomics etc. These databases have given leads for research in human health, disease, drug development, gene therapy and structure-function relationships in biomolecules. Thus, bioinformatics provides different tools to a biotechnologist for access and compilation of data for rigorous sequence similarities, search, data mining and planning for new experimental designs in all areas of biotechnology research.

With the advent of the Human Genome Project, more than 30,000 human genes have already been mapped. A "working draft of the human genome" has already been produced [124]. By the year 2003, a high quality complete genomic sequence will be available, which could identify all the estimated 100,000 or so genes in the human DNA. Thus, the sequence of billions of bases making up the human DNA could be determined and the databases built up [125].

Apart from the human genome, information that exists for other organisms would be added, amassing huge information databases available to active researchers in biotechnology, which are to be managed, absorbed, stored and accessed for future use [126]. These data will keep increasing. As the information is being piled up, it is not only on genes and proteins, but also includes continuous updating of the sequence information for genes and proteins, structure/function annotations, population variations, disease correlations as well as every bit of information that is being generated in life sciences on bacteria, viruses, fungi, protozoa, algae and various specific chromosomes of human and animals [122]. The accumulated valuable data resources can be accessed and there is immediate communication with national and international research groups all over the world at affordable costs. These utilities have made bioinformatics so relevant that biotechnologists cannot do without bioinformatics tools today in research.

9.2 Database and Bioinformatics

Amos Bairoch, at the beginning of the last decade, established a moderate database pertaining to protein sequence and structural correlations of the ""net" [127]. This database was known as PROSITE, and was further strengthened and complemented with a database on sequence analysis and comparison of protein sequences known as SEQQANALREE [128]. From 1991, scientists [128] started depending more on the internet for remote accessing of data, depositing their own sequences and accessing information available on the net. The early 1990s also experienced the development of proteomics-related databases on two-dimensional polyacrylamide gel electrophoresis maps of proteins pertaining to diverse groups of healthy and diseased tissues, known as SWISS-2DPAG [129]. This was the period when compilation of databases in functionality of non-coding sequences, computational protein structures, and protein-peptide characterization based on structure-function analysis started accumulating. Recent developments in this database include cross-references to additional databases. A computer-annotated supplement to SWISS-PROT has improved it significantly. This was possible with a variety of new documentation files and improvement to TrEMBL [130]. The TrEMBL consists of entries in a SWISS-PROT-like format derived from the translation of all coding sequences (CDS) in the EMBL nucleotide sequences database. Some important databases which are commonly used by biotechnologists are those from Incyte, Pangea systems, PE-informatics, GCG, NCBI, PDB, SRS, UDB, SWISS-PROT, EMBLnet, ICCBnet, Medline, FlyBase, Mendelian Inheritance in Man (MIN), USPTO database and SeqWeb. There are specific databases available on the plant, animal or human genomics and proteomics. Practically, a number of URLs for specific databases can be located on the web. There are various programs and tools available for making searches, algorithms analysis, modeling and the computer graphics of the databases on genomics and proteomics like BLAST, Smith-Waterman, ENTERZ, MAGE, CHIME, RasMO, CASP, CAFASPI, PDB-3D Browser, Swiss PDB-Viewer, CHROMAS, CINEMA, EDITVIEW, DNA-SEQUENCER, FACTURA, AUTO ASSEMBLY, GCG software, GENE Explorer 1.4 etc. Some of the tools used in bioinformatics and computational biology have been compiled by Horton after obtaining information from their manufacturers [131].

9.3 Indian Scenario [133]

The impact of bioinformatics on Indian bioscience and biotechnology can be seen both in tangible and non-tangible terms. Research and development activities in these fields grew in quantity as well as quality, as can be seen from research papers published from India. The Department of Biotechnology, Ministry of Science and Technology, has played a key role in the advent of bioinformatics in the country by establishing bioinformatics centers. The initiatives launched by the Indian Government in liberalizing the access to internet and deciding to establish national networks are expected to benefit the program significantly in its attempt to disseminate bioinformatics resources to a large number of scientists in universities and R & D institutions. Presently, the network comprises fifty-two bioinformatics centers with a main center in New Delhi. The Biotechnology Information Centre (BTIC) is responsible for coordinating, organizing and providing information services at a national level covering a wide range of subjects on large sectors of national endeavors in biotechnology. It is coordinating the activities of other centers and plans to provide a nation-wide communication network between the Distributed Informatics Centres (DICs) and other Sub-Centres. These centers are located at Indian Institute of Science, Bangalore, Jawaharlal Nehru University, New Delhi, Madurai Kamaraj

University, Madurai, Bose Institute, Calcutta, University of Poona, Pune, Indian Agricultural Research Institute, New Delhi, Centre for Cellular and Molecular Biology, Hyderabad, National Institute of Immunology, New Delhi, Institute of Microbial Technology, Chandigarh and National Brain Research Centre, New Delhi. Thirty-eight R & D institutions and universities form chains of Distributed Sub-centres set up in the country. The main center also co-ordinates linkages and cooperation with external sources in bioinformatics, including documentation and information centers abroad. The network encourages sharing of knowledge and greater interaction amongst the scientific community irrespective of their geographical locations. It has endeavored in establishing national databases in the country in collaboration with several national and international agencies and has encouraged bilateral and international collaboration in bioiniformatics, e.g., with EMBLnet, ICCBnet etc. The University of Pune has developed several databanks on animal viruses, agricultural pests, biological and medical research, etc. and provides an update and accurate information in the area of biotechnology. The URLs for this important site which has recently been selected to feature on the web, "Pick of the Day" for searching databanks on genome-related sites in India, are http:/bioinformo.ernet.in and http://bioinfor.ernet.in/~sunita/main.html. The other important Indian site on the web is "Gateway Biologics" maintained by the Molecular Biology Group of Tata Institute of Fundamental Research, Mumbai, similarly, MKY, Madurai and IISc, Bangalore have their own databases, which can be accessed by biotechnologists. The important URLs can be found at "Resource of Biology Pages at http://neehow.ym.edu.tw/wonderful/biosites/bioinforg97.html, which is updated periodically and any center can add its URL for access by researchers. Similarly there are a number of such resource sites suitable for providing Indian links (Table 20). Despite the fact that a number of private organizations have excelled in information technology, none has entered in the area of bioinformatics so far, like in developed countries. Soon, multinational biopharmaceutical giants may open this arena in India, like e-commerce for industry, as a business for developing new drugs, delivery systems, geneticeuticals and immunobiologicals.

Table 20. Bioinformatics industry in India [120]

Bioinformatics Service	Major Indian Bioinformatics Companies
Genome of Rice	Avestha Gengraine
Bioinformatics	Biogen Technology, DSQ Biotech, GVK Biotech, Spectramind, Syngene International, Oeium Biosolution
Computational Chemistry	Mahendra British
Genomics	Nagarjuna Group
Pharmacogenomics Informations	Nicholas Parimal
Software for Toxicity & Clinical Study	Syngene International
Software for Target Identification	Incyte
Molecular Modeling	GVK Biotech

Some of the companies are just marketing some of the biotechnology-related information from US companies in India, such as Incyte Genomic etc. Some Indian companies have set up training programs, e.g., GVK Biotech, DSQ Biotech etc.

10 Future Prospects of Biotechnology

During the last decade biotech products have gained considerable ground. More than 95% of biotech products approved for medical uses appeared during 1990s. This number is expected to grow at least 3–5-fold during the present decade. With the Human Genome Project coming to a successful conclusion, several new therapeutic targets are expected to be revealed in the present decade and several new drugs based on the understanding of gene function/protein function are expected to enter clinical trials and the market. Biopharmaceuticals are expected to play a leading role in management of diseases. Conservative estimate suggests that global biotech industry [115] is expected to grow by 17% to $ 40 billion by 2005. US FDA has already approved more than 120 products and over 2500 products are in clinical development and over 350 biopharmaceuticals are either in advanced clinical trial or awaiting FDA approval, which is approximately 30% of new drugs in the pipeline.

In the USA alone, the demand for biopharmaceuticals is growing by almost 13%, and US biotech industry alone spends nearly $ 14 billion annually on R & D. The world market in 2000 for biotech products was in the range of $ US 100 to 110 billion out of which the global vaccine market alone is worth US $ 35 billion.

The European market alone for biotech products is expected to grow to $ 100 billion by 2005. The USA and Europe accounts for almost 5% and 35% of the market share. At present, the in vitro diagnostic market is estimated to be around US $ 5.5 billion and is expected to grow by 10% over the next five years and in certain cases such as cancer, cardiac, infections, genetic disorder, it is expected to grow even more (15–17%) every year. The industrial enzyme market at present is estimated to be US $ 1.8 billion and is further expected to grow by 5–6% every year. The global bioinformatics industry is expected to grow rapidly in the next 5 years with a present estimated market of US $ 2.5 billion increasing to US $ 50–60 billion. It is expected that biopharmaceuticals may get a 15% share of the total worldwide pharmaceutical market by 2003, and in terms of sale it is expected to generate US $ 30–35 billion. With the human genome opening a floodgate of newer ways of drug discovery, several biopharmaceuticals may assume the front-line position in the years to come. Nearly 300–350 biopharmaceutical products are in clinical development across 800 companies world-wide, which may generate US $ 30 - 50 billion in the next 2–3 years.

10.1 Constraints on Biotechnology Development in India

India, being a developing country, has several constraints specially to develop biotechnology in the country.

10.1.1 *Human Resources*

Although, several IITs notably IIT, Delhi, and universities have programs in biochemical engineering, molecular biology and biotechnology with world-class facilities, an acute shortage of trained manpower continues to exist. This is because of the shortage of centers where scientists could be trained in the experimental arts of biological research. Several universities offer training programs, but may not have facilities to give experimental exposures to the students, and therefore, there exists a great gap between the required essential facilities for training and the vision of progress of biotechnology in the country.

Moreover, opportunities in biotechnology globally have increased tremendously, as is universally known. Most of the trained manpower would like to avail of such opportunities available particularly in the United States. Until reasonably good job prospects in the biotechnology industrial sector in India exist, it will be difficult to control this brain drain.

10.2 Biotechnology and Indian Intellectual Property Issues

The rules and regulations governing biotechnology patents are country-specific. Each country has set its own standards for granting biotechnological patents. Some countries place stricter standards than others regarding the patentability of biotechnological inventions.

Exciting inventions in the field of biotechnology have been made in recent years. Following the examples of the owners of the biotech patents, companies with high investments in the field of biotechnology now recognize the advantages of protecting and enforcing their intellectual property rights.

The current controversy in biotechnology is what is patentable and what is not. Generally, an invention is patentable while a discovery is not. While this rule may, in other areas, appear well defined, in biotechnology, it is often the cause of differences in regulations between countries.

Discovery is merely making available what already exists in nature. A substance freely occurring in nature, if merely found or discovered, is not patentable. However, if the substance found in nature has first to be isolated from its surroundings, and a process for obtaining it is developed, that process is patentable.

The granting of a patent is subject to strict criteria of novelty, inventiveness and industrial application. In the case of biotechnology, often involving living entities like microorganisms, there is an additional requirement of sufficient disclosure, for which the invention is required to be deposited at an authorized Depository Authority.

India is a storehouse of biological resources and one of the world's richest biodiversity countries. In the past two years, there has been a rise in the investment in the biotech-oriented industries. It is estimated that over the next five years, biotechnology can offer opportunities for fresh investment of US $ 160 million (Rs 7–8 billion) in India. Biotechnology is poised to take India to a different playing field where it can dominate the world market. However, to achieve its full potential, the biotech sector requires a facilitative environment. The government needs to make a concerted effort to amend its patent laws, strengthen its IPR regime, so as to protect the economic interests of those who innovate. Additionally, India needs to sign the "Budapest Treaty on the International Recognition of the Deposit of Microorganisms for the Purposes of Patent Procedure" to assist in the standardization process of biotechnology patenting.

10.2.1 *What is Patentable in Biotechnology?*

This is the question that India must decide in regards to new biotechnology (genomics, proteomics, bioinformatics, genetic engineering). As per the Indian Patent Act, the following are patentable:

- **Microorganisms**: Under the Indian Patent Act, microbiological processes can be patented. Also patentable are processes for producing new-microorganisms through genetic engineering and the products that result out of this process, such as microorganisms including plasmids and viruses if they are non-living.
- **Cell lines:** A cell line is patentable if artificially produced.
- **γ-DNA, RNA, amino Acids**: If the end result is non-living, it is patentable.
- **Hybridoma technology**: Patents are also allowed on hybridoma technology, but not on protoplast fusion.
- **Expressed sequence tag's or ESTs**: These are small fragments of genetic material obtained by reverse transcriptions of messenger RNA (mRNA) from expressed genes. The gene sequence, or expressed sequence tags (ESTs), can be patented if it has a use, such as if it works as a probe.

10.2.2 *What is not Patentable in Biotechnology?*

Additionally, the Indian Patent Act defines what is not patentable biotechnology: inventability does not apply to plant or animals. Accordingly, a method of producing a new form of a known plant or tissue-culture method for production of plant variety is not patentable, nor is a method of treatment of a human body by surgery or operation for diagnosis. Nor is a method of improving or changing the appearance of the human body or parts of it patentable.

10.2.3 *Patents and Ethics*

Inventions which offend human dignity or encompass the human body at any stage in its formation and development are unpatentable. Patent legislation should prohibit protection for inventions in the following fields:

- Processes for reproductive cloning of human beings.
- Processes for modifying the germ line genetic identity of human beings:
- Use of human embryos for industrial and commercial purpose.

10.3
Funding Institutions in Biotechnology in India

There are two major resources of funding in biotechnology in India. The Department of Biotechnology, Govt. of India was established in 1986. Funding through government grants is quite limited and only available to government institutions and research laboratories. The other source of funding in biotechnology has been research in private research and development setups of pharmaceutical industries.

During the last 10 years or so, some of the private companies have invested money in carrying out research in biotechnology. A few major pharma industries like Wockhardt, Dr. Reddy's Laboratories, Cadila Healthcare, Panecia Biotech, Shanta Biotech, Biocon India, Bharath Biotech, Biological Evans have invested money to create infrastructure and research in biotechnology.

10.3.1
Venture Funding

Venture funding to carry out research in biotechnology is extremely poor in India. Most of the venture capitalists fund research only after ample proof of concept has been carried out. Unlike in the United States, venture capitalists in India do not want to invest money on a startup, biotech company based on research ideas and business plans. ICICI and Tata are among the few groups who have invested money in some of the biotechnology companies, however, the amount of investment is very very low. Other venture capital firms are Morgan Stanley, ILFS, ICF, SIDBI and CDC, but none of them are willing to fund towards conducting basic R & D. These venture capitals are mostly available for commercialization of products that have already been developed.

This funding situation in the country discourages entrepreneurs development in the country.

11
Conclusion

The strength of biotechnology has been steadily growing in the past few decades and several new therapeutic proteins, vaccines, antibodies, diagnostics, etc. have reached the market. With the knowledge of the complete human genome, it is expected that several new targets will be in the limelight for the discovery of new drugs. Gene therapy and DNA vaccines are going to take the front-line in the future management of several metabolic diseases. Biotechnology is going to bring revolution in stem cell research and open a complete new horizon of health management. India, because of its tremendous biodiversity can play a pivotal role in shaping the future of biotechnology at large.

Acknowledgement. I am grateful to Mr. K. Banerjee, Mr. Binu Thomas, Ms. Megha Patel, Mr. S. K. Shah, Dr. (Mrs.) Vidya Lohray for help in literature search, reference and preparation of the manuscript, and Shri. P. R. Patel for encouragement.

12 References

1. Nichols RA (1999) Biotechnology, The science and the business, 2nd edn. Harwood Academic Publishers, Amsterdam, pp 649–657
2. Goeddel DV, Kleid DG, Itakura K (1982) EP Patent No 0055945
3. Ehrlich HA (1989) Polymerase Chain Reaction Technology. Stockton Press, New York
4. Warford A (1998) In-situ hybridisation: a new tool in pathology. Medical Laboratory Science 45:381
5. (a) William JH, John RA (1997) Antibody Therapeutics. CRC Press; (b) Winter G, Milstein C (1991) Nature 349:293
6. Kohler G, Milstein C (1975) Nature 256:495
7. Waldmann T (1991) Science 252:1657
8. (a) Engvall E, Perlman P (1971) Immunochemistry 8:871; (b) Abright LM, Coen DM, Varki A (2001) Current protocols in molecular biology, vol 2, Chapter 11:2.1., Wiley, New York
9. (a) Leary JJ, Brigati DJ, Ward DC (1983) Proc Natl Acad Sci USA 80:4045; (b) Langer PR, Waldrop AA, Ward DC (1981) Proc Natl Acad Sci USA 78:6633
10. Nicholas RA (1999) In: Springham DG (ed), The application of molecular genetics techniques in forensic science in biotechnology, 2nd edn. Harwood Academic Publisher, India
11. Lancini G, Lorenzetti R (1993) Biotechnology of antibiotics and other bioactive microbial metabolites. Plenum Press, New York
12. Dey N C, Dey T K (1998) Medical bacteriology & microbiology, 14th edn. Allied Agency, Calcutta
13. Kaucers A, Crowe ML, Hoy JF (1997) Uses of Antibiotics: A clinical review of antibacterials, antifungal and antiviral drugs, 5th edn. Butterworth-Heinemann, Oxford, UK
14. Billstein S (1994) Agents Chemother 38:2679
15. Hancock R (1997) Lancet 349:418
16. (1983) Quality control of biologicals produced by r-DNA technique, WHO consultation, Bulletin of WHO, 61:897
17. (1996) General Requirements for manufacturing establishments and control laboratories in Requirements for biological substance, Report of WHO Experts Group, WHO, 11. (WHO Technicians Report, Series no. 323)
18. (1987) Acceptability of cell substrates for production of biologicals, Geneva, WHO, (WHO Technical Report Series no. 747)
19. Bristow A (1993) TIBTECH 11:301
20. Fuh G, Cunningham BC, Fukunaga R, Nagata S, Goeddel DV, Wells JA (1992) Receptor Sciences 256:1677
21. Gibson F, Hinds C (1997) Intens Care Med 23:369
22. Olson K, Kenneth C, Fenno J, Lin N, Richard N, Snider C, Kohr WH, Ross MJ, Fodge D, Prender G, Stebbing N (1981) Nature 293:408
23. Koury M, Bondurant M (1992) Eur J Biochem 210:649
24. Roberts D, Smith D (1994) Erythropoietin induction of synthesis to signal transduction. J Mol Endocrinol 12:131
25. (a) Sylvia LH (1984) Proc Natl Acad Sci USA 81:2708; (b) Linfu-Kuen (1999) US Patent no. 5955422; (c) Inoue N, Takeuchi M, Ohashih, Suzukit (1995) Biotechnology Annu. Rev 1:297; (d) Chung Kug I Hsueh K (1997) Hsueh Yaun Hsueh 19:389
26. Fried W (1995) Ann Rev Nature 15:353
27. (a) Joost JO, Marc F (2001) Cytokine reference: A compendium of cytokines and other mediators of host defense, vol 2. Academic Press, New York, pp 899–940; (b) Michael TL, Maurice KG, Giorgio T, Stanley FW (1996) Cellular and Molecular Immunology of an Important Regulatory Cytokines. New York, Academy of Sciences
28. (a) Dranoff G, Crawford AD, Sadelain M, Ream B, Rashid A, Bronson RT, Dickersin GR, Bachurski CJ, Mark EL, Whitsett JA, Mulligan RC (1994) Science 264:713; (b) Cantrell MA, Anderson D, Cerretti DP, Price V, Mckereghan K, Tushinski RJ, Mochizuki DY,

Larsen A, Grabstein K (1985) Cloning, sequence and expression of a human GM-CSF, PNAS, USA, 82:6250; (c) Lee F, Gemmell L, Larson N, Luh J, Rennick D (1985) Isolation of cDNA for a human GM-CSF by functional expression in mammalian cells PNAS, USA 82:4360; (d) Cho JM, Park YW (1955) EP Patent no. 0646643; (e) Arai K, Lee FD (1989) EP Patent no. 299782
29. (a) Kawasaki Ernest, Coyne MY, Koths KE, Ladner MB, Martin GA, Noble Janelle A, Halenbeck RF (2000) US Patent no. 6156300; (b) Robinson AS, Wittrup KD (1998) US Patent no. 5773245; (c) Wong Garden, Clarksteven (1999) US Patent no. 5908763
30. Harousseau J (1997) Biodrugs 7:448
31. (a) Kaufman RJ, Wasky LC (1988) J Biol Chem 263:6352; (b) Pannekoek H, Van Leen RW, Van Ooyen AJ, Verbeet Martinus P (2002) US Patent no. 6346513; (c) Wood WI, Capon DJ, Simonsen CC, Eaton DL, Gitschier J, Keyt B, Seeburg PH, Smith DH, Hollingshead P, Wion KL (1984) Nature 312:330
32. Capon DJ, Lawn RM, Vehar GA (1997) US Patent no. 5668108
33. Martine L, Mayanx JF, Sarminentos P (1992) US Patent no. 5,100,784
34. Latta M, Mayaux J, Sarmietos P (1993) US Patent no. 5,187,261
35. (a) Sreekrishna K, Fuke M (1989) US Patent no. 4,857,467; (b) Kobayashi K, Tomomitsu K, Kuwae S, Ohya T, Ohda T (1997) US Patent no. 5,631,145
36. Boonstra J, Rijken P, Humbel B, Cremers F, Verkleij A (1995) Cell. Biol Int 19:413
37. Burgess W, Maciog T (1989) Annu Rev Biochem 58:575
38. Elliott S, Fagin KD, Narhi LO, Miller JA, Jones M, Koski R, Peters M, Hsieh P, Sachdev R, Rosenfeld RD (1990) J Protein Chem 9:95
39. Bayne ML, Applebaum J, Chicchi GG (1988) Gene 66:235
40. (a) Bell G, Fong NM, Stempien MM, Wormsted MA, Caput Daniel, Ku Lailing, Urdea MS, Rall LB, Sanchez-Pescador R (1986) Nucl Acid Res 14:8427; (b) Brake AJ, Merryweather JP (1988) Biochemistry 27:797
41. Siegel RS, Thill GP, Buckholz RG, Wondrack LM (1992) WO Patent no. 90/10697
42. Greene John M, Rosen CA, Gruber JR (2002) US Patent no. 6,368,822
43. Claesson-Welsh L (1994) J Biol Chem 269:32023
44. Deuel T (1991) Annu Rev Med 42:567
45. Kelly James D, Murray Mark J (1988) US Patent no. 4,769,328
46. Thomason Arlen R (1993) US Patent no. 5,272,064
47. Heldin C, Westernmark B (1989) Eur J Biochem 184:487
48. Wong EY, Seetharam R (1988) Gene 68:193
49. Smith Daniel D (1993) US Patent no. 5,234,639
50. Hunt SMN, Pak SCO (1997) Cytotechnology 24:55
51. Ullrich A, Lee JM, Singh A (2001) US Patent no. 6,331,609
52. Meyer-Ingold W (1993) TIBTECH 11:387
53. Pestka S, Langer J (1987) Annu Rev Biochem 56:727
54. (a) Haria M, Benfield P (1995) Interferon α-2a, Drugs 50:873; (b) Mory Y, Ben-Barak J, Segev D, Cohen B, Novick D, Fischer DG, Rubinstein M, Kargman S, Zilberstein A, Vigneron M (1986) DNA 5:181; (c) Feinstein SI, Chernajovsky Y, Chen L, Maroteaux L, Mory Y (1983) Nucleic Acids Res 11:2927
55. (a) Todd P, Goa K (1992) Interferon gamma 1b Drugs 43:11; (b) Chernajovsky Y, Mory Y, Chen L, Marks Z, Novick D, Rubinstein M, Revel M (1984) DNA 3:297
56. Young H, Hardy K (1995) Role of interferons – α in immune cell regulation. J Leukocyte Bio 58:373
57. Woll P, Pettengell R (1997) Interferon in oncology. Br J Clin Pract 51:111
58. (a) Babu K.R., Swaminathan S. Martin S. Rinas U (2000) Appl Microbial Technol 53:655; (b) Mory Y, Chernajousky Y, Revel M (1981) Eur J Biochem 120:197; (c) Gray PW, Leung DW (1982) Nature 295:503
59. (a) Silva TO, Perez S, Thin JM (1983) Interferon Biotechnol 5:125; (b) Derynck R, Singh A, Goeddel (1983) Nucleic Acids Res 11:1819
60. (a) Rossmall C, Sharp N (1996) Protein Expression and Purification 7:335; (b) Chernajovsky Y, Mory Y (1984) DNA 3:297; (c) Mory Y, Ben-Barak J, Segev D (1986) DNA 5:181

61. Aulitzky W, Schuler M, Peschel C, Huber C (1994) Drugs 48:667
62. Gubler UA, Mizel SB, Lomedico PT, Hoffmann LR. Univ Pennsylvania (1999) US Patent no. 5,936,066
63. (a) Rene D, Geert P, Hilde C (1983) Nucleic Acid Research 11:4307; (b) Tadatsugu T, Hirushi M (1983) Nature 302:305
64. (a) Jeong W, Shin N, Shin H (1997) Biotechnol Lett 19:579; (b) Korn JH, Mory Y, Ziberstein A (1988) Lymphokine Res 7:349; (c) Sreekrishna K, Nelles L (1989) Biochemistry 28:4117
65. Camussi G, Albano E, Tetta C, Bussoline F (1991) Eur J Biochem 202:3
66. Feldmann M, Elliott MJ, Woody JH, Maini Ravinder N (1997) Adv. Immunol 64:283
67. (a) Dobrynin VN, Bercova NP (1988) Bioorgan Chemistry (Russia) 14:1530; (b) Chai H, Vasudevan SG, Porter AG, Chua KL, Oh S, Yap M (1993) Biotechnol Appl Biochem 18:259
68. Verstraete M, Lijnen Henri R, Desire C (1995) Drugs 50:29
69. Emeis J, Verheijen JH, Ronday HK, DeMatt MPM, Brakman P (1997) Proteolysis 11:67
70. Datai R, Cartwright T, Rosen CG (1993) Biotechnology 11:349
71. Pennica D, Holmes WE, Kohr WJ, Warkins RN (1983) Nature 301:214
72. Martegani E, Forlani N, Mauri I (1992) Appl Microbial Biotechnol 37:604
73. Browne MJ, Carrey JE (1988) J Biological Chem 283:1599
74. Castillo P, Palmer CS, Halpern MT, Hatziandreu EJ, Gersh BJ (1997) Ann Pharmacother 31:596
75. (a) Martin S (1994) Biotechnol Adv 12:619; (b) Dunn PM (1996) Dr Edward Jenner (1749–1823) of Berkeley, and vaccination against smallpox. Arch Dis Child Fetal Neonatal Ed 74:F77–8
76. (a) Innis BL, Snitbhan R, Kunasol P (1994) JAMA 271:1328; (b) Clemens R, Safary A, Hepburn A, Roche C, Stanbury WJ, Andre FE (1995) J Infect Dis 171[Suppl 1]:544–549
77. Kang CV, Weiner DB, Boyer JD (2000) Exp Opin Ther Patents 10:1345
78. Lemon S, Thomas D (1997) N Engl J Med 336:196
79. (a) Mi Z, Guo J, Feng F, Chen W, Zhang X, Tong Y (1997) Chin Med Sci. J 12:37; (b) Park SG, Lee JS, Je EY, Kim IJ, Chung JH, Choi IH (2000) Biochem Biophys Res Commun 28:553; (c) Nunez E, Wei X, Delgado C, Rodriguez-Crespo I, Yelamos B, Comez-Gutierrez J, Peterson DL, Gavilanes F (2001) Protein Expr Purif 21:183; (d) Colbere-Garapin F, Horodniceanu F, Kourilsky P, Garapin AC (1983) EMBO J 2:21
80. (a) Talwar GP, Diwan M, Razvi F, Malhotra R (1999) Natl Med J India 12:274; (b) Goldstein KP, Philipson TJ, Joo H, Daum RS (1996) JAMA 276:56
81. Strominger J (1995) Nature Med 1:1140
82. (a) Andino R, Deborah S, Suggett SD, Fichacoso P L, Miller CJ, Baltimore D, Feinberg MB (1994) Science 265:1448; (b) Yokoyama N, Maeda K, Mikami T (1997) J Vet Med Sci 59:311
83. Perkus ME, Tartaglia J, Paoletti E (1995) J Leukocyte Biol 58:1
84. Nilsson C, Makitalo B, Thorstensson R, Norley S, Binninger-Schinzel D, Granagl M, Rud E, Biberfeld G, Putkonnen P (1998) AIDS, pp 2261–2270
85. White DO, Fenner F (1994) Medical Virology, 4th edn. Academic Press, San Diego, pp 306–315
86. Gurwith MJ, Horwith GS, Impellizzeri CA, Davis AR, Lubeck MD, Hung PP (1989) Semin Respir Infect 4:299
87. (a) Cho SP, Lee B, Min MK (2000) Vaccine 18:2878; (b) Hayden FG, Coats T, Kim K, Hassman HA, Blatter MM, Zhang B, Liu S (2002) Antivir Ther 7:53
88. Cohen J (1994) Science 264:1072
89. Esparza J, Osmanov S, Heyward WL (1995) Drugs 50:792
90. Gupta R, Siber G (1995) Vaccine 13:1263
91. Cox J, Coutter A (1997) Vaccine 15:248
92. Donnelly J, Ulmer U, Shiver J, Lui M (1997) Annu Rev Immunol 15:617
93. Krishnan S, Haensler J, Meulien P (1995) Nature Med 1:521
94. Carter P, Kelley RF, Rodrigues ML, Snedecor B, Covarrubias M, Velligan MD, Wong WLT, Rowland AM, Kotts CE, Carver ME, Yang M, Bourell JH, Shepard HM, Henner D (1992) Bio/Technology 10:163

95. Chester K, Hawkins R (1995) TIBTECH 13:294
96. Mountain A, Adair J (1992) Engineering antibodies for therapy. In: Tombs MP (ed), Biotechnology and Genetic Engineering Reviews 10:1
97. Waldman T (1991) Science 252:1657
98. Winter G, Milstein C (1991) Nature 349:293
99. Anderson W (1995) Sci Am September: 96–98b
100. Blaese R (1997) Sci Am June: 91–95
101. Crystal R (1995) Nature Med 1:15
102. Donnelly J (1997) Annu Rev Immunol 15:617
103. Dykes C (1996) Br J Clin Pharmacol 42:683
104. Kay M, Woo S (1994) Trends Genet 10:253
105. Smith A (1995) Annu Rev Microbiol 49:807
106. Morgan R, Anderson W (1993) Annu Rev Biochem 62:191
107. Friedmann T (1997) Sci Am June: 80–85
108. Tolstoshev P (1993) Annu Rev Pharmacol Toxicol 32:573
109. Askari F (1996) N Engl J Med 334:316
110. Putnam D (1996) Am J Health Syst Pharm 53:151
111. Reddy D (1996) Drugs Today 32:113
112. Weintraub H (1990) Sci Am January: 34–40
113. (1993) Research Profile of Biotechnology Activities in India – A Directory. Publications and Information Directorate Dr. K. S. Krishnan Marg New Delhi - 110012
114. Depth Survey of Biotechnology in India by Associated Marketquest Consultants Pvt
115. Coombs J, Campbell PN (1991) Biotechnology Worldwide. COBIOTECH, International Scientific Committee for Biotechnology, CPL press, Science House, Newbury, UK
116. Foo Fatt Kah (2001) Asian Biotechnology Dawn, S. G. Cowen, Hongkong
117. Walsh G (1998) Biopharmaceuticals: Biochemistry and Biotechnology. Wiley, England
118. Bioworld Financial Watch at www.bioworld.com
119. (2001) Focus on Fundamentals: The Biotechnology Report, 15th Annual Review, Ernst and Young Pvt Ltd., India
120. www.biotechsupportindia.com
121. ORG Report, 2001
122. Edelman M (1999) The ICCB Bioinformatics Training Workshop, International Centre for Co-operation in bioinformatics network, Weizmann Institute of Science, Israel, July 11–20
123. Philip HA, Hines PJ (1999) Science 285:367
124. Deloukas P, Schuler GD, Gyapay G, Beasley EM, Myers RM, Cox DR, Weissenbach J, Boguski MS, Bentley DR (1998) Science 282:744
125. Collins FS, Patrinos A, Jordan E, Chakravarti A, Gesteland R, Walters L (1998) Science 282:682
126. Reichardt T (1999) Nature 399:517
127. Bairoch A (1991) Nucleic Acids Res 19:2241
128. Bairoch A (1991) Comput Appl Biosci 7:268
129. Persidis A (1999) Nature Biotechnol 17:828
130. Bairoch A, Boeckmann B (1991) Nucleic Acids Res 19:2247
131. Horton B (1998) Nature 393:603
132. Annual Report 1999–2000, Department of Biotechnology, Ministry of Science and Technology, Government of India, Biotechnology Information System Network (BTISnet), Chapter 8
133. Tripathi KK (2000) Current Science 79:570
134. www.biospace.com

Received: June 2002

Author Index Volumes 51–85

Author Index Volumes 1–50 see Volume 50

Subject Index

Kolibri 360 GmbH
Ferdinand-Jühlke-Straße 7
99095 Erfurt, Deutschland
produktsicherheit@kolibri360.de